GLOBAL FOOD SAFETY

Microbial Interventions and Molecular Advancements

GLOBAL FOOD SAFETY

Microbial Interventions and Molecular Advancements

Edited by

Saher Islam, PhD

Devarajan Thangadurai, PhD

Jeyabalan Sangeetha, PhD

Natália Cruz-Martins, PhD

AAP | APPLE ACADEMIC PRESS

First edition published 2023

Apple Academic Press Inc.
1265 Goldenrod Circle, NE,
Palm Bay, FL 32905 USA

760 Laurentian Drive, Unit 19,
Burlington, ON L7N 0A4, CANADA

CRC Press
6000 Broken Sound Parkway NW,
Suite 300, Boca Raton, FL 33487-2742 USA

4 Park Square, Milton Park,
Abingdon, Oxon, OX14 4RN UK

Library and Archives Canada Cataloguing in Publication

Title: Global food safety : microbial interventions and molecular advancements / edited by: Saher Islam, PhD, Devarajan Thangadurai, PhD, Jeyabalan Sangeetha, PhD, Natália Cruz-Martins, PhD.
Names: Islam, Saher, editor. | Thangadurai, D. (Devarajan), 1976- editor. | Sangeetha, Jeyabalan, editor. | Cruz-Martins, Natália, editor.
Description: First edition. | Includes bibliographical references and index.
Identifiers: Canadiana (print) 20220281394 | Canadiana (ebook) 20220281483 | ISBN 9781774910108 (hardcover) | ISBN 9781774910115 (softcover) | ISBN 9781003283140 (ebook)
Subjects: LCSH: Food—Microbiology. | LCSH: Pathogenic microorganisms—Detection. | LCSH: Foodborne diseases—Prevention.
Classification: LCC QR115 .G56 2023 | DDC 664.001/579—dc23

Library of Congress Cataloging-in-Publication Data

Names: Islam, Saher, editor. | Thangadurai, Devarajan, 1976- editor. | Sangeetha, Jeyabalan, editor. | Cruz-Martins, Natália, editor.

Title: Global food safety : microbial interventions and molecular advancements / edited by Saher Islam, PhD, Devarajan Thangadurai, PhD, Jeyabalan Sangeetha, PhD, Nata¡lia Cruz-Martins, PhD.

Description: First edition. | Palm Bay, FL : Apple Academic Press, 2023. | Includes bibliographical references and index. | Summary: "Foodborne microbial outbreaks are a serious food safety and public health concern worldwide. One of the most challenging issues in food safety is the identification and characterization of foodborne microbial communities, which is a core objective of research by food scientists and food microbiologists. This book reviews the molecular advances in food science related to the safety and quality of food along with recent diagnostic tools for the detection of emerging pathogens based on the food commodities. It presents a wide selection of methods for the identification and characterization of foodborne infectious agents. The volume provides an overview of foodborne pathogens, diseases, and outbreaks and then proceeds to delve into techniques for characterizing foodborne pathogens using molecular approaches. It reviews state-of-the-art methods for detecting and tracing foodborne pathogens using next-generation sequencing and whole-genome sequencing for controlling foodborne illnesses as well as the application of microorganisms in food production for preventing foodborne illnesses. The volume ends with a chapter that provides an overview of management systems and analytical tools for public health protection from foodborne illnesses. Key features: Reviews the detection of foodborne pathogens using recently developed molecular technologies Provides deep insights into the fundamental and functional aspects of microbial interventions in food science Addresses current challenges for the management of food safety Includes updated studies on microbial safety for processed food items Discusses advanced management systems for foodborne pathogens This book will be an important reference source for food scientists and food safety supervisors from production to processing and from transport to retail. This book will also be vital source for teachers, students, and researchers for exploring recent developments for food safety by employing advanced microbial and molecular methods"-- Provided by publisher.

Identifiers: LCCN 2022031058 (print) | LCCN 2022031059 (ebook) | ISBN 9781774910108 (hardcover) | ISBN 9781774910115 (paperback) | ISBN 9781003283140 (ebook)

Subjects: LCSH: Food--Microbiology. | Food--Safety measures.

Classification: LCC QR115 .G56 2023 (print) | LCC QR115 (ebook) | DDC 664.001/579--dc23/eng/20220706

LC record available at https://lccn.loc.gov/2022031058

LC ebook record available at https://lccn.loc.gov/2022031059

ISBN: 978-1-77491-010-8 (hbk)
ISBN: 978-1-77491-011-5 (pbk)
ISBN: 978-1-00328-314-0 (ebk)

About the Editors

Saher Islam, PhD
Visiting Lecturer, Lahore College for Women University, Pakistan

Saher Islam, PhD, is a Visiting Lecturer at the Department of Biotechnology of Lahore College for Women University, Pakistan. She received her PhD in Molecular Biology and Biotechnology from the University of Veterinary and Animal Sciences, Lahore, Pakistan. She was an IRSIP Scholar at Cornell University, New York, and Visiting Scholar at West Virginia State University, West Virginia, USA. She has keen research interests in genetics, molecular biology, biotechnology, and bioinformatics and has ample hands-on experience in molecular marker analysis, whole-genome sequencing (WGS), and RNA sequencing. She has edited five books, including Sequencing Technologies in Microbial Food Safety and Quality and Beneficial Microbes for Sustainable Agriculture and Environmental Management. She has visited the USA, UK, Singapore, Germany, Italy, and Russia for academic and scientific trainings, courses, and meetings. She is the recipient of the 2016 Boehringer Ingelheim Fonds Travel Grant from European Molecular Biology Laboratory, Germany.

Devarajan Thangadurai, PhD
Associate Professor, Karnatak University, Dharwad, India

Devarajan Thangadurai, PhD, is an Associate Professor at Karnatak University, Dharwad in India, and did his postdoctoral research at the University of Madeira, Portugal, University of Delhi, India, and ICAR National Research Center for Banana, India. He is the recipient of the Best Young Scientist Award with Gold Medal from Acharya Nagarjuna University, India, and the VGST-SMYSR Young Scientist Award of the Government of Karnataka, India. He has a keen interest and expertise in the fields of biodiversity and biotechnology, genetics, and genomics of food crops and beneficial microbes for crop productivity and food safety towards sustainable agricultural development. He has authored and edited more than 35 books, including Genetic Resources and Biotechnology (3 Vols.), Genes, Genomes, and Genomics (2 Vols.), Genomics and Proteomics, Biotechnology of Microorganisms, and Sequencing Technologies in Microbial Food Safety and Quality with international scientific publishers in USA, UK, Canada, Switzerland, and India. He has extensively traveled to many universities and institutes in Africa, Europe, and Asia for academic works, scientific meetings, and international collaborations. He is also a reviewer for several journals, including Comprehensive Reviews in Food Science and Food Safety, Food Science and Technology International, Journal of Agricultural and Food Chemistry, Journal of Food Science, Journal of the Science of Food and Agriculture, and International Journal of Plant Production.

Jeyabalan Sangeetha, PhD
Assistant Professor, Central University of Kerala, Kasaragod, India

Jeyabalan Sangeetha, PhD, is an Assistant Professor at the Central University of Kerala, Kasaragod, India. She earned her BSc in Microbiology (2001) and PhD in Environmental Sciences (2010) from Bharathidasan University, Tiruchirappalli, Tamil Nadu, India. She holds an MSc in Environmental Sciences (2003) from Bharathiar University, Coimbatore, Tamil Nadu, India. Between 2004 and 2008, she was the recipient of the Tamil Nadu Government Scholarship and the Rajiv Gandhi National Fellowship of University Grants Commission, Government of India, for doctoral studies. She served as Dr. D. S. Kothari Postdoctoral Fellow and UGC Postdoctoral Fellow at Karnatak University, Dharwad, South India, during 2012–2016 with funding from the University Grants Commission, Government of India, New Delhi. She has carried out her doctoral and postdoctoral research in the fields of microbiology and metagenomics, respectively. She has authored a textbook on Molecular Mycology and edited 10 books, including Genomics and Proteomics, Industrial Biotechnology, Biotechnology of Microorganisms, and Sequencing Technologies in Microbial Food Safety and Quality. She has got trained in next-generation sequencing (NGS) and data analysis at Malaysia Genome Institute, Malaysia.

Natália Cruz-Martins, PhD
Invited Professor, Faculty of Medicine, Researcher at the Laboratory of Neuropsychophysiology, Faculty of Psychology and Education Sciences (FPCEUP); and Institute for Research and Innovation in Health (i3S), University of Porto, Portugal

Natália Cruz-Martins, PhD, is an invited Professor at the Faculty of Psychology and Education Sciences and at the Institute for Research and Innovation in Health (i3S) at the University of Porto, Portugal. She has academic and research background in dietetics, nutrition, natural product chemistry, biochemistry, drug discovery, phytochemistry, phytopharmacology, functional foods, and nutraceuticals. She has been increasingly focused on the use of naturally occurring bioactives for human health, not only from a health promotion and disease prevention point of view, but also from a treatment point of view. She has worked as university professor since 2017, was an advisor of several MSc and PhD theses, and is a member of the evaluation panel of the College of Nutritionists (Porto, Portugal). She has participated in various research projects, received several grants and awards, published more than 130 articles in peer-reviewed international journals (h-index: 25) and several book chapters, and presented more than 40 communications at national and international conferences. She is a member of the Council for Nutritional and Environmental Medicine (CONEM, Norway), reviewer for more than 50 highly reputed international journals, invited reviewer for several book publishers, and editorial board member of several international journals. She also edited several special issues on research topics in highly reputed journals and is currently editing several books for renowned publishers.

Contents

Contributors

Kayode Adebisi Arowora
Department of Biochemistry, Federal University Wukari, PMB 1020, Taraba, Nigeria

Charles Oluwaseun Adetunji
Applied Microbiology, Biotechnology and Nanotechnology Laboratory,
Department of Microbiology, Edo University Iyamho, Edo, Nigeria

Juliana Bunmi Adetunji
Nutritional and Toxicological Research Laboratory, Department of Biochemistry,
Osun State University, Osogbo, Osun, Nigeria

Muhammad Akram
Department of Eastern Medicine, Government College University, Faisalabad, Pakistan

Saadia Andleeb
Atta-Ur-Rahman School of Applied Biosciences, National University of Sciences and Technology,
Islamabad, Pakistan

Hina Anwar
Department of Eastern Medicine, Government College University, Faisalabad, Pakistan

Kayode Adebisi Arowora
Department of Biochemistry, Federal University Wukari, P.M.B 1020, Taraba State, Nigeria

P. Ashwini
Division of Microbiology and Tissue Culture, School of Life Sciences,
JSS Academy of Higher Education and Research, Mysuru, Karnataka, India

Ajith Athira
Department of Food Process Engineering, School of Bioengineering,
College of Engineering and Technology, SRM Institute of Science and Technology,
Kattankulathur, Tamil Nadu – 603203, India

Ayodele Eugene Ayeni
Department of Biological Sciences, Microbiology Unit, Landmark University, PMB 1001,
Omu-Aran, Kwara, Nigeria

Eugene Ayodele Ayeni
Department of Biological Sciences, Microbiology Unit, Landmark University, PMB 1001,
Omu-Aran, Kwara, Nigeria

Manoj Manjunath Bongale
Department of Studies in Biotechnology, Davangere University, Shivagangotri, Davangere,
Karnataka, India

Sayed Muhammad Ata Ullah Shah Bukhari
Department of Microbiology, Abdul Wali Khan University, Garden Campus, Mardan, Pakistan

Manjunath Chavadi
VRDL Laboratory, Department of Microbiology, Hassan Institute of Medical Science (HIMS), Hassan, Karnataka, India

Jayabrata Das
Department of Biotechnology, School of Bioengineering, SRM Institute of Science and Technology, Kattankulathur, Chennai, Tamil Nadu, India

Atul Dhiman
Department of Food Science and Technology, National Institute of Food Technology Entrepreneurship and Management (NIFTEM), Sonipat, Haryana, India

Manjunatha Hanumanthappa
Department of Biochemistry, Bangalore University, Bangalore, Karnataka, India

Daniel Ingo Hefft
Edgbaston Campus, University of Birmingham, School of Chemical Engineering, Birmingham B15 2TT, United Kingdom

Saher Islam
Institute of Biochemistry and Biotechnology, Faculty of Biosciences, University of Veterinary and Animal Sciences, Lahore – 54000, Pakistan

Muhsin Jamal
Department of Microbiology, Abdul Wali Khan University, Garden Campus, Mardan, Pakistan

Naveen Kumar Kudure Jayanna
Department of P.G. Studies and Research in Microbiology, Kuvempu University, Shimoga, Karnataka, India

Fahad Said Khan
Department of Eastern Medicine, Government College University, Faisalabad, Pakistan

Smitha Chandrashekar Kollarahithlu
SMart and Innovative Laboratory for Energy Devices (SMILE), Indian Institute of Information Technology Design and Manufacturing (IIITDM), Kancheepuram, Chennai, Tamil Nadu, India; Department of Chemical Engineering, National Institute of Technology Karnataka Surathkal, Mangalore, Karnataka, India

Arvind Kumar
Department of Dairy Science and Food Technology, Institute of Agricultural Sciences, Banaras Hindu University, Varanasi – 221005, Uttar Pradesh, India

Umme Laila
Department of Biological Sciences, Microbiology Unit, Landmark University, PMB 1001, Omu-Aran, Kwara, Nigeria

Murali Mahadevamurthy
Department of Studies in Botany, University of Mysore, Manasagangothri, Mysuru, Karnataka, India

Olugbenga Samuel Michael
Cardiometabolic Research Unit, Department of Physiology, College of Health Sciences, Bowen University, Iwo, Osun, Nigeria

Sadhna Mishra
Faculty of Agricultural Sciences, GLA University, Mathura 281406, Uttar Pradesh, India

Muhammad Asif Nawaz
Department of Biotechnology, Shaheed Benazir Bhutto University, Sheringal, Dir (Upper), Pakistan

Olugbemi Tope Olaniyan
Laboratory for Reproductive Biology and Developmental Programming, Department of Physiology, Edo University, Iyahmo, Edo, Nigeria

Wadzani Dauda Palnam
Crop Science Unit, Department of Agronomy, Federal University, Gashua, Yobe, Nigeria

Shikha Pandhi
Department of Dairy Science and Food Technology, Institute of Agricultural Sciences, Banaras Hindu University, Varanasi – 221005, Uttar Pradesh, India

Gollapalli Pavan
Central Research Lab, K.S. Hegde Medical Academy, Nitte University (Deemed to be University), Mangalore, Karnataka, India

Kollur Shiva Prasad
Department of Sciences, Amrita School of Arts and Sciences, Amrita Vishwa Vidyapeetham, Mysuru Campus, Karnataka, India

Radhakrishnan Preetha
Department of Food Process Engineering, School of Bioengineering, College of Engineering and Technology, SRM Institute of Science and Technology, Kattankulathur, Tamil Nadu – 603203, India

Santosh Kumar Shimoga Rajanna
Department of Studies in Food Technology, Davangere University, Shivagangotri, Davangere, Karnataka, India

Manjula Ramu
National Center for Biological Sciences (NCBS-TIFR), Bangalore, Karnataka, India

Sana Raza
Department of Microbiology, Abdul Wali Khan University, Garden Campus, Mardan, Pakistan

Redaina
Department of Microbiology, Abdul Wali Khan University, Garden Campus, Mardan, Pakistan

Suhasini Sathiyamoorthy
Department of Electronics and Communication Engineering, SRM Institute of Science and Technology, Kattankulathur – 603203, Tamil Nadu, India

Liloma Shah
Department of Microbiology, Abdul Wali Khan University, Garden Campus, Mardan, Pakistan

Saranya Thiruvottriyur Shanmugam
AXES Research Group and NANOlab Center of Excellence, University of Antwerp, Antwerp, Belgium

Chandan Shivamallu
Division of Biotechnology and Bioinformatics, School of Life Sciences, JSS Academy of Higher Education and Research, Mysuru, Karnataka, India

Amna Siddique
Department of Eastern Medicine, Government College University, Faisalabad, Pakistan

Anurag Singh
Food Technology Department, Harcourt Butler Technical University, Nawabganj, Kanpur 208002, Uttar Pradesh, India

Chandrashekar Srinivasa
Department of Studies in Biotechnology, Davangere University, Shivagangotri, Davangere, Karnataka, India

Rajat Suhag
Department of Food Science and Technology, National Institute of Food Technology Entrepreneurship and Management (NIFTEM), Sonipat, Haryana, India

Ajayi Kolawole Temidayo
Department of Microbiology, University of Ilorin, Ilorin, Kwara, Nigeria

Dhruv Thakur
Department of Food Science and Technology, National Institute of Food Technology Entrepreneurship and Management (NIFTEM), Sonipat, Haryana, India

Devarajan Thangadurai
Department of Botany, Karnatak University, Dharwad – 580003, Karnataka, India

Benjamin Ewa Ubi
Department of Biotechnology, Ebonyi State University, P.M.B. 053 Abakaliki, Ebonyi, Nigeria

Benssan K. Varghese
Department of Food Process Engineering, School of Bioengineering, College of Engineering and Technology, SRM Institute of Science and Technology, Kattankulathur, Tamil Nadu – 603203, India

Pandiyarasan Veluswamy
SMart and Innovative Laboratory for Energy Devices (SMILE) and School of Interdisciplinary Design and Innovation (SIDI), Indian Institute of Information Technology Design and Manufacturing (IIITDM), Kancheepuram, Chennai, Tamil Nadu, India

Karolien De Wael
AXES Research Group, University of Antwerp, Antwerp, Belgium; NANO Lab Center of Excellence, University of Antwerp, Groenenborgerlaan – 171, 2020, Antwerp, Belgium

Abbreviations

AAB	acetic acid bacteria
ABR	antibiotic-resistant
AFLP	amplified fragment length polymorphism
AGE	acute gastroenteritis
AMR	antimicrobial resistance
AMV	avian myeloblastosis virus
API	analytical profile index
CA-MRSA	community-associated MRSA
CCs	clonal complexes
CDC	Centers for Disease Control
CDCP	Centers for Disease Control and Prevention
cDNA	complementary DNA
CDT	cytolethal distending toxin
CGST	core-genome sequence typer
CN	copy number
CNS	central nervous system
CPAD	*C. perfringens*-associated disease
CPE	*C. perfringens* enterotoxin
CRISPR	clustered regularly interspaced short palindromic repeat
CSF	cerebrospinal fluid
DAEC	diffusely adhering *E. coli*
ddNTPs	dideoxynucleotide triphosphates
DON	deoxynivalenol
EAEC	enteroaggregative *E. coli*
EFSA	European Food Safety Authority
EHEC	enterohemorrhagic *E. coli*
EIEC	enteroinvasive *E. coli*
EMA	ethidium monoazide
EPEC	enteropathogenic *E. coli*
ETEC	enterotoxigenic *E. coli*
FAFLP	fluorescent amplified fragment length polymorphism
FBI	foodborne illness

GA	genetic analyzer
GBS	Guillain–Barré syndrome
GC	gas chromatography
GFSI	global food safety initiative
GTM	gene trait matching
HA	hemagglutinin
HACCP	hazard analysis and critical control point
HA-MRSA	health-associated MRSA
HAV	hepatitis A virus
HC	hemorrhagic colitis
HCC	high-confidence core
HEV	hepatitis E virus
HPAI	highly pathogenic avian influenza
HPLC	high-performance liquid chromatography
HU	hemolytic uremic
IMBs	immunomagnetic beads
ISO	International Organization of Standardization
ITS	internal transcribed spacer
LAB	lactic acid bacteria
LAMP	loop-mediated isothermal amplification
LCA	lifecycle assessment
LmCGST	*L. monocytogenes* core-genome sequence typing
LPS	lipopolysaccharide
MALDI-TOF	matrix-assisted laser desorption ionization-time of flight
MAP	*Mycobacterium avium* subsp. *paratuberculosis*
MHLW	Ministry of Health, Labor, and Welfare
miRNA	micro-RNA
MLST	multilocus sequence typing
MLVA	multilocus variable-number tandem repeat analysis
MOST	metric oriented sequence typing
mRNA	messenger RNA
MRSA	methicillin-resistant *S. aureus*
MS	mass spectroscopy
NA	neuraminidase
NASBA	nucleic acid sequence-based amplification
NCBI	National Center for Biotechnology Information
NGS	next-generation sequencing
NM	non-motile

NMR	nuclear magnetic resonance
OTUs	operational taxonomic units
PAIs	pathogenicity islands
PCR	polymerase chain reaction
PFGE	pulsed-field gel electrophoresis
PMA	propidium monoazide
PPi	pyrophosphate
QIIME	quantitative insights into microbial ecology
qPCR	quantitative PCR
RAPD	random amplified polymorphic DNA
RDP	ribosomal database project
REA	restriction enzyme analysis
RFLP	restriction fragment length polymorphism
RNA	ribonucleic acid
RTE	ready-to-eat
SARS-CoV-2	severe acute respiratory syndrome coronavirus-2
SE	staphylococcal enterotoxins
SFD	*Staphylococcus* foodborne disease
siRNA	small interfering RNA
SNP	single nucleotide polymorphism
snRNA	small nuclear RNA
SRST	short read sequence typing
ssRNA	single-stranded RNA
STEC	Shiga toxigenic *Escherichia coli*
TDH	thermostable direct hemolysin
TMP-SMX	trimethoprim-sulfamethoxazole
tRNA	transfer RNA
TTM	transcriptome-trait matching
USFDA	US Food and Drug Administration
VBNC	viable but non-cultivable
VNTR	variable number tandem repeats
WGS	whole-genome sequencing
WHO	World Health Organization
ZMW	zero-mode waveguides

Preface

Consumers' demand for safe and healthy food and fostered economic standards have instigated the necessity for emergent technologies over traditional analytic processes. Foodborne microbial outbreaks are serious food safety and public health concerns worldwide. Therefore, food safety and food quality is the core objective of research for food scientists and food microbiologists. A present challenging issue for food safety is the identification and characterization of foodborne microbial communities. As the infectious amount of pathogens in food matter is as less as few cells, the sensitivity of diagnostic tools has become highly essential. In this manner, high-throughput and sensitive diagnostic methods with microbial interventions and molecular advancements are of utmost importance to monitor microbial pathogens present in food. Hence, appropriate strategies need to be applied to monitor the occurrence of foodborne pathogens along the food chain, and their persistence of them has to be cautiously prevented.

Cultivation-dependent traditional methods for pathogens detection have been considered standard tools; however, these methods demand about a week to infer results. Moreover, these traditional tools are not always suitable to explore the complex microbial populations as they often fail to correctly profile the prevailing diversity. Thus, studies on formulation, optimization, application, and advancements of molecular technologies in food science are of great interest for the development of powerful tools accessible for the enhancement of food safety. Molecular assays, including PCR, fluorescent *in situ* hybridization, NGS, and microarray technologies, have become well-established protocols for the detection of microbial pathogens present in food. Among these assays, PCR is a widely applied assay in the food industry for its high-level sensitivity. But a major shortcoming of this assay is its incapability to differentiate DNA from viable and dead cells, and that is the most critical aspect for regulatory agencies and the food industry. To solve this issue, researchers are designing biological dyes to intercalate with DNA of dead cells present in the food samples before setting up DNA extraction. DNA testing is also feasible now with single-molecule, and high-throughput assays enable thousands of reactions to be accomplished simultaneously and rapidly for detection. These diagnostic methods based on molecular advances are offering better sensitivity, accuracy, and reproducibility to improve the safety of food.

This book presents recently developed molecular techniques that are used in food science to ensure food safety. The book addresses the current challenges for the management of food safety and also provides strategic insights in order to avoid them. Each chapter highlights its subject and their potential applications to food safety. This book will serve as a reference source for food scientists and food safety supervisors from production to processing, transport, and retail. This book will also be a vital source for teachers, students, and researchers to explore recent developments towards food safety by employing advanced microbial and molecular methods.

We sincerely thank all the contributing authors for their keen interest in submitting potential chapters to this book. We also thank Ashish Kumar and Sandra Sickels for their confidence, support, and encouragement during the preparation and publication of this book.

—***Editors***

Part I

Foodborne Pathogens, Diseases, and Outbreaks

CHAPTER 1

Foodborne Pathogens and Their Associated Infections: An Introduction

MUHSIN JAMAL, SAYED MUHAMMAD ATA ULLAH SHAH BUKHARI, SAADIA ANDLEEB, MUHAMMAD ASIF NAWAZ, LILOMA SHAH, SANA RAZA, and REDAINA

ABSTRACT

According to World Health Organization (WHO), foodborne illness (FBI) is the diseased condition caused by ingesting contaminated food and water. FBI is characterized into two broad groups: intoxication and infection. Foodborne intoxication is caused when the preformed toxin is consumed, which is made by pathogens within the food, while the foodborne infection is triggered through the uptake of food which contains the viable microbes. There are several causes of FBIs (i.e., bacteria, viruses, parasites). Several critical foodborne pathogens are: *Clostridium* spp., *Salmonella*, *E. coli*, *Shigella*, *campylobacter*, *Bacillus*, and *Staphylococcus* food poisoning are the major food illness caused by bacteria. Several foodborne viruses (i.e., hepatitis A virus (HAV), Norovirus) and parasites (i.e., *Trichinella spiralis*, *Toxoplasma gondii,* and *Cyclospora cayetanensis*) are responsible for FBI. The common symptoms of FBI are nausea, abdominal cramps, diarrhea, headache, and vomiting. The prevention of FBI depends on the safety and control measurement of food during production and processing. The ingredients used in the food must be of excellent microbiological quality, so the foods have to be safe from any hazardous pathogen which is the major cause of FBI. This suggests the need to implement strict hygienic control

Global Food Safety: Microbial Interventions and Molecular Advancements. Saher Islam, PhD, Devarajan Thangadurai, PhD, Jeyabalan Sangeetha, PhD, & Natália Cruz-Martins, PhD (Editors)

measures during handling, storage, manufacturing, and commercialization of foods. This chapter determines the characteristics of the most significant foodborne microbes which are involved in food spoilage and FBIs.

1.1 INTRODUCTION

It was early recognized by Hippocrates (460 BC) that there exists relation among the consumption of food and human diseases (Hutt, 1984). The chief reasons of foodborne disease are parasites, virus, and bacteria. The two main kinds of foodborne illnesses (FBIs) are infection and intoxication. Intoxication happens when food poisoning is caused by the toxins which are produced via the microbes, while infection occurs by ingesting food that contains viable microbes (Adams and Moss, 2008).

Foodborne microbes are accountable for food intoxication (ingestion of preformed toxin) and toxico-infection. Numerous bacterial species, i.e., Gram-positive, mainly *Clostridium perfringens*, *Staphylococcus aureus*, *Bacillus cereus*, *Clostridium botulinum,* and *Listeria monocytogenes* are thought to be major foodborne pathogens and are accountable for foodborne illnesses outbreaks worldwide. Apart from these Gram-positive bacteria, numerous Gram-negative bacterial species can also cause foodborne diseases, including *Aeromonas* spp., *Campylobacter* spp., *Vibrio* spp., *Salmonella* spp., *Shigella* spp., and *Escherichia coli.* About 100 types of enteric viruses cause foodborne illnesses; most common foodborne viruses are Noroviruses and Hepatitis A. Several other parasites can also cause foodborne illnesses. The common types of foodborne parasites are *Cyclospora cayetanensis*, *Toxoplasma gondii*, and *Trichinella spiralis*. Infections caused by protozoan are most commonly related to fresh produced foods like vegetables, fruits, and water. However, the countries where food processing and production are not achieved under hygienic conditions, where they import larger proportion of such commodities which might contributes for the high incidences of FBI (Bhunia, 2018).

Food is used by several microbes as a nutrient source for their growth. Microorganisms grow within the food and produce toxic metabolites, so not just the food is made unsafe to eat but also consumption of such food caused health problems. Numerous foods favor the growth of disease-causing microbes or serves as transmission vector of microbes. Food contamination might occur from water, soil, air, animals, plant surfaces, sewage or from handlers of food during the processing and handling (Bean and Griffins, 1990).

Foodborne diseases are a global problem, so a joint and united strategy by the entire nations and related international organizations would be necessary for the identification and control of the overall food-borne problems that threaten human health. The public health agencies, food industry, and regulatory agencies need to make constant efforts so that to prevent food contamination during processing in the farm, restaurants, and homes. With proper education systems associated with the safety of food for overall involved people can reduce the prevalence of foodborne illnesses. This chapter determines the characteristics of the most significant foodborne microbes mostly involved in food spoilage and FBIs (Figure 1.1).

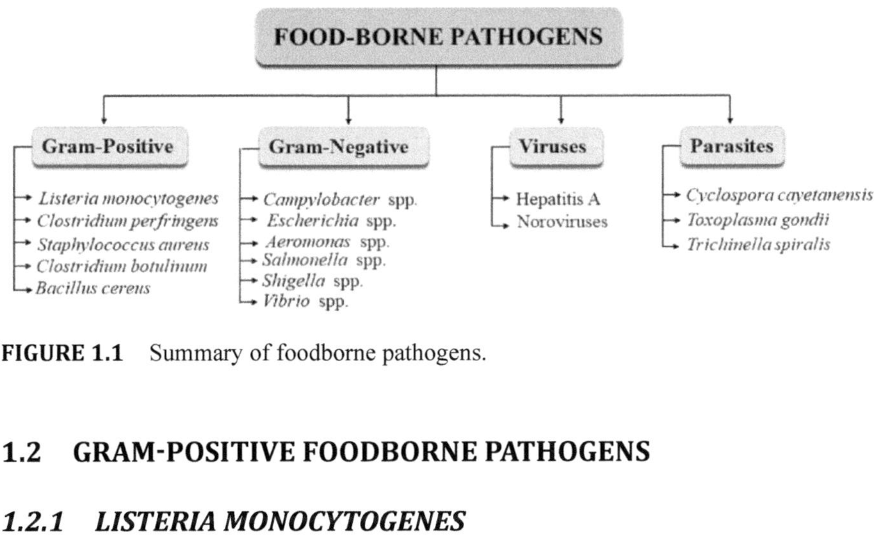

FIGURE 1.1 Summary of foodborne pathogens.

1.2 GRAM-POSITIVE FOODBORNE PATHOGENS

1.2.1 LISTERIA MONOCYTOGENES

Listeria spp. are facultative anaerobic, non-spore formers, oxidase-negative, catalase-positive, smaller Gram-positive rod-shaped (0.5–4 μm in diameter and 0.5–2 μm in length) organisms. Tumbling motility is shown by the *Listeria* on 20–25°C because of its peritrichous flagellum. On basis of flagellar (H) and somatic (O) antigens, 13 serotypes are being recognized in *Listeria monocytogenes* (*L. monocytogenes*) comprising "1/2a, 1/2b, 1/2c, 3a, 3b, 3c, 4a, 4ab, 4b, 4c, 4d, 4e, and 7" (Meloni, 2014).

The optimal growth temperature for *L. monocytogenes* is 30–37°C, but it could persist at temperature varying from 0 to 45°C. At freezing temperature *L. monocytogenes* could multiply, show resistance to disinfectants, and attaches to numerous surfaces (Meloni, 2014). Once it is subjected to the processing plants, it gets adapted to continue and stay for a longer period underneath unfavorable situations (Meloni, 2014). Within the food industry, biofilm can be formed by the *L. monocytogenes* which could be a contamination source (Colagiorgi et al., 2017). In nature, *L. monocytogenes* is found everywhere and the main reservoir is soil. Furthermore, from animals, healthy humans, infected wild or domestic animals its isolation is possible (Liu et al., 2007). It is also found in different types of uncooked foods such as raw meats and vegetables. It can also be found in foods which are contaminated after processing or cooking.

When a food contaminated with *L. monocytogenes* is eaten, so listeriosis is caused, which is a serious infection. Even though, its comparatively rare illness having higher mortality rate (20–30%) that makes it one of the lethal foodborne infections (Jemmi and Stephen, 2006). Unlike numerous other foodborne microbes, *L. monocytogenes* survives within colder environments like refrigerators (Ghandhi and Chikindas, 2007). Once, *L. monocytogenes* gets into a factory where food is processed, so it could survive over there for many years and occasionally contaminate different food products (Jemmi and Stephen, 2006). *L. monocytogenes* gains entry to the host cells by using its surface proteins called internalins. These proteins are particularly inlB and InlA. Furthermore, InlJ, and InlC also show involvement within the post-intestinal phases of *L. monocytogenes* infection (Liu et al., 2007). Three chief clinical features are shown by the foodborne listeriosis; specifically, abortion, meningitis, and septicemia. It can cause febrile gastroenteritis in healthy humans, but in susceptible groups (like elderly, children, pregnant women, and immune-compromised persons) it might lead to meningitis and septicemia. The incubation period for gastrointestinal symptoms is about 9–48 hours while established invasive disease in 2–6 weeks (Meloni, 2014). Within Asian countries, there exist rare reports of listeriosis because of detection failure. Though, *L. monocytogenes* is considered as one of etiologic reasons of spontaneous abortions or stillbirth within India (WHO, 2013). Higher infection rates with *L. monocytogenes* have been shown by the individuals more than 65 years or neonates (Denny and McLauchlin, 2008). *L. monocytogenes* isolation from various types of ready to eat foods makes it a notable foodborne microbe (Osaili et al., 2011). One of the key vehicles for transmission of *L. monocytogenes* is poultry flocks that could transmit the organism into the poultry carcass and environment because of unhygienic practices (Dhama et al., 2013). *Listeria* is rarely isolated from the feces of chicken and poultry. In different poultry products, the *Listeria* spp. has been detected. Uncooked poultry meat products normally have much contamination of *L. monocytogenes* as compared to cooked ones (Jamali and Thong, 2014).

1.2.2 BACILLUS CEREUS

B. cereus is aerobic, motile, facultative, spore-forming, Gram-positive, and rod-shaped bacterium (Blackburn and McClure, 2005). There are several species of *Bacillus* which are closely related and shares genetic resemblance to *B. cereus* (Blackburn and McClure, 2005). Group of *B. cereus* includes the furthermost significant hazardous species such as *B. weihenstephanensis*, *B. thuringiensis*, *B. mycoides*, *B. cereus*, *B. pseudomycoides* and *B. anthracis* (Lechner et al., 1998; Nakamura, 1998). Soil is usual locality for many of species, therefore contaminate agriculture product directly and cause foodborne intoxication, infection, and food spoilage. The most significant species of *Bacillus* relating to food is *B. cereus* (Nakamura, 1998). It is also present within numerous additives and spices, and due to such reason, the meat contamination with *B. cereus* rises with each supplementary phase within the processing of raw meat (Volkova, 1971). An important role is played by the meat additives in enhancing the *B. cereus* load in final products (Cantoni and Bresciani, 1987).

B. cereus not only undergoes survival at room temperature but in heat-treated foods it also undergo survival. In a study, *B. cereus* was isolated from 28% of the meat products (heat-treated) and such a food samples, 48% were positive for *B. cereus* (Schlegelova et al., 2003). This is due to the reason that such bacteria show resistance to heat stress. Bolstad has stated a food poisoning case including two people within Norway, who had taken prepared grilled chicken, kept at room temperature within the local shop of food and such food showed the occurrence of *B. cereus* (10,000/g) (Bolstad, 1990). *B. cereus* is regarded as a common raw milk contaminant (Ahmed et al., 1983). Contamination related to *B. cereus* resulted from raw milk within which bacterium is existing in spores and capable to show resistant at pasteurization temperature.

B. cereus causes both emetic and diarrheal type diseases which are self-limiting (24–48 h). Apart from the foodborne diseases, the non-gastrointestinal diseases like endophthalmitis and endocarditis are also caused by *B. cereus* (Logan and Rodrigez-Diaz, 2006). Diarrheal condition mainly results when heat-labile enterotoxins are produced in the host small intestine during growth of vegetative cells (10^4–10^9) is the infective dosage per gram of food (Logan and Rodrigez-Diaz, 2006). Such diarrheal condition is milder and diarrhea and abdominal cramps are common in this condition. It having incubation period of 8–16 hours which lasts for 6–12 hours (Murray et al., 2007). Diarrhea might be watery, profuse, or mild. This kind is mentioned as "long-incubation or diarrheal form" of the illness and it shows resemblance to food poisoning of *Clostridium perfringens* (Drobniewski, 1993).

While the emetic condition is much severe and acute as compared to the diarrheal syndrome characterized for short incubation. Symptoms of emetic syndrome include abdominal cramps, nausea, and vomiting. The toxin accountable for such emetic syndrome is a smaller cyclic heat-stable peptide that results in vomiting afterwards the 1–6 hours (average 2–5 hour) of its ingestion (Mortimer and McCann, 1974). Toxin is already present in ingested food. In emetic disease, dosage is around (10^5–10^8 cells) per gram so that to make enough toxin (Logan and Rodrigez-Diaz, 2006). It shows resemblance to *S. aureus* food poisoning in its incubation period and symptoms. The amount of microbes required for producing such syndrome appears to be high in number as compared to diarrheal syndrome. Both syndromes happen because of the reason that spores of *B. cereus* could survive ordinary conditions of cooking. Underneath inappropriate conditions of storage after cooking, the spores follow germination and vegetative cells are multiplied (Logan and Rodrigez-Diaz, 2006). In order to keep a number of *B. cereus* to be lowest the temperature of storage is the key factor. Apart from this, food poisoning usually happens because of food handling practice and/or poor hygiene. Therefore, it's significant to teach food handlers regarding their duties in maintaining food safety and adequate training is also required.

1.2.3 *CLOSTRIDIUM BOTULINUM*

Clostridium botulinum (and several isolates of *C. argentinense*, and rare isolates of *C. baratii* and *C. butyricum*) are a various classes of rod-shaped, anaerobic, spore-forming, Gram-positive bacteria have the ability to produce a potent neuro-toxin, i.e., botulinum toxin. Neurotoxin producing *Clostridium* spp. produces the botulinum toxin and it causes 4 types of botulism within human beings: wound, foodborne, infant, and adult colonization, which have been reported within the US to the Centers for Disease Control and Prevention (CDCP). There are 7 serotypes exist for botulinum toxin, also mentioned as toxin types, and designated as; "A, B, C (occasionally referred to as C1), D, E, F, and G" (Hatheway, 1995). The type C and D strains of *C. botulinum* might also yield C2 and C3 toxins, but such toxins are not neuro-toxins having no involvement in botulism (Koepke et al., 2008).

C. botulinum exists within marine sediments, freshwater, soils, and animals intestinal tracts. *C. botulinum* has an incubation period of nearby 12–72 hours and a large number of foods for instance green beans, asparagus, spinach, canned corn, mushrooms, peppers, luncheon meats, soups, beets, chicken livers, ripe olives, tuna fish, stuffed eggplant, chicken, liver pate, ham, sausage, lobster, salted, and smoked fish are linked with botulinum toxin. Foodborne botulism might also happen because of ingestion of a food product deliberately contaminated with botulinum toxin (FDA, 2012).

The clinical symptoms of foodborne botulism are symmetric descending paralysis, without any fever, and unchanged mental status. Initial neurological symptoms comprise diplopia, blurred vision, and dry mouth. Such symptoms are followed by dysphagia, dysphonia, peripheral muscle weakness, dysarthria, and finally respiratory failure; development of symptoms, including progression rate, is reliant on the degree of toxin ingested. Ventilator support for 2 weeks-7 months is required for affected patients (CDCP, 1998).

1.2.4 *CLOSTRIDIUM PERFRINGENS*

Clostridium perfringens, which was formerly recognized as *Clostridium welchii*, fit in the family Bacillaceae and is one of the important reasons for foodborne illness. These bacteria are encapsulated, rod-shaped, and non-motile (NM) cells. They produce protein toxins and spores which show resistance to certain environmental stresses like heat, desiccation, and radiation (Bacon and Sofos, 2003). Amongst the species of *Clostridium*, *C. perfringens* produces the largest amount of toxin, and moreover it is the most common humans and animals microbiota. It is also found within the soil. Such bacterium is categorized into 5 types on the basis of production of 4 main toxins, i.e., beta (β), alpha (α), iota (ι) and epsilon (ε). Besides these chief toxins, more than 15 other toxins are produced by it. Several of these virulence-factors like beta-2 toxin, enterotoxin, and necrotic enteritis B-like toxin (NetB) have gained much consideration as compared to others due to the reason that they show part in the pathogenesis of "*C. perfringens*-associated disease (CPAD)" within animals and humans (Monma et al., 2015). Food poisoning because of *C. perfringens* occurs when the inappropriately stored and cooked food is eaten. Generally, bacteria are present on food surface afterward cooking, and such bacteria could start multiplication and trigger *C. perfringens* food poisoning. Commonly infected foodstuffs comprise gravy, meat, and meat products (Havelaar et al., 2015). *C. perfringens* food poisoning symptoms comprise watery diarrhea and extreme abdominal cramps. Following the food ingestion comprising a high amount of *C. perfringens* results in the frequent appearing of symptoms after 8–16 hours (Eriksen et al., 2010).

Food poisoning could be triggered via *C. perfringens* enterotoxin (CPE) formed via the *C. perfringens* spores within the small intestine, that could germinate within foodstuffs like poultry and meat. Within US consuming a larger number of *C. perfringens* is thought out to be a significant reason for watery diarrhea. Chief symptoms of such an illness include diarrhea, nausea, and abdominal pain. Illness is frequently self-limiting and mild within healthy individuals and the symptoms resolve in 24–48 hours (Miki et al., 2008). Clostridial myonecrosis is commonly caused by *C. perfringens*. In this disease, the muscle tissue is breakdown because of the action of certain powerful exotoxins, i.e., alpha and beta; which are formed via the bacterium. It involves edema, pallor, intense pain, and tenderness followed through hemorrhagic bullae, discoloration, and at wound site gas is produced. Hypotension, renal failure, shock, and bacteremia along with intravascular hemolysis, which finally leads to coma and death are systemic manifestations of this disease (Jay, 2000).

C. perfringens type C causes the Enteritis necroticans which is a life-threatening infection and it involves the ischemic necrosis of the jejunum. It involves the hemorrhagic, ischemic, or inflammatory necrosis of the jejunum (Acheson, 2011). The growth of this organism can be prevented by quickly chilling ready to eat foods within shallower containers and when cold foodstuff is kept cold and hot foodstuff as hot. When all working area is kept clean and proper sanitary conditions are followed so level of contamination could be reduced (Quinn et al., 2001).

1.2.5 *STAPHYLOCOCCUS AUREUS*

Staphylococcus aureus, a NM cocci, Gram-positive appears as single or in pairs form or short chains, tetrads, or typical grape-like clusters. *Staphylococci* are facultative anaerobes but with few exceptions, such as *S. aureus* subsp. *anaerobius* and *S. saccharolyticus*. These strains multiply in faster rate in aerobic conditions (Bacon and Sofos, 2003).

S. aureus is an opportunistic microbe which could cause a variety of diseases like superficial skin infections to intense, lethal, and invasive illness (Lowy, 1998). It is a major microbe for both community-acquired and nosocomial infections. Spores are not formed but it could cause contamination of foodstuffs throughout the preparation and processing stage. *S. aureus* could undergo multiplication within a varied temperatures range, i.e., "7°C to 48.5°C; optimum 30 to 37°C," pH (4.2 to 9.3; optimum 7 to 7.5), and in the presence about 15% sodium chloride. *S. aureus* is tolerant to drying and has capability to subsist possibly within stressful and dry surroundings. It can survive on skin, nose of human and inanimate surfaces like clothing (Chaibenjawong and Foster, 2011). These properties support its growth within numerous food products (Kusumaningrum et al., 2002).

Several toxins are produced by *S. aureus*. Staphylococcal enterotoxins (SEs) belongs to a family of 9 main serological kinds of heat-stable enterotoxins (such as SEA, SEB, SEC, SED, SEE, SEG, SEH, SEI, and SEJ) that fit in larger family of pyrogenic toxin superantigens (Balaban and Rasooly, 2000) and can activate non-specific T-cell proliferation and immunosuppression (Le Loir et al., 2003).

Foodborne disease produced by *S. aureus* has a fast onset following consumption of contaminated foods (frequently 3 to 5 hours). This occurs when one or more than one toxin is produced via the bacterium throughout its multiplication at suitable temperatures (Le Loir et al., 2003). Though, *Staphylococcus* foodborne disease (SFD) incubation period is dependent on quantity of toxin consumed or ingested (Murray, 2005). Very small dosage is enough to trigger SFD. For instance, it was indicated in one report that about 0.5 ng/mL levels of SEs contaminated with chocolate milk triggered a large outbreak (Murray, 2005).

Onset of the SFD is rapid. Symptoms comprise cramping of the abdomen with or without diarrhea, hypersalivation, vomiting, and nausea. If there is loss of fluid, then the physical examination might show signs of hypotension and dehydration (Balaban and Rasooly, 2000). Vomiting, nausea, and abdominal cramps are most common symptoms. Though SFD is usually self-limiting and resolves in 24–48 hours of onset, it could be serious predominantly in immunocompromised patients who are infants and elderly. The incubation period for *S. aureus* is 1–6 hours. There is no use of antibiotics for therapy (Scallan et al., 2006).

For *S. aureus* certain foodstuffs serve as optimal growth medium. Foodstuffs involved in SFD are poultry, milk, egg products, dairy products, meat, meat products, salads, bakery products such as cream-filled cakes, sandwich fillings, and pastries (Tamarapu et al., 2001). Foods involved in SFD differ from one state to another, mainly because of changes in food eating manners (Le Loir et al., 2003). About 13,000 cases of SFD happened within Japan in 2000 because of contamination of milk at a dairy-food production plant (Asao et al., 2003). It was investigated during the outbreak that inappropriate handling of processed or cooked food was the chief cause of contamination. Though *S. aureus* could be eradicated via heat treatment and through competition with other flora within fermented and pasteurized foodstuffs but still SEs formed through *S. aureus* can trigger SFD due to their heat tolerance capability.

1.3 GRAM-NEGATIVE FOODBORNE PATHOGENS

1.3.1 AEROMONAS SPP.

Aeromonas are glucose-fermenting, oxidase-positive, Gram-negative, facultative anaerobic rods, and most of them are motile having polar flagellum. The word *Aeromonas* was originated from the Greek words "*aer*, meaning air or gas" and "*monas*, meaning unit or monad." The size of cell is particularly from 1.0 to 4.4 mm. Aeromonads have ability to ferment maltose, fructose, glucose, and trehalose to acid and gas or only to acid. Glycerol dextrin and starch are hydrolyzed by it (Khardori and Fainstein, 1988).

The role played by aeromonads and the toxins which they yield and their association to virulence in human gastroenteritis is not completely clear. Mostly *Aeromonas* causes gastroenteritis in immunocompromised, children, and elderly people. Several putative virulence factors are recognized; comprising capsules, siderophores, hemolysins, adhesins, proteases, invasins, pili, endotoxin, lipopolysaccharide (LPS), fimbriae, S-layers, and other different types of extracellular enzymes (Janda and Abbott, 1999). Numerous hemolysins might be produced, but among them, aerolysin is the well-categorized (Bernheimer and Avigad, 1974).

Clinical signs of infection related to *Aeromonas* frequently rely on severity and the site of infection (Taher et al., 2000). Wound infections often outcome in necrotizing fasciitis and cellulitis. Septicemia might occur along with wound infection, or there might be secondary systemic illnesses for instance, biliary disease, diabetes, cirrhosis, cancer or illnesses resultant in gastrointestinal perforations, endocarditis, or meningitis (Mukhopadhyay et al., 2008). Pneumonia is uncommon and it is frequently linked with aspiration (Qu et al., 2003). Gastroenteritis symptoms differ from minor self-limiting to dysentery or cholera-like disease (Bravo et al., 2003).

Aeromonas species also trigger meningitis, gastroenteritis, peritonitis, cellulitis, pneumonia, and disseminated infections within immunocompromised individuals (von Graevenitz, 2007). *A. hydrophila* is responsible for diarrhea within children and travelers in the developing states (Jiang et al., 2002). Recognized risk factors which expose humans to the illness comprise intake of contaminated water or swimming in the contaminated water and likewise contaminated food consumption. *Aeromonas* species has an incubation period of 1–48 hours and the illness duration is about 24–48 hrs (Jiang et al., 2002).

Aeromonas species are documented as enteric microbes and the mechanism of their pathogenicity in unclear. *A. caviae* attaches to mucosal epithelial cells (Figueras et al., 2005), and majority of aeromonads linked to gastroenteritis comprise such as *A. trota* (HG-13); *A. veronii*, *A. hydrophila* (HG-1), biovar *sobria* (HG-8/10), *A. jandaei* (HG-9), *A.*

caviae (HG-4) and *A. veronii* biovar *veronii* (HG-8/10) (Bravo et al., 2003). Gastroenteritis caused by *A. sobria* is categorized by vomiting, abdominal pain, acute watery, fever, and diarrhea (Taher et al., 2000). Goldsweig and Pacheco (2001) stated that infectious colitis is triggered through *Aeromonas* species. Species of *Aeromonas* are described as a reason for diarrhea within 2% of travelers to Asia, Latin America, and Africa (Yamada et al., 1997). *Aeromonas* is not remarkably resistant to traditional techniques of food processing but is quite common in seafood, meats, and fresh products.

1.3.2 CAMPYLOBACTER SPP.

Campylobacter are small spiral-shaped or curved, Gram-negative bacilli which show corkscrew motility by means of polar flagellum. *Campylobacter* multiply optimally amongst 37–42°C. Due to phenotypic, metabolic, and high genetic diversity in their population *Campylobacter* spp. could be present within varied environmental circumstances (Labbé and García, 2013). Though numerous *Campylobacter* species, such as *Campylobacter jejuni*, *C. hyointestinalis*, *C. concisus*, *C. sputorum*, *C. lari*, *C. coli*, *C. upsaliensis*, *C. rectus*, *C. insulaenigrae*, *C. helveticus*, *C. fetus*, *C. ureolyticus* and *C. mucosalis* are stated to trigger gastroenteritis (Kaakoush et al., 2015). The *C. jejuni* species are commonly isolated from retail poultry and man and *C. coli* was the 2nd most frequently isolated species. Though, the ratio of *C. coli* to *C. jejuni* was significantly changed in diverse states like South Africa and Thailand, where *C. coli* isolation was dominantly from retail poultry (Suzuki and Yamamoto, 2009).

Because of lack of pathogenic resemblance among *Campylobacter* and other microbes, specific mechanisms of virulence are not obviously explained for *Campylobacter* spp. (Dastia et al., 2010). Adherence of bacteria to mucosa of intestine, flagella-mediated motility, invasive ability, and capability to yield toxins are recognized as virulence factors (Dastia et al., 2010). Cyto-lethal distending toxin (CDT) is extensively produced by Gram-negative bacterial strains (Ge et al., 2008). It is defined as a vital virulence factor for *Campylobacter* spp. (Asakura et al., 2008).

The disease produced by these bacteria is termed as campylobacteriosis and symptoms of this disease include abdominal pain, acute onset of diarrhea (i.e., bloody) and fever is mostly self-limiting. *Campylobacter* has an incubation period of 2–5 days and the illness duration is about 2–10 days (Kaakoush et al., 2015). Though, a variety of further conditions in the gastrointestinal tract has been described such as functional gastrointestinal disorders, cholecystitis, intestinal bloody diarrhea, esophageal diseases, celiac disease, colon cancer and periodontitis. Guillaine Barre Syndrome is severe demyelinating neuropathy and is developed by about 3 out of 10,000 cases of campylobacteriosis (Skarp et al., 2016).

The issue is becoming the worst because the amount of campylobacteriosis cases has intensely increased within Australia, Europe, and North America. Similarly, reports from numerous Asian, Middle East, and African countries show that the illness is endemic, particularly within children (Kaakoush et al., 2015). Each year globally, about 400–500 cases of infection occur due to *Campylobacter* (Labbé and García, 2013), and along with *Salmonella*, it is the frequently isolated foodborne microbe (Kaakoush et al., 2015). It is stated that poultry products, unpasteurized milk and water are the chief vehicles for *C. coli* and *C. jejuni* infections (Butzler, 2004). Poultry can transmit the food associated with *Campylobacter* species to human beings (Kaakoush et al., 2015). Consumption, handling, and preparation of broiler meat might be responsible for 20–30% cases of human campylobacteriosis, whereas 50–80% might be accredited to the chicken reservoir (Kaakoush et al., 2015).

1.3.3 SHIGELLA SPP.

The genus *Shigella* belongs to the family Enterobacteriaceae and has 4 serogroups (serogroup B, i.e., *Shigella flexneri*, serogroup A, i.e., *Shigella dysenteriae*, serogroup D, i.e., *Shigella sonnei* and serogroup C, i.e., *Shigella boydii*). About 38 serotypes are found in the serogroups A, B, and C and just 1 in serogroup D (Bacon and Sofos, 2003). *Shigella* are non-spore-formers, Gram-negative rods, facultative anaerobes and NM. These bacteria can multiply at temperatures ranging from 6–8°C, but preferably at 37°C. *S. sonnei* seems to be tolerating at lower temperature as compared to the other serogroups. Optimum pH range for their growth is 6.0 and 8.0, although their growth has also been reported at pH 4.8 and 9.3 (ICMSF, 1996).

Shigella adheres to and penetrates in the cell walls of the small intestine by producing toxins, which might promote the diarrhea. Bacteria are capable of penetrating the epithelial lining of intestines by Shiga toxin, which collapse epithelial lining leading to hemorrhage. *Shigella* moreover possess adhesions which promotes its adherence to epithelial cell surfaces and also carry invasion plasmid antigens which enable bacteria to enter into the target cells, hence increasing its virulence (http://www.cdc.gov/).

Shigella spp. are present most commonly within environment of poor hygiene, and though the chief route of transmission is via person-person contact. Shigellosis could happen afterward the intake of contaminated water or food (Bacon and Sofos, 2003). *Shigella* spp. are not linked with exact one kind of food poisoning. Various foods linked with shigellosis outbreaks include shellfish, milk, chicken, salads, and other fresh foods (ICMSF, 1996). Around 20% of all shigellosis cases in the United States are linked with international travel (i.e., traveler's diarrhea), S. *sonnei* is most predominant, and *S. flexneri* is the 2nd most common in developed nations (Vargas et al., 1999). Though, within developing states, S. *dysenteriae* type 1 and *S. flexneri* are the commonest serogroups, and *S. dysenteviae* type-1 have been involved in a prolonged epidemic within southern Africa, Central America and in Asia (NCBI, 2017). High mortality and morbidity rates were resulted because of these epidemics, particularly in elderly, immunocompromised individuals, and malnourished children (NCBI, 2017).

Incubation period of *Shigella* is about 24–48 hours and duration of illness is about 4–7 days. Gastrointestinal infections could be caused by the entire serogroups of *Shigella* followed by watery diarrhea in combination with fatigue, fever, abdominal cramps, and malaise is experienced by the affected persons. Although all four serogroups of *Shigella* could cause dysentery but *S. dysenteriae* type 1 most frequently trigger epidemic dysentery, which is associated with a particularly severe form of the disease that might be accompanied by further problems such as HUS (Vargas et al., 1999). *Shigella* infections could be prevented when food is properly cooked and handled. No vaccine is available for preventing shigellosis, but there is recent research going on which seems to be hopeful. The best method of prevention is regular and proper handwashing with warm water (Mead et al., 1999).

1.3.4 *SALMONELLA SPP.*

Salmonella is rod-shaped, Gram-negative, facultative anaerobe and flagellated bacterium. Its size is approximately 2–3×0.4–0.6 μm (Montville and Matthews, 2008). *Salmonella* is non-fastidious and could multiply in numerous environmental settings. Sodium chloride is not needed for their growth. Most serotypes of *Salmonella* multiply on temperature range of 5–47°C but 35 to 37°C is the optimal temperature for growth. Several could multiply on temperature as high as 54°C or as low as 2–4°C (Montville and Matthews, 2008). They are sensitive to heat, and frequently a temperature of 70°C or greater is sufficient to kill them. *Salmonella* multiplication could happen at a pH range of 4–9 but 6.5–7.5 is the optimal pH for their growth. A high-water activity (aw) between 0.99 and 0.94 (pure water aw=1.0) is required for their sufficient growth.

The genus *Salmonella* is categorized into two species that could trigger disease in humans: *S. bongori* and *S. enterica* (FDA, 2012). For instance, "*Salmonella enterica* subsp. *enterica*" is additionally classified into several serotypes for instance *S. enteritidis* and *S. typhimurium* (FDA, 2012). Several *Salmonella enterica* serovars are accountable for many serious illnesses such as typhoid fever. Several virulence factors enable *Salmonella* spp. to infect and colonize the host and such virulence factors are encoded by numerous pathogenicity islands (PAIs). There are two PAIs, i.e., *Salmonella* pathogenicity island 1 and 2 (SPI-1 and SPI-2), which encodes for 2 diverse type III secretion systems for delivering effector molecules into the cell of the host which results in internalization of bacteria that afterward progresses to systemic spread.

Salmonella spp. can survive within the intestines of numerous wild animals and most of the livestock. The food which is contaminated by the animal or human feces that carries bacteria and when such a contaminated food is eaten by the person leading to *Salmonella* spp. infection. Outbreaks of *Salmonella* are generally linked with poultry, eggs, and meat, but other foods like vegetables and fruits are also contaminated by these bacteria (CDCP, 2012).

Clinical pattern of salmonellosis in humans could be categorized into three disease patterns such as bacteremia, enteric fever and gastroenteritis. *S. typhi* causes typhoid fever, whereas *S. paratyphi* A, B, and C causes paratyphoid fever with lower mortality rate and mild symptoms. Both serotypes are merely pathogens of humans. When the water or food is contaminated with human waste and is ingested that usually results in infections (Scherer and Miller, 2001).

Enterocolitis or non-typhoidal salmonellosis is triggered by approximately 150 serotypes of *Salmonella*. *S. enteritidis* and *S. typhimurium* are the commonest serotypes within the US. When food which is contaminated with animal waste rather than human waste and is ingested so infection happens. The emergence of *S. typhimurium* DT104 strain (multidrug-resistant) is linked with outbreaks associated with beef contamination resulted in double hospitalization rates as compared to other foodborne salmonellosis (Yousef and Carlstrom, 2003).

Bacteremia is a state in which bacterial specie gains entry into bloodstream afterwards crossing the barrier of the intestine. The incubation period of *Salmonella* is about 1–3 days and the duration of illness is about 4–7 days. It is the commonest foodborne microbe linked with vegetables and fresh fruits. Annually there are about 16 million cases of typhoid fever and 1.3 billion of gastroenteritis. About 3 million deaths occur globally because of *Salmonella* (Bhunia, 2018). The information regarding *Salmonella* and its evolution are very significant so that to guarantee the quality and safety of food. Intervention approaches are therefore needed for controlling *Salmonella* from farm to fork.

1.3.5 ESCHERICHIA SPP.

The genus *Escherichia* includes facultative anaerobic bacilli, Gram-negative, common resident of the mammal's gastrointestinal tract and belongs to the family Enterobacteriaceae. They are non-fastidious; lactose fermenters and their growth occur under the mesophilic temperatures. The β-glucuronidase enzyme is found in most of *E. coli* which have ability to breakdown the complex carbohydrates (Kaper, 2005).

Severe illness and mortality are caused by the Enterohemorrhagic *Escherichia coli* (EHEC). Different enteropathogenic groups of *E. coli* are reasons for numerous kinds of gastrointestinal infections. Six key *E. coli* types could be distinguished; diffusely adhering *E. coli* (DAEC), enteropathogenic *E. coli* (EPEC), enterotoxigenic *E. coli* (ETEC), enteroaggregative *E. coli* (EAEC), Enterohemorrhagic *Escherichia coli* (EHEC), and enteroinvasive *E. coli* (EIEC). All pathotypes of *E. coli* use multistep systems of pathogenesis, involved in general mucosal site colonization, host defenses evasion, multiplication, and damage to the host (Kaper, 2005). After the first outbreak in 1982, *E. coli* O157:H7 turn out to be the utmost commonly known strain of EHEC. It was shown by the reports that *E. coli* O157:H7 strain was involved in hemorrhagic colitis (HC) outbreak within the US which produced Shiga-toxins. The Shiga toxins of EHEC are categorized in two different sub-groups (Stx1 and Stx2). Pathogens produce such toxins in the colon and cause local damage. These toxins also have the capability to pass through the bloodstream to the kidney triggering hemolytic uremic (HU) syndrome (HUS) and HC. Two diverse kinds of Shiga toxins could be produced by *E. coli* O157. Stx1 shows similarity to type-1 toxin of *S. dysenteriae* whereas Stx2 is genetically and immunologically distinct having 55–60% resemblance in genetic sequences and amino acid. The expression and possession of the "Stx2 gene" and the variant Stx2c correlate strongly with the causation of HU syndrome and bloody diarrhea (Persson et al., 2007).

E. coli is transmitted when water or food is contaminated with humans or animals feces and such foods are ingested. Product contamination occurs during the slaughtering and processing of animals. *E. coli* could continue for the longer time period and might multiply within several foods and vegetables. *E. coli* is found in unpasteurized juices and milk, undercooked beef, contaminated water and raw fruits and vegetables. *E. coli* has an incubation period of about 1–8 days while illness duration is about 5–10 days. Symptoms of the illness include abdominal pain, severe diarrhea (often bloody), vomiting, and slight or no fever (Garcia et al., 2010).

The most common source of *E. coli* O157:H7 outbreaks is ground beef and accounts for about 75% of outbreaks (Vugia et al., 2006). Under-cooked crushed beef and dairy products could be contaminated directly through the feces of cattle during either slaughtering processes or milking processes. Consequences from a study of Snedeker et al. (2009) showed that the source of transmission was food in almost 42.2% of the outbreaks, animal contact in 7.8%, environmental in 2.2%, dairy products in 12.2%, and water in 6.7% (Snedeker et al., 2009). Frequently used approaches for controlling bacteria such as boiling could not eradicate the Shiga toxin produced by the *E. coli*. Preventive measures are the best alternative to control transmission of these microbes. Furthermore, there is need of hygienic practices to be applied by the food handlers.

1.3.6 *VIBRIO SPP.*

The genus *Vibrio* belongs to the Vibrionaceae family and includes more than 35 species (Bacon and Sofos, 2003). In this genus, the organisms are mainly motile having a single polar-sheathed flagellum. These are non-spore formers, facultative anaerobic and Gram-negative straight or curved rods. All pathogenic species of *Vibrio* such as *Vibrio vulnificus*, *Vibrio parahaemolyticus* and *Vibrio cholerae* need sodium for their optimum growth. They are present mainly in marine or brackish habitats located in temperate or tropical areas and their occurrence decline as the temperature of water drops below 20°C (Bacon and Sofos, 2003). *V. cholerae* is also isolated from those areas that are not related with a brackish or marine water supply including rivers, freshwater lakes, herbivores, and birds (Bacon and Sofos, 2003).

V. cholerae requires an optimal temperature between 30–37°C for its growth, however it can also grow at temperature between 10–43°C. On a pH range of 5.0–9.6, *V. cholerae* multiplication occurs but the optimum pH is 7.6. The optimum growth of *V. cholerae* happens when environment having 0.5% concentration of sodium chloride (ICMSF, 1996). *V. cholerae* characteristically gain entrance into the humans by ingesting contaminated food, for example eating raw crustaceans or mollusks (raw oysters), eating undercooked food, or food which is contaminated after cooking, or open wound exposure to a contaminated source of water.

The illness having duration from few hours-5 days, reliant on size of inoculum and the quantity of food is ingested. Disease symptoms comprise muscle cramps because of severe dehydration, vomiting, enhanced peristalsis, which is followed by loose stools that progresses to watery stools, and mucus flecked diarrhea (distinguishing feature of cholera) (Wu et al., 2014). In addition to dehydration, other problems might include metabolic acidosis, hypoglycemia, and hypovolemic shock (Wu et al., 2014). Food sources which act as a transmission vehicle for *Vibrio parahaemolyticus* include oysters, clams, scallops, crabs, prawns, and seaweed (Bacon and Sofos, 2003). The growth of *V. parahaemolyticus* occurs at temperatures range between 5–44°C, but their optimal growth temperature is approximately 30–37°C. The optimum pH required for their growth is nearby 7.6 to 8.6 (ICMSF, 1996).

In US, *V. parahaemolyticus* is mostly isolated from clinical samples (Bacon and Sofos, 2003). Gastroenteritis is characteristically related with consumption of improperly cooked food or raw food consumption. The incubation period is about 4–96 hours, *V. parahaemolyticus* symptoms include headache, chills, abdominal cramps, nausea, slight fever, watery diarrhea (which is occasionally bloody), and vomiting (Bacon and Sofos, 2003). Symptoms are generally self-limiting and last for about 2–3 days but severe cases might outcome in primary septicemia, dysentery, or cholera-like illness (Janda et al., 1988). The presence of a pre-existing condition (e.g., diabetes mellitus, liver disease, peptic ulcer disease, alcoholism, immune disorder, and antacid medication) significantly increases the possibility to develop a clinical syndrome, i.e., septicemia, wound infection, or gastroenteritis (Bacon and Sofos, 2003). Four hemolytic components are present in *V. parahaemolyticus* such as a lysophospholipase, a thermolabile direct hemolysin, thermostable direct hemolysin (TDH) and phospholipase A (Bacon and Sofos, 2003). *V. parahaemolyticus* are invasive in nature and could penetrate the lamina propria followed by entering into circulation and mainly found in pancreas, heart, liver, and spleens (Mossel et al., 1995).

Infection-related to *V. parahaemolyticus* is thought out to be the important reason of bacterial diseases mainly related with consumption of seafood in Guangdong province of China (Chen et al., 2017). Outbreaks most commonly happened in cafeteria (12.7%), private residents (21.1%) and restaurants (50.7%) (Chen et al., 2017). So as to avoid outbreaks of *V. parahaemolyticus* triggered by inappropriate cooking, cross-contamination, and inappropriate storage within high-temperature seasons, guidelines for seafood safety from the production stage to the consumption stage must be strengthened (Chen et al., 2017). Raw shellfish consumption, chiefly oysters, has been associated to numerous outbreaks of *V. parahaemolyticus* illness within the US (Wu et al., 2014). *Vibrio* infections can be controlled by the hygienic handling and proper cooking of marine and other aquatic food products.

1.4 FOODBORNE VIRUSES

Viruses are obligate intracellular parasites and could multiply only in living cells. Hence, outside the host they are unable to survive for a longer period of time. Foodborne illnesses are triggered by more than 100 kinds of enteric viruses. Common foodborne viruses are noroviruses and hepatitis A. Such viruses are commonly transmitted through food; for example, mussels, clams, oysters, and cockles are particularly vulnerable to spread viruses. The waters within which their growth occurs are progressively subject to fecal contamination of humans, occasionally from sewage discharges and from diseased shellfish reapers. Viruses are collected by the shellfish in the course of their filter-feeding action. These species are not infected by human viruses, but for days or weeks they are present inside the digestive tract of shellfish and it is apparently much hard to remove it as compared to bacteria during processes planned to clean the shellfish (for example, depuration) (Power and Collins, 1989).

1.4.1 NOROVIRUS

Noroviruses fit in the family Caliciviridae which comprises about 5 genera: Vesivirus (VeV), Nebovirus (NeV), Sapovirus (SaV), Norovirus (NoV), and Lagovirus (LaV) (Glass et al., 2009). Viruses which belong to this family are icosahedral having diameter of 23 to 40 nm without non-envelope. The capsid comprises the minor VP2 protein and the main VP1-protein about 60 k Dalton. The genome of this virus is, non-segmented, +ve sense ssRNA, which is about 7–8 Kb in length (Kapikian et al., 1972). Norovirus of human is mainly hard to study. Unlike from other viruses, it is not probable to grow the human Noroviruses by means of cultured cells. Only humans are infected by the human Norovirus and these viruses multiply only within the intestines of human. Even within the cultured cells, this virus is unable to proliferate. Hence, more elaborate research is required to further explore human Norovirus (Farkas et al., 2008).

When norovirus consumed orally into the body, so, it multiplies within the upper intestinal tract (upper jejunum and duodenum). It causes blunting and widening of intestinal villi. When intestinal cells are infected by the norovirus infection so, enzyme activity of such infected intestinal cells is altered (Flynn et al., 1988). Norovirus shedding occurs from the infected cells and the virus then gain entry to the intestinal lumen, but RNA of Norovirus is detected within cerebrospinal fluid (CSF) and serum also (Ito et al., 2006).

Transmission of noroviruses mainly occur through the fecal-oral route, whichever via person-to-person direct spread (88%) or via ingesting food which is contaminated by Norovirus (10%) (Kroneman et al., 2008). Noroviruses could moreover be transmitted through a droplet from vomitus. Though, in children, almost all the infections are produced because of unknown reasons. Norovirus outbreaks are usually reported in winter. Infection transmission occurs by person-to-person contact in a family and formerly spread to playschool or further public organizations to trigger outbreak.

Noroviruses are extremely infectious in their nature: and it is thought that very few (about 10) living viral particles are enough to cause infection. The symptoms might range from asymptomatic to severe vomiting, diarrhea, dehydration, fever, headache, and myalgia. Norovirus has an incubation period of 12–48 hours and the illness duration is 12–60 hrs (Green, 2013). Significant mortality and morbidity are caused by foodborne viruses. These could be controlled only when good agricultural practices, good personal and food hygiene practices are adopted and when human sewage is properly managed so it can prevent further transmission (Tuladhar et al., 2015). Proper handwashing by using soap is the best way to remove Norovirus from fingers (Tuladhar et al., 2015). Though, formulations of hand sanitizers complemented with citric acid and urea might be much effective against Noroviruses (Ionidis et al., 2016).

1.4.2 HEPATITIS A

Hepatitis A virus (HAV) has ssRNA genome of about 7.5 kilobases. HAV shares some main features with genera of the Picornavirus family but also shows a sufficient diverse property which helps it categorizing it as the only species within the genus Hepatovirus (Cuthbert, 2001). RNA sequence analysis has shown that there are about 6 genotypes of Hepatitis A (I-VI). I, II, and III genotypes of Hepatitis A comprise strains which are linked with infections of humans, and the majority of human strains are placed within genotypes I and III. HAV is non-enveloped (contains no lipid envelope), hydrophobic virus which is about 22–30 nm in its size (FDA, 2012). It is mainly transmitted through the fecal-oral route. Gastric acid does not neutralize the HAV and is supposed to be moved across the epithelium of the intestine and then hepatocytes take it (Taylor et al., 1992). Target organ for HAV is liver. Viral replication occurs in the cytoplasm of the diseased hepatocytes (Karayiannis et al., 1986). Within US, about 10%

of reported cases of hepatitis A are linked with waterborne transmission or suspected food (CDCP, 2006). Recent estimations of foodborne disease in the US indicate that hepatitis-A is 4th important reason of viral food-related infection (Mead et al., 1999). In 1988 in Shanghai, the largest foodborne outbreak of hepatitis A was happened, and approximately 300,000 persons got affected due to ingesting clams which were harvested from contaminated waters (Halliday et al., 1991).

HAV can be spread by contaminated water, food, surfaces (e.g., cooking utensils and polluted tabletops) and by indirect or direct person-person contact (Cuthbert, 2001). HAV cannot multiply in the environment, though; they are very stable in an extensive range of ecological conditions, involving desiccation, heat, freezing, and chemicals (Cuthbert, 2001). Salad items, contaminated water, uncooked foods, sandwiches, and fresh fruit made by diseased food handlers are involved in outbreaks of hepatitis A (Levy et al., 1975).

Hepatitis A is a self-limited illness in most of individuals. The incubation period of HAV is 28 days average. Jaundice is the distinctive symptom of hepatitis A, which develops in 1–2 weeks after the beginning of prodromal symptoms. Jaundice is accompanied by tenderness, pruritus, and minor liver enlargement. Further clinical symptoms might comprise anorexia, diarrhea, arthralgia, fever, and abdominal pain. Generally, the recovery is complete, even though affected individuals might feel not well for numerous weeks. Children under the age of six years have mild or asymptomatic infections and general symptoms include vomiting, nausea, diarrhea, and malaise (Smith et al., 1997).

Personal hygienic practices are required in order to prevent hepatitis A. Public sanitation, comprising suitable water-treatment systems and disposal of human waste are significant in maintaining a lower incidence rate of hepatitis A within industrialized countries (Shapiro and Margolis, 1993). Transmission of HAV can be prevented if hygienic practices are adopted and proper handwashing is followed in surroundings where there is contact with human waste.

1.5 FOODBORNE PARASITES

Parasites are single-celled microbes. They do not have a rigid cell wall around but have an organized nucleus. As compared to bacteria, they are larger in size. They are transmissible in the form of cyst. Parasites depend on the other living organism for their nourishment and protection. Parasite transmission might occur from human-human, human-animals, and animals-human. Several parasites cause important foodborne and waterborne diseases. Parasites survive and multiply within the tissues and organs of infected individuals. Most common foodborne parasites are *Trichinella spiralis*, *Toxoplasma gondii* and *Cyclospora cayetanensis*.

1.5.1 CYCLOSPORA CAYETANENSIS

Cyclospora cayetanensis is a protozoan parasite and belongs to the family Eimeriidae. *C. cayetanensis* resides within small intestine, where this parasite spends the intermediate life cycle phase in cytoplasm of enterocytes and then oocysts are produced which contains two sporocysts that encapsulates four sporozoites (Bacon and Sofos, 2003). *C. cayetanensis* is identified as a reason for foodborne illness. Numerous outbreaks occurred within the 1990s and were related with imported raspberries consumption. Further outbreaks involved basil, salads, or berries as a vehicle of infections. It is supposed that such foodborne outbreaks are caused by using contaminated water in irrigation. When oocysts are 1st eradicated in their host feces, they are non-sporulated and are not immediately able to cause infection. Hence, it is believed that the diseased food handlers are possibly not able to transmit the disease. Though, the environmental circumstances needed to bring sporulation are unknown, and it is possible that under some conditions the sporulation may happen rapidly (Bern et al., 2002).

Cyclospora seems to be endemic within numerous developing countries because of the deficiency of sanitary facilities and absence of clean water which enable the *Cyclospora* transmission via the fecal-oral route. It has been reported by several epidemiological investigations that there was a relation between cyclosporiasis and domestic animals. Though, human is the only identified host for *C. cayetanensis*. Immunity to *Cyclospora* has been developed by the people who live in endemic areas and eat local food. Individuals might pass oocysts but no gastrointestinal symptoms can be experienced (Bern et al., 2002). *C. cayetanensis* is having an incubation period of about 1–14 days. Prolonged illness (six weeks or longer) is caused by *C. cayetanensis* in both immunocompetent and immunocompromised persons, with typical symptoms comprising vomiting, malaise, anorexia, non-bloody diarrhea, nausea, fatigue, abdominal cramping, fever, and bloating (Bacon and Sofos, 2003).

1.5.2 TOXOPLASMA GONDII

Toxoplasma gondii is a protozoan parasite and belongs to the phylum *Apicomplexa.* It is an obligate intracellular microbe which is responsible for toxoplasmosis within human. Cats are the primary reservoir of *T. gondii* and other warm-blooded animals can act as transitional hosts for *T. gondii* (Bacon and Sofos, 2003). This protozoan might be in the form of sporozoites, bradyzoites or tachyzoites, which are the three stages of its life cycle. Bradyzoites and tachyzoites occur in body tissues where the tachyzoites multiply and cause damage to infected host cells and bradyzoites proliferate in tissue cysts. Sporozoites are shedded, in oocysts, within feces or cats where they undergo sporulation after 1 to 5 days and survive for months by using their capability to resist drying, freezing, and disinfectants (Bacon and Sofos, 2003). According to NCBI, there are about 17 genomes of *T. gondii* completed until now. The intermediate entire length of the genome is about 64.1936 Mb (NCBI, 2017).

Humans could acquire the *T. gondii* in numerous ways, for example, the intake of contaminated food or water which contain the oocysts, contaminated transfusion of blood or transplantation of organ, accidental inoculation of tachyzoite or transplacental transmission. *T. gondii* infections characteristically occur when cysts are ingested in undercooked or raw. Similarly, beef and fresh pork appear to be the chief source (Bacon and Sofos, 2003). Toxoplasmosis occurs when about 100 oocysts or tissue cysts are ingested, followed by rupturing of walls of cyst releasing sporozoites or bradyzoites, which then pass through the intestinal epithelium and circulates in the body (Bacon and Sofos, 2003). Bradyzoites and sporozoites are converted into tachyzoites and start intracellular multiplication causing the death of cells inside the host. The adjacent cells are invaded the tachyzoites, and then they repeated process of the reproduction. Host immune response transforms these tachyzoites back into bradyzoites and cysts are formed within the local tissue, where they could stay in the host organism for entire life (Bacon and Sofos, 2003). The incubation period of *T. gondii* is about 15–23 days and symptoms of the toxoplasmosis comprise headache, rash, muscle aches, fever, pain, and lymph nodes swelling. *T. gondii* infection is associated with accidental intake of contaminated ingredients (e.g., soil, fruits, vegetables), raw, and partially cooked meat (FDA, 2012).

T. gondii is commonest parasites within the world. Although cats are the only hosts of these parasites where the parasite might complete its life cycle but nearly all the warm-blooded vertebrates are used by this parasite, e.g., humans, as their hosts. In US each year about 87,000 illnesses, 330 deaths and 4,400 hospitalizations are caused by *T. gondii*, which makes it as an important reason of foodborne mortality and third important reason of foodborne hospitalizations in the US (Scallan et al., 2011). Most important *Toxoplasma* sources are animal feces, raw meat, and transmission from mother to unborn child.

1.5.3 TRICHINELLA SPIRALIS

Trichinella spiralis is a roundworm parasite belongs to the phylum Nematoda responsible for most of trichinosis infections in humans. Apart from humans, carnivorous mammals are also infected by *T. spiralis*. According to data regained from NCBI, two genomes of *T. spiralis* are completed until now. The intermediate entire genome length is about 56.7757 Mb (NCBI, 2017).

Adult worms are about 1.4–1.8 mm in their size and are embedded within the epithelium of small intestine inside the host, where males and females mate (Bacon and Sofos, 2003). Larvae is passed by the female adults into the bloodstream, and these then arrive muscle fibers wherever they are encysted; the larvae which is encysted within muscle stay viable for an extended time period (Bacon and Sofos, 2003). The pathogenicity and symptoms are mostly because of the encystment and migrating process which trigger edema, pain, neurological disorders, fever, and even it causes death. Adult nematodes survive within the jejunal and duodenal mucosal epithelium, where they could live for about eight weeks; throughout this transient stage, about 1,500 larvae are released by the adult female nematodes into the bloodstream which travel in the body and then reach muscle tissue, where they could live for numerous years

(Bacon and Sofos, 2003). Within the skeletal muscle, larvae undergo maturation and encapsulated within a calcified wall after 6 to 18 months. The encysted larvae can stay viable for about 10 years (Bacon and Sofos, 2003).

T. spiralis is associated with raw and undercooked contaminated meat. Incubation period of *T. spiralis* is about 3–14 days and symptoms comprise vomiting, visual deficiencies, nausea, headaches, fever, night sweating, nonspecific gastroenteritis, breathing difficulty, circumorbital edema, eosinophilia, myalgia, and chills (Bacon and Sofos, 2003). Thermal process can inactivate this nematode, and hence it is recommended by the USDA that pork products ought to be cooked to an internal temperature of nearby 76.7°C (Bacon and Sofos, 2003). In US, 52 cases of foodborne disease are annually produced via *T. spiralis* with a fatality rate of the cases as 0.003 (Mead et al., 1999) (Table 1.1).

1.6 CONCLUSION

In the recent times, the bacterial foodborne illnesses are amongst the most prevalent global public health issues. Though, the accurate prevalence of bacterial food-related diseases are unidentified due to some reasons such as victims' poor response during discussions with health officials, improper diagnosis of the infection, improper examination by the laboratory and inadequate samples collection for laboratory analysis. The occurrence of numerous pathogenic bacteria within different variety of foods poses a health-related risks and increase concerns related to food products safety. Apart from bacterial FBIs, parasites, viruses, and fungi are also involved in food borne illness. To avoid foodborne diseases, strict hygienic practices are required to be implemented in the storage, handling, selling, and manufacturing of food so as to guarantee the safety and quality of these foodstuffs in order to eliminate or lessen FBI risks. There is a need for further extensive research to be carried out so that to develop effective approaches against bacterial species, viruses, and parasites that cause foodborne illnesses.

TABLE 1.1 List of Hazardous Foodborne Pathogens

Pathogen	Incubation Period	Sign and Symptoms	Duration of Illness	Associated Food	References
Bacterial Foodborne Illnesses (Gram-Positive)					
Listeria monocytogenes	9–48 hrs for gastrointestinal symptoms, 2–6 weeks for invasive disease	Fever, nausea, vomiting, diarrhea, pregnant women may have flu-like illness, infection may lead to stillbirth, immunocompromised patient may have bacteremia or meningitis	Variable	Ready to eat food, uncooked vegetables, uncooked poultry meat products	Jamali and Thong (2014); Meloni (2014)
Bacillus cereus	10–16 hrs	Abdominal cramps, watery diarrhea, nausea	24–48 hrs	Raw milk, meat, and agriculture products	Logan and Rodrigez-Diaz (2006); Murray et al. (2007)
Clostridium botulinum	12–72 hrs	Vomiting, diarrhea, blurred vision, diplopia, dysphagia, muscle weakness	variable	Home-canned foods, chicken, liver pate, ham, sausage, lobster, and salted and smoked fish	CDCP (1998); FDA (2012)
Clostridium perfringens	8–16 hrs	Watery diarrhea, nausea, abdominal cramps, rare fever	24–48 hrs	Meat, poultry, many ready to eat food and stored foods	Miki et al. (2008); Eriksen et al. (2010)
Staphylococcus aureus	1–6 hrs	Sudden onset of severe nausea and vomiting, abdominal cramps, diarrhea, and fever may be present	24–48 hrs	Unrefrigerated or improperly refrigerated meat, potato, eggs, salads	Tamarapu et al. (2001); Scallan et al. (2006)
Bacterial Foodborne Illnesses (Gram-Negative)					
Aeromonas spp.	1–48 hrs	Septicemia, gastrointestinal perforations, endocarditis or meningitis, pneumonia is uncommon	24–48 hrs	Contaminated water, fish, and shellfish, meats, and fresh vegetables	Jiang et al. (2002); Qu et al. (2003)

TABLE 1.1 *(Continued)*

Pathogen	Incubation Period	Sign and Symptoms	Duration of Illness	Associated Food	References
Campylobacter spp.	2–5 days	Bloody diarrhea, fever, cramps, and vomiting	2–10 days	Raw and undercooked poultry, unpasteurized milk, and contaminated water	Butzler (2004); Kaakoush et al. (2015)
Shigella spp.	24–48 hrs	Abdominal cramps, fever, and diarrhea, stool may contain blood and mucus	4–7 days	Food and water contaminated with fecal material, ready to eat foods, salads, milk, shellfish, vegetables, chicken	ICMSF (1996); Vargas et al. (1999)
Salmonella spp.	1–3 days	Abdominal cramps, fever, and diarrhea, vomiting, *S. typhi* and *paratyphi* produce typhoid with the onset of headache, fever, muscle weakness, diarrhea	4–7 days	Contaminated eggs, unpasteurized milk, poultry, contaminated foods, and vegetables	Bhunia (2008); CDCP (2012)
Escherichia spp.	1–8 days	Severe diarrhea that is often bloody, abdominal pain and vomiting, little or no fever is present	5–10 days	Undercooked beef, unpasteurized milk and juices, raw fruits and vegetables and contaminated water	Garcia et al. (2010)
Vibrio spp.	24–72 hrs	Water diarrhea, abdominal cramps, vomiting, severe dehydration death within hours	3–7 days	Contaminated water, fish, undercooked or raw seafoods	Wu et al. (2014)
Viral Foodborne Illnesses					
Norovirus	12–48 hrs	Nausea, vomiting, abdominal cramps, diarrhea, fever, myalgia, and headache	12–60 hrs	Shellfish, contaminated foods, ready to eat foods, salads, cookies, and fruits	Cotterelle et al. (2005); Green (2013)

TABLE 1.1 *(Continued)*

Pathogen	Incubation Period	Sign and Symptoms	Duration of Illness	Associated Food	References
Hepatitis A	28 days average	Diarrhea, dark urine, jaundice, flu-like symptoms, fever, headache, nausea, and vomiting	Variable, 1–2 weeks	Shellfish harvested from contaminated water, contaminated drinking water and uncooked foods	Levy et al. (1975); Smith et al. (1997)
Parasitic Foodborne Illnesses					
Cyclospora cayetanensis	1–14 days	Watery diarrhea, loss of appetite, loss of weight, stomach cramps, nausea, vomiting, and fatigue	Weeks to months	Various types of fresh produce (basil, salad greens, or berries)	Bern et al. (2002); Bacon and Sofos (2003)
Toxoplasma gondii	15–23 days	Headache, rash, muscle aches, fever, pain, and lymph nodes swelling	Months	Accidental ingestion of contaminated substances (soil, fruits, vegetables) raw and partial cooked meat	FDA (2012)
Trichinella spiralis	3 to 14 days	Vomiting, nausea, visual deficiencies, fever, headaches, night sweating, nonspecific gastroenteritis, difficulty breathing, circumorbital edema, eosinophilia, myalgia, and chills	Months	Raw and undercooked contaminated meat	Bacon and Sofos (2003)

Note: hrs: hours.

KEYWORDS

- **foodborne illness**
- **hazardous pathogens**
- **intoxication**
- **microbes**
- **safety**
- **spoilage**
- **toxin**

REFERENCES

Acheson, D., (2004). *Differential Diagnosis of Microbial Foodborne Disease*. Up to Date.

Adams, M., & Moss, M., (2008). *Food Microbiology*. UK: RSC Press.

Ahmed, A. A., Moustafa, M. K., & Marth, E. H., (1983). Incidence of *Bacillus cereus* in milk and some milk products. *Journal of Food Protection, 46*(2), 126–128.

Asakura, M., Samosornsuk, W., Hinenoya, A., et al., (2008). Development of a cytolethal distending toxin (cdt) gene-based species-specific multiplex PCR assay for the detection and identification of *Campylobacter jejuni, Campylobacter coli* and *Campylobacter fetus*. *FEMS Immunology and Medical Microbiology, 52*(2), 260–266.

Asao, T., Kumeda, Y., Kawai, T., Shibata, T., Oda, H., Haruki, K., & Kozaki, S., (2003). An extensive outbreak of staphylococcal food poisoning due to low-fat milk in Japan: Estimation of enterotoxin A in the incriminated milk and powdered skim milk. *Epidemiology and Infection, 130*(1), 33.

Bacon, R. T., & Sofos, J. N., (2003). Characteristics of biological hazards in foods. *Food Safety Handbook*, 157–195.

Balaban, N., & Rasooly, A., (2000). *Staphylococcal* enterotoxins. *International Journal of Food Microbiology, 61*(1), 1–10.

Bean, N. H., & Griffin, P. M., (1990). Foodborne disease outbreaks in the United States, 1973–1987: Pathogens, vehicles, and trends. *Journal of Food Protection, 53*(9), 804–817.

Bern, C., Ortega, Y., Checkley, W., et al., (2002). Epidemiologic differences between cyclosporiasis and cryptosporidiosis in Peruvian children. *Emerging Infectious Diseases, 8*(6), 581.

Bernheimer, A. W., & Avigad, L. S., (1974). Partial characterization of aerolysin, a lytic exotoxin from *Aeromonas hydrophila*. *Infection and Immunity, 9*(6), 1016–1021.

Bhunia, A. K., (2018). *Foodborne Microbial Pathogens: Mechanisms and Pathogenesis*: Springer.

Blackburn, C. W., & McClure, P. J., (2005). *Foodborne Pathogens - Hazards, Risk Analysis and Control* (pp. 423–425). Woodhead Publishing Limited, Cambridge.

Bolstad, I., (1990). Food poisoning caused by *B. cereus* - chicken. *Nor. Vet., 102*, 39.

Bravo, L., Morier, L., Castañeda, N., Ramírez, M., Silva, M., & Castro-Escarpulli, G., (2003). *Aeromonas*: An emerging pathogen associated with extraintestinal infection in Cuba. *Revista Cubana de Medicina Tropical, 55*(3), 208, 209.

Butzler, J., (2004). *Campylobacter*, from obscurity to celebrity. *Clinical Microbiology Infection, 10*, 868–876.

Cantoni, C., & Bresciani, C. M., (1987). Occurrence of *Bacillus cereus* in foods. *Industrie Alimentari, 26*, 7, 8.

Centers for Disease Control and Prevention, (1998). Botulism in the United States, 1899–1996. *Handbook for Epidemiologists, Clinicians, and Laboratory Workers*.

Centers for Disease Control and Prevention, (2006). Prevention of hepatitis A through active or passive immunization. recommendations of the advisory committee on immunization practices. *MMWR, 55*, 1–23.

Centers for Disease Control and Prevention, (2012). Multistate outbreak of *Salmonella* Bareilly and *Salmonella* Nchanga infections associated with a raw scraped ground tuna product. http://www.cdc.gov/salmonella/bareilly-04-12/index.html?s_cid=cs_654 (accessed on 16 January 2022).

Chaibenjawong, P., & Foster, S. J., (2011). Desiccation tolerance in *Staphylococcus aureus*. *Archives of Microbiology, 193*(2), 125–135.

Chen, J., Zhang, R., Qi, X., et al., (2017). Epidemiology of foodborne disease outbreaks caused by *Vibrio parahaemolyticus* during 2010–2014 in Zhejiang Province, China. *Food Control, 77*, 110–115.

Colagiorgi, A., Bruini, I., Di Ciccio, P. A., Zanardi, E., Ghidini, S., & Ianieri, A., (2017). *Listeria monocytogenes* biofilms in the wonderland of food industry. *Pathogens, 6*(3), 41.

Cuthbert, J., (2001). Erratum: Hepatitis A: Old and new. *Clinical Microbiology Reviews, 14*(3), 642.

Dasti, J. I., Tareen, A. M., Lugert, R., Zautner, A. E., & Groß, U., (2010). *Campylobacter jejuni*: A brief overview on pathogenicity-associated factors and disease-mediating mechanisms. *International Journal of Medical Microbiology, 300*(4), 205–211.

Denny, J., & McLauchlin, J., (2008). Human *Listeria monocytogenes* infections in Europe - an opportunity for improved European surveillance. *Eurosurveillance, 13*(13), 9, 10.

Dhama, K., Verma, A. K., Rajagunalan, S., et al., (2013). *Listeria monocytogenes* infection in poultry and its public health importance with special reference to foodborne zoonoses. *Pakistan Journal of Biological Sciences, 16*(7), 301–308.

Drobniewski, F. A., (1993). *Bacillus cereus* and related species. *Clinical Microbiology Reviews, 6*(4), 324–338.

Eriksen, J., Zenner, D., Anderson, S. R., Grant, K., & Kumar, D., (2010). *Clostridium perfringens* in London, July 2009: Two weddings and an outbreak. *Eurosurveillance, 15*(25), 19598.

Farkas, T., Sestak, K., Wei, C., & Jiang, X., (2008). Characterization of a rhesus monkey calicivirus representing a new genus of *Caliciviridae*. *Journal of Virology, 82*(11), 5408–5416.

Figueras, M. J., Suarez-Franquet, A., Chacon, M. R., et al., (2005). First record of the rare species *Aeromonas culicicola* from a drinking water supply. *Applied and Environmental Microbiology, 71*(1), 538–541.

Flynn, W. T., Saif, L. J., & Moorhead, P. D., (1988). Pathogenesis of porcine enteric calicivirus-like virus in four-day-old gnotobiotic pigs. *American Journal of Veterinary Research, 49*(6), 819.

Food and Drug Administration, (2012). In: Lampel, K., Al-Khaldi, S., & Cahill, S., (eds.), *Bad Bug Book, Foodborne Pathogenic Microorganisms and Natural Toxins. Gram-positive bacteria.* Silver Spring: Center for Food Safety and Applied Nutrition of the Food and Drug Administration (FDA), US Department of Health and Human Services.

Gandhi, M., & Chikindas, M. L., (2007). *Listeria*: A foodborne pathogen that knows how to survive. *International Journal of Food Microbiology, 113*(1), 1–15.

García, A., Fox, J. G., & Besser, T. E., (2010). Zoonotic enterohemorrhagic *Escherichia coli*: A one health perspective. *ILAR Journal, 51*(3), 221–232.

Ge, Z., Schauer, D. B., & Fox, J. G., (2008). In vivo virulence properties of bacterial cytolethal-distending toxin. *Cellular Microbiology, 10*(8), 1599–1607.

Glass, R. I., Parashar, U. D., & Estes, M. K., (2009). Norovirus gastroenteritis. *New England Journal of Medicine, 361*(18), 1776–1785.

Goldsweig, C. D., & Pacheco, P. A., (2001). Infectious colitis excluding *E. coli* O157: H7 and *C. difficile*. *Gastroenterology Clinics, 30*(3), 709–733.

Green, K. Y., (2013). *Caliciviridae*: The noroviruses. *Field Virology, 1*, 582–608.

Halliday, M. L., Kang, L. Y., Zhou, T. K., et al., (1991). An epidemic of hepatitis A attributable to the ingestion of raw clams in Shanghai, China. *Journal of Infectious Diseases, 164*(5), 852–859.

Hatheway, C. L., (1995). Botulism: The present status of the disease. In: *Clostridial Neurotoxins* (pp. 55–75). Springer, Berlin.

Havelaar, A. H., Kirk, M. D., Torgerson, P. R., et al., (2015). World health organization global estimates and regional comparisons of the burden of foodborne disease in 2010. *PLoS Medicine, 12*(12), e1001923.

Hutt, P. B., & Hutt, P. B. I., (1984). A history of government regulation of adulteration and misbranding of food. *Food Drug Cosmetic Law Journal, 39*, 2.

International Commission on Microbiological Specifications for Foods, (1996). *Microorganisms in Foods 5: Characteristics of Microbial Pathogens* (Vol. 5). Springer Science and Business Media.

Ionidis, G., Hübscher, J., Jack, T., et al., (2016). Development and virucidal activity of a novel alcohol-based hand disinfectant supplemented with urea and citric acid. *BMC Infectious Diseases, 16*(1), 77.

Ito, S., Takeshita, S., Nezu, A., Aihara, Y., Usuku, S., Noguchi, Y., & Yokota, S., (2006). Norovirus-associated encephalopathy. *The Pediatric Infectious Disease Journal, 25*(7), 651, 652.

Jamali, H., & Thong, K. L., (2014). Genotypic characterization and antimicrobial resistance of *Listeria monocytogenes* from ready-to-eat foods. *Food Control, 44*, 1–6.

Janda, J. M., & Abbott, S. L., (1999). Unusual foodborne pathogens: *Listeria monocytogenes, Aeromonas, Plesiomonas*, and *Edwardsiella* species. *Clinics in Laboratory Medicine, 19*(3), 553–582.

Janda, J. M., Powers, C., Bryant, R. G., & Abbott, S. L., (1988). Current perspectives on the epidemiology and pathogenesis of clinically significant *Vibrio* spp. *Clinical Microbiology Reviews, 1*(3), 245–267.

Jay, J. M., (2000). *Modern Food Microbiology*. Aspen Publication Inc., Gaithersburg, Maryland.

Jemmi, T., & Stephan, R., (2006). *Listeria monocytogenes*: Foodborne pathogen and hygiene indicator. *Rev. Sci. Tech., 25*(2), 571–580.

Jiang, Z. D., Lowe, B., Verenkar, M. P., et al., (2002). Prevalence of enteric pathogens among international travelers with diarrhea acquired in Kenya (Mombasa), India (Goa), or Jamaica (Montego Bay). *The Journal of Infectious Diseases, 185*(4), 497–502.

Kaakoush, N. O., Castaño-Rodríguez, N., Mitchell, H. M., & Man, S. M., (2015). Global epidemiology of *Campylobacter* infection. *Clinical Microbiology Reviews, 28*(3), 687–720.

Kaper, J. B., (2005). Pathogenic *Escherichia coli. International Journal of Medical Microbiology, 295*, 355–356.

Kapikian, A. Z., Wyatt, R. G., Dolin, R., Thornhill, T. S., Kalica, A. R., & Chanock, R. M., (1972). Visualization by immune electron microscopy of a 27-nm particle associated with acute infectious nonbacterial gastroenteritis. *Journal of Virology, 10*(5), 1075–1081.

Karayiannis, P., Jowett, T., Enticott, M., et al., (1986). Hepatitis A virus replication in tamarins and host immune response in relation to pathogenesis of liver cell damage. *Journal of Medical Virology, 18*(3), 261–276.

Khardori, N., & Fainstein, V., (1988). *Aeromonas* and *Plesiomonas* as etiological agents. *Annual Reviews in Microbiology, 42*(1), 395–419.

Koepke, R., Sobel, J., & Arnon, S. S., (2008). Global occurrence of infant botulism, 1976–2006. *Pediatrics, 122*(1), e73–82.

Kroneman, A., Verhoef, L., Harris, J., et al., (2008). Analysis of integrated virological and epidemiological reports of norovirus outbreaks collected within the foodborne viruses in Europe network from 1 July 2001 to 30 June 2006. *Journal of Clinical Microbiology, 46*(9), 2959–2965.

Kusumaningrum, H. D., Van, P. M. M., Rombouts, F. M., & Beumer, R. R., (2002). Effects of antibacterial dishwashing liquid on foodborne pathogens and competitive microorganisms in kitchen sponges. *Journal of Food Protection, 65*(1), 61–65.

Labbé, R. G., & García, S., (2013). *Guide to Foodborne Pathogens*. Wiley Blackwell.

Le Loir, Y., Baron, F., & Gautier, M., (2003). *Staphylococcus aureus* and food poisoning. *Genet. Mol. Res., 2*(1), 63–76.

Lechner, S., Mayr, R., Francis, K. P., et al., (1998). *Bacillus weihenstephanensis* sp. nov. is a new psychrotolerant species of the *Bacillus cereus* group. *International Journal of Systematic and Evolutionary Microbiology, 48*(4), 1373–1382.

Levy, B. S., Fontaine, R. E., Smith, C. A., et al., (1975). A large foodborne outbreak of hepatitis A: Possible transmission via oropharyngeal secretions. *JAMA, 234*(3), 289–294.

Liu, D., Lawrence, M. L., Ainsworth, A. J., & Austin, F. W., (2007). Toward an improved laboratory definition of *Listeria monocytogenes* virulence. *International Journal of Food Microbiology, 118*(2), 101–115.

Liu, D., Lawrence, M. L., Austin, F. W., & Ainsworth, A. J., (2007). A multiplex PCR for species- and virulence-specific determination of *Listeria monocytogenes*. *Journal of Microbiological Methods, 71*(2), 133–140.

Logan, N. A., & Rodrigez-Diaz, M., (2006). *Bacillus* spp. and related genera. *Principles and Practice of Clinical Bacteriology, 2*, 139–158.

Lowy, F. D., (1998). *Staphylococcus aureus* infections. *New England Journal of Medicine, 339*(8), 520–532.

Mead, P. S., Slutsker, L., Dietz, V., et al., (1999). Food-related illness and death in the United States. *Emerging Infectious Diseases, 5*(5), 607.

Meloni, D., (2014). Focusing on the main morphological and physiological characteristics of the foodborne pathogen *Listeria monocytogenes. J. Vet. Sci. Res., 1*, 1, 2.

Miki, Y., Miyamoto, K., Kaneko-Hirano, I., Fujiuchi, K., & Akimoto, S., (2008). Prevalence and characterization of enterotoxin gene-carrying *Clostridium perfringens* isolates from retail meat products in Japan. *Applied and Environmental Microbiology, 74*(17), 5366–5372.

Monma, C., Hatakeyama, K., Obata, H., et al., (2015). Four foodborne disease outbreaks caused by a new type of enterotoxin-producing *Clostridium perfringens. Journal of Clinical Microbiology, 53*(3), 859–867.

Montville, T. J., & Matthews, K. R., (2008). *Food Microbiology: An Introduction.* ASM Press, Washington, DC.

Mortimer, P. R., & McCann, G., (1974). Food-poisoning episodes associated with *Bacillus cereus* in fried rice. *The Lancet, 303*(7865), 1043–1045.

Mossel, D. A. A., Corry, J. E., Struijk, C. B., & Baird, R. M., (1995). *Essentials of the Microbiology of Foods: A Textbook for Advanced Studies*. John Wiley and Sons.

Mukhopadhyay, C., Chawla, K., Sharma, Y., & Bairy, I., (2008). Emerging extra-intestinal infections with *Aeromonas hydrophila* in coastal region of southern Karnataka. *Journal of Postgraduate Medicine, 54*(3), 199.

Murray, P. R., Baron, E. J., Jorgensen, J. H., Landry, M. L., & Pfaller, M. A., (2007). *Manual of Clinical Microbiology*. ASM Press, Washington, DC.

Murray, R., (2005). Recognition and management of *Staphylococcus aureus* toxin-mediated disease. *Intern. Med. J., 35*(S2), S106–119.

Nakamura, L. K., (1998). *Bacillus pseudomycoides* sp. nov. *International Journal of Systematic and Evolutionary Microbiology, 48*(3), 1031–1035.

NCBI, National Centre for Biotechnology Information, (2017). https://www.ncbi.nlm.nih.gov/genome (accessed on 20 December 2021).

Osaili, T. M., Alaboudi, A. R., & Nesiar, E. A., (2011). Prevalence of *Listeria* spp. and antibiotic susceptibility of *Listeria monocytogenes* isolated from raw chicken and ready-to-eat chicken products in Jordan. *Food Control, 22*(3, 4), 586–590.

Persson, S., Olsen, K. E., Ethelberg, S., & Scheutz, F., (2007). Subtyping method for *Escherichia coli* Shiga toxin (verocytotoxin) 2 variants and correlations to clinical manifestations. *Journal of Clinical Microbiology, 45*(6), 2020–2024.

Power, U. F., & Collins, J. K., (1989). Differential depuration of poliovirus, *Escherichia coli*, and a coliphage by the common mussel, *Mytilus edulis*. *Applied and Environmental Microbiology, 55*(6), 1386–1390.

Qu, F., Cui, E. B., Xia, G. M., et al., (2003). The clinical features and prognosis of *Aeromonas* septicaemia in hepatic cirrhosis: A report of 50 cases. *Zhonghua Nei Ke Za Zhi, 42*(12), 840.

Quinn, P. J., Markey, B. K., Carter, M. E., Demnelly, W. J., & Leonard, F. C., (2001). *Veterinary Microbiology and Microbial Disease*. Blackwell Publishing, Oxford, UK.

Scallan, E., Hoekstra, R. M., Angulo, F. J., et al., (2011). Foodborne illness acquired in the United States - major pathogens. *Emerging Infectious Diseases, 17*(1), 7.

Scallan, E., Jones, T. F., Cronquist, A., et al., (2006). Factors associated with seeking medical care and submitting a stool sample in estimating the burden of foodborne illness. *Foodborne Pathogens and Disease, 3*(4), 432–438.

Scherer, C. A., & Miller, S. I., (2001). *Molecular Pathogenesis of Salmonellae*: *Principles of Bacterial Pathogenesis* (pp. 265–316). USA: Academic Press.

Schlegelova, J., Babak, V., Brychta, J., Klimova, E., & Napravnikova, E., (2003). The prevalence of and resistance to antimicrobial agents of *Bacillus cereus* isolates from foodstuffs. *Veterinarni Medicina-UZPI (Czech Republic), 11*, 331–338.

Shapiro, C. N., & Margolis, H. S., (1993). Worldwide epidemiology of hepatitis A virus infection. *Journal of Hepatology, 18*, S11–14.

Silva, J., Leite, D., Fernandes, M., Mena, C., Gibbs, P. A., & Teixeira, P., (2011). *Campylobacter* spp. as a foodborne pathogen: A review. *Frontiers in Microbiology, 2*, 200.

Skarp, C. P. A., Hänninen, M. L., & Rautelin, H. I. K., (2016). Campylobacteriosis: The role of poultry meat. *Clinical Microbiology and Infection, 22*(2), 103–109.

Smith, P. F., Grabau, J. C., Werzberger, A., et al., (1997). The role of young children in a community-wide outbreak of hepatitis A. *Epidemiology and Infection, 118*(3), 243–252.

Snedeker, K. G., Shaw, D. J., Locking, M. E., & Prescott, R. J., (2009). Primary and secondary cases in *Escherichia coli* O157 outbreaks: A statistical analysis. *BMC Infectious Diseases, 9*(1), 144.

Suzuki, H., & Yamamoto, S., (2009). *Campylobacter* contamination in retail poultry meats and by-products in the world: A literature survey. *Journal of Veterinary Medical Science, 71*(3), 255–261.

Taher, A. A. I., Rao, B. N., Alganay, K. G., & El-Arabi, M. B., (2000). An outbreak of acute gastroenteritis due to *Aeromonas sobria* in Benghazi, Libyan Arab Jamahiriya. *Eastern Mediterranean Health Journal, 6*(1), 497–499.

Tamarapu, S., Mckillip, J. L., & Drake, M., (2001). Development of a multiplex polymerase chain reaction assay for detection and differentiation of *Staphylococcus aureus* in dairy products. *Journal of Food Protection, 64*(5), 664–668.

Taylor, G. M., Goldin, R. D., Karayiannis, P., & Thomas, H. C., (1992). *In situ* hybridization studies in hepatitis A infection. *Hepatology, 16*(3), 642–648.

Tuladhar, E., Hazeleger, W. C., Koopmans, M., Zwietering, M. H., Duizer, E., & Beumer, R. R., (2015). Reducing viral contamination from finger pads: Handwashing is more effective than alcohol-based hand disinfectants. *Journal of Hospital Infection, 90*(3), 226–234.

Vargas, M., Gascon, J., De Anta, M. T. J., & Vila, J., (1999). Prevalence of *Shigella* enterotoxins 1 and 2 among *Shigella* strains isolated from patients with traveler's diarrhea. *Journal of Clinical Microbiology, 37*(11), 3608–3611.

Volkova, R. S., (1971). *Bacillus cereus* contamination of foods and environment at institutional feeding points. *Gigienai-Sanitariya, 36*(2), 108, 109.

Von, G. A., (2007). The role of *Aeromonas* in diarrhea: A review. *Infection, 35*(2), 59.

Vugia, D., Cronquist, A., Hadler, J., et al., (2006). Preliminary FoodNet data on the incidence of infection with pathogens transmitted commonly through food in 10 states, United States, (2005). *Morb. Mortal. Wkly. Rep., 55*, 392–395.

WHO, (2013). *Basic Information on Emerging Infectious Diseases (EIDs): Listeriosis: What we Should Know*. http://www.searo.who.int/entity/emerging diseases/ (accessed on 20 December 2021).

Wu, Y., Wen, J., Ma, Y., Ma, X., & Chen, Y., (2014). Epidemiology of foodborne disease outbreaks caused by *Vibrio parahaemolyticus*, China, 2003–2008. *Food Control, 46*, 197–202.

Yamada, S., Matsushita, S., Dejsirilert, S., & Kudoh, Y., (1997). Incidence and clinical symptoms of *Aeromonas*-associated travellers' diarrhoea in Tokyo. *Epidemiology and Infection, 119*(2), 121–126.

Yousef, A. E., & Carlstrom, C., (2003). *Food Microbiology: A Laboratory Manual*. John Wiley and Sons.

CHAPTER 2

Foodborne Diseases: Causative Agents and Related Microorganisms

SHIKHA PANDHI, ARVIND KUMAR, and SADHNA MISHRA

ABSTRACT

Foodborne diseases are amongst the most widespread public health concern globally. The ability of various foodborne pathogens to cause foodborne diseases in consumers have highlighted the need for effective monitoring and prevention strategies. A foodborne disease or illness is commonly associated with ingestion of food contaminated with pathogens like bacteria, fungi, protozoa, or parasites. They are broadly categorized as "food infection," "food intoxication," and "food toxicoinfection." The emergence of new pathogens or re-emergence of already controlled ones in a new way has further escalated the associated public health concern. This chapter delivers an updated review of various foodborne diseases and its related causative microorganisms in brief.

2.1 INTRODUCTION

The science of food microbiology deals with the study of microorganisms related to food intended consumption by a human. Microorganisms utilize food as an efficient source of nutrients for their growth and development. Also, they use food as a source for their transmission to human hosts

Global Food Safety: Microbial Interventions and Molecular Advancements. Saher Islam, PhD, Devarajan Thangadurai, PhD, Jeyabalan Sangeetha, PhD, & Natália Cruz-Martins, PhD (Editors)

(Odeyemi, 2016). Microorganisms in food can be broadly divided into three categories such as (Bhunia, 2018):

- Beneficial microorganisms
- Spoilage-causing microorganisms, and
- Pathogenic microorganisms

Beneficial microorganisms are used in the preparation of various fermented products such as cheese, curd, yogurt, sauerkraut, idli, alcoholic beverages, etc. (Elshafei, 2017). During fermentation, these microorganisms' breakdown complex substances into simpler ones to give products like alcohol, acids, and antibacterial substances that increase product stability as well as provide desirable characteristics. Further, probiotics are also considered to be one of the most important categories of beneficial microorganisms that provide a health benefit to the host too by modulating its immune system. These beneficial microorganisms are generally safe for human consumption and do not cause any illness in the host unless certain unfavorable factor disseminates their growth. Spoilage causing microorganism grows on food to give products that result in product quality deterioration by altering its physical, chemical, and sensorial properties. Various food processing industries employ various preservation techniques to prevent food from spoilage. Food spoilage microorganisms possess less risk to public health as compared to pathogenic microorganisms.

Food pathogens are organisms which when grown on food, alter its sensorial as well as aesthetic quality and produce toxins that may cause foodborne illness. Foodborne illness is caused by either ingestion of pathogen itself or toxin produced by pathogen along with the food (Bintsis, 2017). A pathogen leads to damage to cells by colonizing in tissue leading to either morbidity (defined by general suffering) or mortality (death). These pathogens can replicate in the host by continuously destroying all the cellular barriers that restrict their entry and by conveying particular virulence factors that facilitate microbe establishment with a host that may be transmitted to another host (Bhunia, 2018) (Figure 2.1).

Bacteria, fungi, viruses, and parasites at any stage of food production can come in contact with food and may serve as a cause of various foodborne illnesses and are called food pathogens. Furthermore, some bacteria and fungi produce toxins under a certain favorable condition that may possess serious concerns and may lead to death (Martinović et al., 2016). Advancement in the microbiological safety of food has been essentially motivated by increasing public demand for safe and wholesome food. Various factors such as changing eating habits, improved transport conditions that promote microbe survival, rapid population growth, and an increasingly global market has greatly contributed towards the varying patterns in foodborne illness. Foodborne illness has become a solemn public health concern as it possesses constant invariable risk that not only affects the wellbeing of human beings but also causes significant obstruction to socio-economic expansion universally (Ellis et al., 2018). The type and number of microorganisms present in the end product are affected by the microbiological quality of the initial raw material, hygienic conditions maintained during processing, packaging, storage, distribution to ultimate consumption by the consumer. Rising concerns associated with the emergence of new pathogens and re-emergence of the existing ones together with their low infectious doses and increased multidrug-resistance has been extensively reported as a major challenge among the sectors dealing with the food domain (Gangiredla et al., 2017).

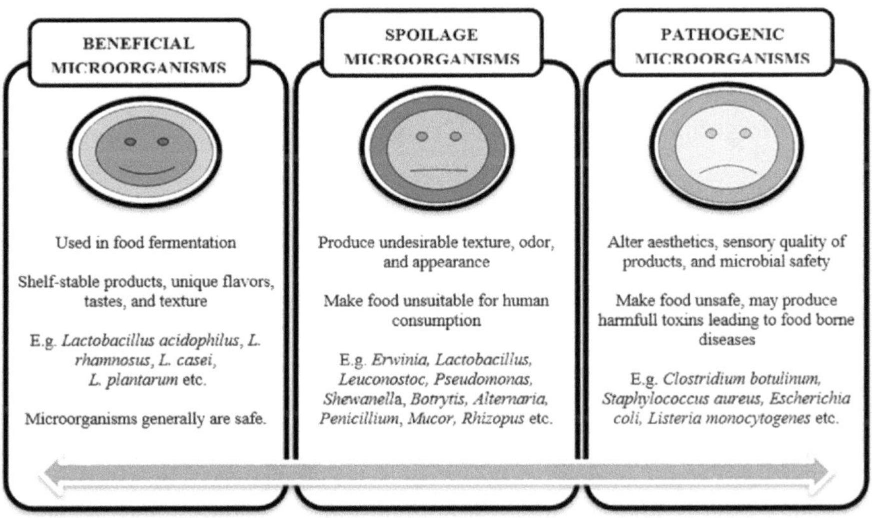

FIGURE 2.1 Three domains of food microbiology.

Improvements in the microbiological safety of foods have been largely driven by public demand in response to disease outbreaks. Critical monitoring and screening of foodborne diseases and pathogens stand essential in reducing the global burden of this foodborne illness. Implementation of a farm-to-fork approach has been acquired by all the stages of food production to perk up product of hygienic and wholesome products and vigorously integrate structured approaches to ensure food safety such as HACCP (Hazard Analysis Critical Control Point) (Newell et al., 2010). This chapter provides comprehensive information regarding various food-related microorganisms with the main emphasis on foodborne pathogens and related public health concerns.

2.2 FOODBORNE DISEASES

A foodborne disease or illness is generally associated with the ingestion of food contaminated with a pathogenic agent such as bacteria, fungi, viruses, protozoa, or parasites. WHO defines foodborne diseases as "any disease of a transferable or lethal nature caused by or considered to be caused by consumption of food or water." The disease-causing organisms such as bacteria, fungi, viruses, and parasites from diverse genera and classes may or may not be associated with toxin production and may acquire varying morbidity-mortality rates (Majumdar et al., 2018). Based on the nature of the disease evoked by pathogens are mainly classified into three types such as food infection, food intoxication, and food toxic infection.

2.2.1 FOOD INFECTION

In this case, foodborne illness occurs as a result of the ingestion of live microorganisms itself through contaminated food or water in the relative quantity of suggested levels causing brutal tissue injury. They are mainly enteropathogenic. These cells are generally alive at the time of consumption and have the potential to multiply within the host to increase their number in the gut to exert a harmful effect on the host. These microorganisms usually invade the intestinal epithelial cell lining and further try to spread to other organs and tissues. Food infection is predominantly associated with bacterial pathogens such as *Salmonella enterica*, *Campylobacter jejuni*, *Escherichia coli* (*E. coli* O157:H7), *Shigella* spp., *Yersinia enterocolitica*, and *Listeria monocytogenes* (Bhunia, 2018; Majumdar et al., 2018; Noor, 2019).

2.2.2 FOOD INTOXICATION

Food intoxication occurs as a result of the ingestion of food and water with a pre-formed bacterial or fungal toxins or secondary metabolites of toxic nature throughout their intensification in the food. The onset of symptoms is very quick in the case of foodborne intoxication. The presences of toxic metabolites in the food are the main contributor to foodborne illness and are the major cause of virulence instead of microorganisms itself. After the release of toxins in the food, there is no need for the presence of the microorganism in the food to further cause foodborne illness. Bacterial toxin-mediated foodborne illness has been principally associated with pathogenic species such as *Clostridium perfringens*, *Clostridium botulinum*, *Staphylococcus aureus*, and *Bacillus cereus*. Preformed toxins are produced by *Staphylococcus aureus*, and *Bacillus cereus* whereas, some strains of *Clostridium perfringens* and *Bacillus cereus* produce in vivo toxins. Botulin is a potent neurotoxin produced by *Clostridium botulinum* causing fatal sickness called botulism. Fungal toxins such as fumonisins, aflatoxin B and ochratoxin A are extensively recognized to be the etiological agents for the foodborne illnesses (Bhunia, 2018; Majumdar et al., 2018; Noor, 2019).

2.2.3 FOOD TOXICOINFECTION

This type of illness results from the ingestion of a large number of viable pathogens through contaminated food and water which produce toxins inside the host body by either sporulating or releasing toxin to further manifest to give related symptoms of illness. These toxins interact with the host epithelium cells and show gastrointestinal symptoms. *Clostridium perfringens*, enterotoxigenic *Escherichia coli* (ETEC), and *Vibrio cholera* are common examples (Majumdar et al., 2018) (Figure 2.2).

It is essential to acquire adequate knowledge regarding the pathogenesis mechanism and transmission sources. Further, it is essential to formulate effective prevention and control strategies. Increased demand for rapid and precise diagnostic tools for microbial pathogens in the food samples has necessitated the adoption of various molecular techniques with enriched laboratory settings for the detection of microorganisms in foodstuff (Bhunia, 2018).

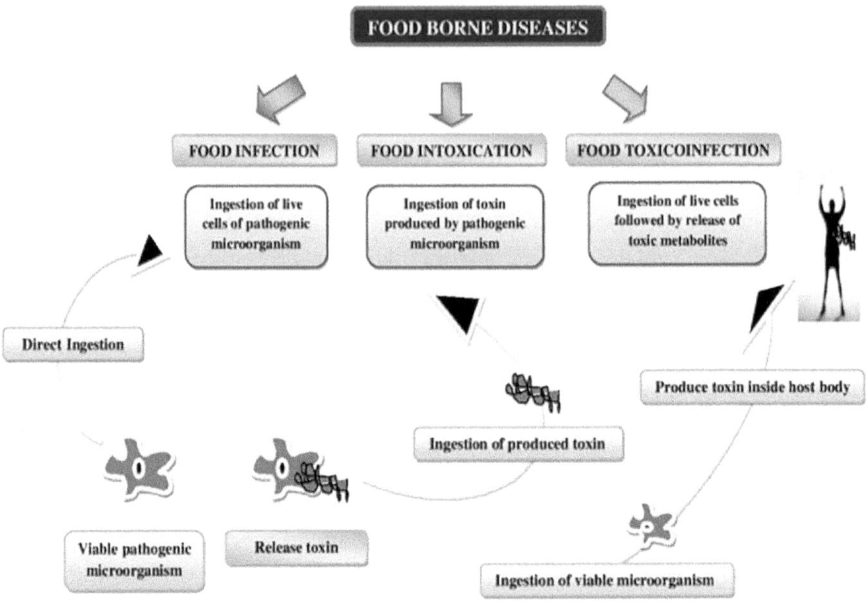

FIGURE 2.2 Flow diagram showing different types of foodborne diseases (Schmidt et al., 2003).

2.3 FOODBORNE PATHOGENS

A foodborne illness occurrence can be brought by contamination of food with any biological agents or pathogens (e.g., bacteria, fungi, viruses, and parasites), chemical agents (e.g., pesticides, metals) or physical agents (e.g., glass fragment). Among the various known foodborne diseases, the diseases caused by biological agent or pathogen is the most prevalent one (Schmidt et al., 2003). A pathogen is described as an organism capable of causing cellular damage to various tissues in the host and leads to outcomes like morbidity or mortality. They can colonize into host cells by continuous replication and ultimately serve as a virulence factor. Bacteria, viruses, fungi, and parasites are mostly associated with foodborne diseases. Pathogens can be designated into two categories; one is "primary pathogens," which are primarily related to foodborne diseases, and the other category is "opportunistic pathogens" that includes those pathogens that mainly attacks immune-compromised or susceptible host. Both these types of pathogens share the same attributes like (1) entry and survival in the host, (2) evade host defense, (3) multiply to significant numbers, and (4) to show desirable attributes within the host. Various factors such as pH, nutrient availability, presence/absence of oxygen, mucus composition, bile salt, and the balance of normal microflora influence pathogen growth and endurance within a host. Pathogens can be introduced into the food chain from various sources such as raw materials, humans, air, water, and equipment that contribute to foodborne illness (Figure 2.3).

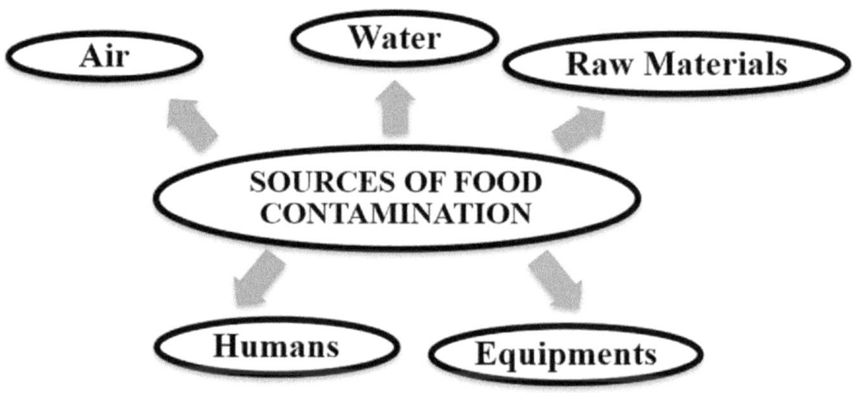

FIGURE 2.3 Sources of pathogen contamination within food chain.

Pathogenic microorganisms adopt different strategies to persist on food such as by forming a biofilm, spore formation, stress response/adaption to food inherent conditions, the viable but not culturable condition. From the food safety point of view, it is essential to know the type, number, and concentration of microorganism or its toxin to assure microbiological quality throughout the food chain by eliminating these biological hazards (Bhunia, 2008). Foodborne illness mediated via different biological agents like bacteria, fungi, viruses, and parasites have been summarized below (Table 2.1).

TABLE 2.1 Common Foodborne Pathogens

Pathogenic Agent	Pathogenic Genera	References
Foodborne bacterial agents	Gram-positive: *Bacillus*, *Staphylococcus*, *Mycobacterium*, and *Clostridium*	Bintsis (2017)
	Gram-negative: *Aeromonas*, *Brucella*, *Salmonella*, *Shigella*, *Escherichia*, *Yersinia*	
Foodborne fungal agents	Mold and its mycotoxin: *Aspergillus flavus*, *Aspergillus parasiticus* (Aflatoxin)	
	Aspergillus ochraceu, *Aspergillus carbonarius* (Ochratoxin)	
	Fusarium verticillioides (Fuminosin)	
	Penicillium expansum (Patulin)	
	Claviceps purpurea (Ergot)	
	Fusarium graminearum (Zearalenone)	
Foodborne parasites	*Giardia lamblia*, *Entamoeba histolytica*, *Trichinella spiralis*, *Toxoplasma gondii*, *Trypanosoma cruzi*, *Cyclospora cayetanensis*, *Cryptosporidium parvum*	
Foodborne viral agents	Norovirus, Adenovirus, Rotavirus, Astrovirus	

2.3.1 FOODBORNE BACTERIAL AGENTS

Bacteria are the most predominant among the various foodborne pathogenic agents and exist in varying shapes, types, and characteristics. Some have the characteristic of spore formation in the case of highly resistant *Clostridium botulinum*, *Clostridium perfringens*, *Bacillus subtilus*, and *Bacillus cereus* whereas some can produce heat-resistant toxins like *Staphylococcus aureus* and *Clostridium botulinum*. These bacterial pathogens mostly require mesophilic optimal growth temperature ranging from 20°C to 45°C. Some of the bacteria are psychrotrophic such as *Listeria monocytogenes* and *Yersinia enterocolitica* that can grow below 10°C under refrigerated conditions. The major types associated with foodborne illnesses are (1) *Salmonella* (contaminating raw and undercooked meat, poultry, dairy products, and seafoods), (2) *Campylobacter jejuni* (contaminating the raw or undercooked chicken and unpasteurized milk), (3) *Shigella* (contaminating the water used for food processing), (4) *Escherichia* (contaminating raw or undercooked meat, unpasteurized fruit juices, and milk and fresh vegetables and fruits), (5) *Listeria monocytogenes* (contaminating raw and undercooked meats, unpasteurized milk, soft cheeses, etc.), (6) *Clostridium botulinum* (contaminating improperly canned foods and smoked and salted fish), (7) *Bacillus cereus* from infant foods, and (8) Shiga toxin-producing *Escherichia coli*, including O157 and many nonO157 serogroups (Noor, 2019). All of these bacteria vary in their susceptibility to various parameters such as high salt, sugar, or total solids level as well as the acidity of food products. A clear understanding of the effects of these parameters on the growth of pathogens can be utilized for designing the preservation method ensuring food safety (Schmidt et al., 2003). Apart from this, as a control system effect surveillance procedures must be adopted to prevent foodborne disease outbreaks. These surveillance systems are utilized to reveal the interaction of food and associated pathogen. Determination of genotype and subtype information of various foodborne pathogenic strains stand necessary to sketch all required details such as transmission source, characteristics, and similarity with other strains. Whole-genome sequencing is an essential tool to serve this purpose (Bergholz et al., 2014).

2.3.2 FOODBORNE FUNGAL AGENTS

Fungi or molds are ubiquitous and cause considerable spoilage of foods such as grains, fruits, vegetables, and nuts. Some molds are capable of producing toxins called mycotoxins, which may possess carcinogenic and teratogenic effects and significantly colonize food crops (Bhunia, 2018). Aflatoxins are amongst the most important mycotoxin associated mainly with crops that cause liver cancer and impairment in child growth. Other important mycotoxins are fumonisins (causing esophageal cancer and neural defects), deoxynivalenol (DON) (causing gastroenteritis) and ochratoxin A (related with renal diseases). These common mycotoxins are produced by food crops infecting fungal genera *Aspergillus*, *Fusarium*, and *Penicillium*. Mycotoxins exert acute health effects targeting kidney and liver and chronic effects that may result in liver cancer, neurotoxicity, birth defects, and ultimately to death by three principal modes of action such as mutagenic, carcinogenic, and teratogenic. These mycotoxins are amongst the most critical foodborne biological contaminants that need to control vigilantly to protect the health of the public on a global scale. Diverse assortments of measures are adopted to mitigate the risks associated with these mycotoxins. These measures mainly include interventions such as pre-harvest, post-harvest, dietary, and clinical methods. The pre-harvest methods include the adoption of good agricultural practices, biocontrol, and pest management. Post-harvest interventions emphasize hygienic procedures followed during post-harvest operations to eliminate any chance of fungal contamination, whereas dietary methods include the use of edible antifungal agents in the food formulations. Clinical intervention ensures the mitigation of various health risks associated with the consumption of these mycotoxins (Wu et al., 2014).

2.3.3 FOODBORNE VIRAL AGENTS

Virology as science has shown a rapid expansion due to the principal discovery of new pathogens at an escalating pace. Over the years the emergence of SARS coronavirus and highly pathogenic avian influenza viruses (HPAI) of subtypes H7N7 and H5N1 that have affected humans have acquired considerable attention.

The viruses have the capability of causing major illness and human mortality. It is essential to understand the elementary properties of foodborne viruses as these properties are necessary for establishing microbiological control criteria as well as valid procedures to eliminate the occurrence of foodborne disease outbreaks. Virus mediated foodborne diseases are difficult to be controlled as these viruses do not grow in food, spread at a very high pace, and lack systematic surveillance facilities. Norovirus has been recognized as one of the most widespread causes of gastroenteritis among the population. It possesses great concern as the symptoms are relatively milder, difficult to notice except when leading to major outbreaks. Detection of the foodborne virus is another important hurdle in controlling foodborne disease outbreaks. Diagnosing these viral agents largely utilizes molecular techniques as these agents are either difficult to culture or even impracticable, as in the case of Noroviruses. It is more relevant to develop effective preventive strategies to eliminate any chance of contamination at early stages in the food chain. The emergence of new viral diseases over the years has made the development of foodborne viral disease control measures a global priority (Newell et al., 2010).

2.3.4 FOOD PARASITES

Parasitism refers to the association between two microorganisms in which one is benefited (parasite) and the other is harmed. The parasite develops this association to promote its growth and development resulting in harm to the host (Nelluri and Thota, 2018). Parasitism is carried out by various microorganisms, amongst which protozoa and helminths are the most common ones. The most common foodborne parasitic pathogens are *Toxoplasma gondii*, *Entamoeba histolytica*, and *Cryptosporidium*. Most of the parasitic infections of enteric type are transmitted through the oral-fecal route by ingestion of food contaminated with free-living parasites (eggs, cysts, and oocysts). The potential sources of this type of contamination are feces, sewage, soil, human handling, or irrigation water.

Parasites generally differ from bacterial-food pathogens in the fact that they do not replicate outside the host, and are generally not susceptible to antibiotics. For these reasons, prevention, and control of foodborne-parasitic diseases are tedious (Newell et al., 2010).

2.3.5 EMERGING FOODBORNE PATHOGENS

The spectrum of foodborne illness has widened over the years and has shown the establishment of pathogens that have been eliminated, controlled, or emerged as newer ones. The emergence of new pathogens or re-emergence of already controlled ones in a new way that has been associated with increasing health concerns in the population (Smith and Fratamico, 2018). An emerging foodborne pathogen has been defined as the "contributory agent of a communicable disease whose occurrence is increasing after its appearance in a new host population or whose occurrence is escalating in the present population as a result of long-term changes in its underlying epidemiology" (Cleaveland et al., 2007). The emergence of new or unanticipated pathogens in food has aroused critical concerns about food safety. These emerging pathogens may include strains that have exhibited improved resistance to stresses as well as enhanced adaption to new environments.

Emergence and re-emergence is a complex process and is occurred as a result of interventions like (1) advancement in agricultural practices, (2) evolution and adaptation of microbes, (3) climate change, (4) changing lifestyle, (5) technological advancement and (6) global trade. Various emerging species and strains of foodborne pathogens have been identified over the years. Amongst the emerging bacterial pathogens *Aeromonas*, *Arcobacter*, *Clostridium difficile*, *Cronobacter*, *Mycobacterium paratuberculosis*, *Streptococcus suis* as well as non-O157 *Shiga* toxin-producing *Escherichia coli* are the most important ones. Apart from this *Campylobacter jejuni* has been identified as one of the most widespread and severe emerging foodborne bacterial pathogens transmitted through meat and poultry products. It is mainly associated with sporadic bacterial gastroenteritis. Viruses such as Hepatitis E virus (HEV) and Coronavirus have emerged as a serious concern raising serious health issues in the population. *Cyclospora cayetanensis*, *Toxoplasma gondii*, *Trichinella*, *Cryptosporidium*, and *Giardia* are amongst the newer emerging foodborne parasites (Koutsoumanis et al., 2014). Development and adoption of robust and effectual surveillance programs stand necessary for early diagnosis and control of challenging foodborne emerging pathogens.

2.4 MECHANISMS OF FOODBORNE PATHOGENS ACTION

Foodborne pathogens have been associated with three different forms of foodborne illness such as foodborne infection, intoxication, and toxicoin-fection. The principal route for the occurrence of these illnesses is oral using food or water as vehicles and primarily targets the gastrointestinal tract. Some of these pathogens or toxins brings about localized damage whereas some may expand to other sites for deeper tissue damage (Bhunia, 2018). Host characteristics greatly affect the probable mechanisms and efficacy of pathogens to cause illness. Different pathogens differ in the way they colonize in the host, pathogenesis mechanism, their adequate numbers, and the quantity of toxin required to cause illness. The host can be broadly categorized into two types, one with the low risk and the other one includes individuals at high risks such as immune-compromised/suppressed, elderly as well as small children (Schmidt et al., 2003).

The pathogens that are accountable for food infection colonize the gut by various mechanisms such as adhesion and cohesion of pathogens in the gastrointestinal tract by producing various adhesion factors such as fimbriae, curli, adhesion proteins, and extracellular matrices that facilitate the formation of biofilm. Other mechanisms of foodborne infection pathogenesis include quorum sensing, natural as well as induced phagocytosis, evasion of immune system and intracellular residence, etc.

Foodborne illness mediated by toxin can be either intoxication or toxicoinfection. Pathogens present in food produce toxins under favorable conditions. These toxins are primarily of two types: exotoxin and endotoxin. Exotoxins are toxins that are generally released outside after cell lysis via an active transport system (e.g., botulinum toxin, cholera toxin, Shiga toxin, and so on) whereas, endotoxins mainly form the part of cellular structure which is liberated on cell disruption. In the case of foodborne intoxication, the illness is mainly mediated via exotoxin which is produced by a pathogen in the food while in foodborne toxicoinfection, both exotoxins and endotoxins can be produced inside the host body after ingestion of pathogenic microbial cells (Figure 2.4).

The preformed or in vivo toxins are absorbed through gastrointestinal lining resulting in local tissue damage and inflammation showing visible symptoms such as vomiting and diarrhea. Whereas, in some cases, toxins may circulate through the blood, lymphatic or distant organs such as liver, kidney, central nervous system (CNS), and cause their damage. These toxins act by either cell death via necrosis or apoptosis or by induction of second messenger pathways that destabilized epithelial cell architecture increases membrane permeability and fluid loss (Bhunia, 2008). Understanding of these probable mechanisms stands essential to know the extent of virulence caused by these pathogens.

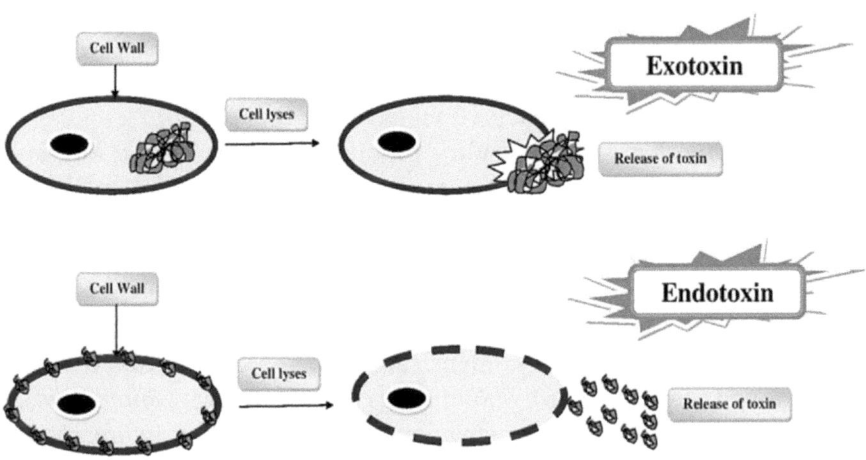

FIGURE 2.4 Exotoxin versus endotoxin.

2.5 FOOD-RELATED MICROORGANISMS

Microorganisms are everywhere and can stay alive and rise in extreme atmospheric conditions as in food, and human hosts. They vary in their optimal conditions necessary for their growth and development. These parameters that affect their growth and propagation in food mainly include factors like pH, temperature, water activity, presence or absence of oxygen, osmotic pressure, inherent constituents of the food itself, and so on. Food consists of a diverse assortment of components like salts, acids, aldehydes, flavoring compounds, ions, and antimicrobial substances. Food serves as a great substrate for the growth of various pathogenic microorganisms due to its high nutrient value. These microorganisms are mostly able to survive in the harsh environmental factors in food. Different food groups are infected by different pathogenic microorganisms depending on various growth-promoting factors in food that particularly regulate their growth.

A source of human pathogen contamination in fruits and vegetables mainly involves pathogens attachment to the surface of fruits and vegetables as well as internalization into the fresh food produced by processes like vacuum or hydro cooling or washing. The most common bacterial food spoilage organisms or pathogens associated with fruits and vegetables are *Pseudomonas syringae*, *Pseudomonas fluorescens*, *Agrobacterium*, *Erwinia*, *Clostridium*, *Bacillus*, *Klebsiella*, etc. (Bhunia, 2018). Milk and dairy products provide an excellent substrate for the growth of pathogenic microorganisms due to their exclusive composition and properties. *Staphylococcus aureus*, *Salmonella*, *Listeria monocytogenes*, *Campylobacter*, and *Escherichia coli* O157:H7 are the most common probable pathogens related to milk or milk products (Cancino-Padilla et al., 2017).

The safety of meat and poultry products has also been a key issue amongst the population. The most common bacterial pathogens associated with meat and meat products are *Listeria monocytogenes*, *Campylobacter*, and *Clostridium perfringens*. *Salmonella* species are predominantly associated with beef. *Salmonella typhimurium* and *S. enteritidis* are amongst the most non-typhoidal *Salmonella* infection leading to gastroenteritis in humans can often be isolated from raw meat (Biswas and Mandal, 2016). *Salmonella enterica* serovars *enteritidis* and *typhimurium*, *Clostridium perfringens*, *Campylobacter jejuni*, and *E. coli* are the most common species associated with raw poultry. Fish and fish products can be heavily contaminated with fecal microorganisms as their quality is largely affected by the quality of water in which they are cultivated. Contaminated water may contaminate fish and fish products with fecal coliforms, fecal *Streptococci*, and *Staphylococcus aureus*. Fish and shellfish can be contaminated with *Pseudomonas*, *Clostridium perfringens*, *Listeria monocytogenes*, *Vibrio parahaemolyticus*, *Vibrio vulnificus*, *Salmonella enterica* serovar *enteritidis* and *typhimurium*, *Campylobacter jejuni*, *Yersinia enterocolitica*, and Enteroviruses (Hepatitis A) (Bhunia, 2018).

Apart from effective control strategies, it is important to develop efficient detection technologies for improved diagnoses of various pathogenic species or strains in food products. Novel approaches such as biosensors, PCR, immunoassays, whole-genome sequencing have been designed to continuously reveal the characteristics and information about pathogens based on various molecular and analytical procedures. These techniques provide advantages over the conventional techniques that are less efficient and give unreliable results.

2.6 CONCLUSION

Food microbiology generally deals with the interaction of microorganisms with the food. These interactions can be desirable as in the case of beneficial microorganisms or undesirable in the case of pathogenic microorganisms. Foodborne pathogen contamination in food possess serious health concern amongst the population. These pathogens may differ in the source of origin, mechanism of pathogenesis, the extent of virulence, and associated factors. Based on the nature of pathogenesis these foodborne diseases are classified as foodborne infection, intoxication, and toxicoinfection. Any form of these foodborne illnesses poses health risks to the host they infect. Ingestion of live pathogenic micro or their toxin has shown a wide range of serious effects putting the health or life of people in danger. Hence, control, and critical understanding of these foodborne diseases are necessary to mitigate issues of food safety concerns. Further, the emergence and re-emergence of foodborne pathogens have led to various serious foodborne outbreaks. Critical screenings, as well as the establishment of effective management and surveillance procedure, stand essential to ensure food safety for all.

KEYWORDS

- **foodborne diseases**
- **mechanism**
- **pathogens**
- **toxicoinfection**
- ***Vibrio vulnificus***

REFERENCES

Bergholz, T. M., Switt, A. I. M., & Wiedmann, M., (2014). Omics approaches in food safety: Fulfilling the promise? *Trends in Microbiology, 22*, 275–281.

Bhunia, A. K., (2008). General mechanism of pathogenesis for foodborne pathogens. *Foodborne Microbial Pathogens: Mechanisms and Pathogenesis*, 93–112.

Bhunia, A. K., (2018). *Foodborne Microbial Pathogens: Mechanisms and Pathogenesis* (pp. 1–351). Springer, New York.

Bintsis, T., (2017). Foodborne pathogens. *AIMS Microbiology, 3*(3), 529–563.

Biswas, A. K., & Mandal, P. K., (2016). Meat-borne pathogens and use of natural antimicrobials for food safety. *Foodborne Pathogens and Antibiotic Resistance*, 225–245.

Cancino-Padilla, N., Fellenberg, M. A., Franco, W., et al., (2017). Foodborne bacteria in dairy products: Detection by molecular techniques. *Ciencia e Investigación Agraria, 44*(3), 215–229.

Cleaveland, S., Haydon, D. T., & Taylor, L., (2007). Overviews of pathogen emergence: Which pathogens emerge, when and why? In: *Wildlife and Emerging Zoonotic Diseases: The Biology, Circumstances and Consequences of Cross-species Transmission* (pp. 85–111). Springer, Berlin.

Ellis, D. I., Muhamadali, H., Chisanga, M., et al., (2018). Omics methods for the detection of foodborne pathogens. In: Melton, L., Shahidi, F., & Varelis, P., (eds.), *Encyclopedia of Food Chemistry* (pp. 364–370).

Elshafei, A. M., (2017). Role of microorganisms in food contamination, processing and safety. *Journal of Food Microbiology, 1*(1), 1–2.

Gangiredla, J., Yan, X., Patel, I. R., et al., (2017). Application of omics technologies and computational approaches for control of foodborne pathogens in foods. In: *Microbial Control and Food Preservation* (pp. 41–54). Springer, New York.

Koutsoumanis, K. P., Lianou, A., & Sofos, J. N., (2014). Food safety: Emerging pathogens. *Reference Module in Food Science, Encyclopedia of Agriculture and Food Systems*, 250–272.

Majumdar, A., Pradhan, N., Sadasivan, J., et al., (2018). Food degradation and foodborne diseases: A microbial approach. In: *Microbial Contamination and Food Degradation* (pp. 109–148). Academic Press.

Martinović, T., Andjelković, U., Gajdošik, M. Š., et al., (2016). Foodborne pathogens and their toxins. *Journal of Proteomics, 147*, 226–235.

Nelluri, K. D., & Thota, N. S., (2018). Challenges in emerging foodborne diseases. In: *Food Safety and Preservation* (pp. 231–268). Academic Press.

Newell, D. G., Koopmans, M., Verhoef, L., et al., (2010). Foodborne diseases – the challenges of 20 years ago still persist while new ones continue to emerge. *International Journal of Food Microbiology, 139*, 3–15.

Noor, R., (2019). Insight to foodborne diseases: Proposed models for infections and intoxications. *Biomedical and Biotechnology Research Journal, 3*, 135–139.

Odeyemi, O. A., (2016). Public health implications of microbial food safety and foodborne diseases in developing countries. *Food and Nutrition Research*, 60.

Schmidt, R. H., Goodrich, R. M., Archer, D. L., et al., (2003). General overview of the causative agents of foodborne illness. *EDIS, 6*, 1–5.

Smith, J. L., & Fratamico, P. M., (2018). Emerging and re-emerging foodborne pathogens. *Foodborne Pathogens and Disease, 15*(12), 737–757.

Wu, F., Groopman, J. D., & Pestka, J. J., (2014). Public health impacts of foodborne mycotoxins. *Annual Review of Food Science and Technology, 5*, 351–372.

CHAPTER 3

Role of Foodborne Pathogens and Microorganisms in Food Safety

SANTOSH KUMAR SHIMOGA RAJANNA,
NAVEEN KUMAR KUDURE JAYANNA, MANOJ MANJUNATH BONGALE,
and CHANDRASHEKAR SRINIVASA

ABSTRACT

Foodborne pathogens have been regarded as one of the chief concerns in food security due to their adverse effect on human metabolism after consumption. Microorganisms like bacteria, viruses, and few fungi and parasitic species are considered as concerning source of food contamination in the 21st Century, as they contaminate both food and sources of food at various levels. The resultant health aberrations are diverse, ranging from gastrointestinal diseases like cholera, diarrhea to neurological diseases like cerebritis and meningitis. These diseases occur due to the various pathogenic metabolites produced from the microbes known as toxins. Despite several precautionary measures taken, morbidities continue to occur, resulting in the deaths of several thousands of lives. This chapter provides a comprehensive overview of foodborne pathogens, including bacteria, viruses, and parasites that are responsible for causing several foodborne illnesses through highlighting their epidemiological profile, mode of infection, types of foods susceptible, and effect on humans. On the other hand, we also report a few bacteria, yeast, and molds that are also considered as beneficiaries. We principally focus on the pathogenesis of these foodborne pathogens that may help in finding different diagnostic

Global Food Safety: Microbial Interventions and Molecular Advancements. Saher Islam, PhD, Devarajan Thangadurai, PhD, Jeyabalan Sangeetha, PhD, & Natália Cruz-Martins, PhD (Editors)

methods to detect and diagnose these pathogens for the betterment of mankind.

3.1 INTRODUCTION

The first identification of the association between food and illness dates back to 460 BC, during the period of Hippocrates who indicated the correlation between human illnesses and the food consumption habits (Hutt and Hutt, 1984). Microbial agents (such as bacteria, fungi, and viruses) and parasites that cause any undesirable condition are classified under foodborne pathogens. The disease outbreak due to foodborne pathogens is mainly defined as the incidence of more than two cases of similar illness as a result of the ingestion of a certain food (CDC, 2012).

The pathogen itself or the toxins released from the pathogen are both prospective of causing diseases, and therefore, such illnesses are classified as foodborne infections and foodborne intoxications. In that foodborne intoxications are seen very early during the prognosis whereas infections from foodborne pathogens are usually observed after a period of time, primarily due to the time required for the pathogen to establish itself within the host tissues. Infections related to foods can likely affect either the younger or older generations, whereas the mid-aged population are less likely to get affected unless they are exposed to a very high dose of the pathogen (CDC, 2012).

Preliminary survey data released from the Center for Disease Control (CDC) suggest that foodborne illnesses peaked in the year 2018 with high incidence from the *Campylobacter* (~10% of the overall outbreaks), *Salmonella* (~22% of the overall outbreaks) and *Cyclospora* (~10% of the overall outbreaks) pathogens. Even the parasites, especially histamines were reported in close to 3% of the cases. This elevation in the numbers was mainly because of the advancement in the diagnostic methods with the advent of culture-independent diagnostics. In the European Union alone, 2015 saw an outbreak affecting over 45,000 patients with almost one-fourth hospitalizations and one in 100 deaths. This increased in the due course until 2019. Most of the foodborne pathogens were transmitted from food sources of animal origin such as egg, pig meat (close to 10% of the outbreak), fish, and fish products (7%), dairy products (5%), and others. *Salmonella* tops the list of the highest outbreak causing pathogen in terms of illness and hospitalization. The advancement in the food processing and distribution and the demand for newer variety of foods have challenged the production industries time and again to increase the supply as per the demand. These changes have resulted in the emergence of newer pathogens as well as the re-emergence of the existing ones with greater virulence (EFSA, 2016).

Although the very first glance of the presence of microorganisms in food sources seems alarming, a large number of microorganisms are very useful for food processing and preparation industry (Hofstra et al., 1994; Bisen et al., 2012). Right from the dawn of civilization, fermentation of milk forming a variety of milk products, vegetables, and fruits forming large number delicacies are known to mankind (Rezac et al., 2018). With no scientific backup, these productions were only artisanal with little or no appreciation of the involvement of microorganisms. However, with the advancement in technologies, a lot of research has been focused on the fermentation industry for the production of superior quality products that include fermented foods, beverages, and medicines such as antibiotics (Mishra et al., 2017; Sahu and Panda, 2018). The mid-19th Century saw significant events that changed the scenario of food industries. Industrialization peaked up only during this period owing to the accumulation of large population within the towns and cities, which created a demand for the mass production of these fermented and processed foods (Steinkraus, 2004). Additionally, this period also witnessed the advent of microbiology as a branch of science and the importance of bacteria and molds for the process of fermentation (Erkmen and Bozoglu, 2016). Luckily for the industrialization, there was a strong backing from the scientific standpoint mainly because the understanding of the pathways, the key regulatory enzymes and methods of mass-production were being understood with greater detail in this century (Mishra et al., 2017). Although the traditional backslapping is still in practice in many of the European countries for the production of cheese, fermented meat, and vegetables (owing to the better retention of flavor and aroma in these products), large-scale production demands the factory set-up for meeting the growing demands of the population (Tamang et al., 2020). This chapter discusses the various microorganisms involved in food fermentation and processing and as foodborne pathogens.

3.2 FOODBORNE PATHOGENS

3.2.1 BACTERIAL PATHOGENS

The most commonly found foodborne diseases are caused by different species of bacteria. While some of the pathogenic bacteria are spore-forming and therefore highly heat-resistant (e.g., *Clostridium botulinum*, *Bacillus subtilis*, *Bacillus cereus*, *C. perfringens*), the others produce toxins that can withstand high temperatures (*Staphylococcus aureus*, *Clostridium botulinum*). In the following sections, various bacteria, their optimal growth conditions, and the diseases caused by them are discussed (Bacon and Sofos, 2003).

3.2.1.1 BACILLUS CEREUS

They are the members of *Bacillaceae* family of Gram-positive, motile rod-shaped bacteria with spore-forming ability. They are present all over including soil, water, and air. They produce spores which possess pili that are the structures used for motility (Bacon and Sofos, 2003). They resist its elimination during sanitation because of the presence of this structures that enable them to adhere to the surfaces (Rajkowski and Smith, 2001). They grow optimally at a range of temperatures from 4–15°C to 35–55°C and exist in an even wider range of pH such as 4.9–9.3 (Bacon and Sofos, 2003; FDA, 2012). Yet, they are optimally active between 30–40°C (Andersson et al., 1995). There are two types of toxins produced by *B. cereus* namely the emetic (produced during the growth phase within food products) and diarrhea-causing (produced during the growth phase in the small intestine) ones. These toxins are produced by the endospore-forming types of *B. cereus* that survive the harsh cooking processes and enter the proliferation stage during the course of storage (ICMSF, 1996; Rajkowski and Smith, 2001). Foods such as meat, soups, vegetables, sauces, and dairy products are implicated in diarrheic food poisoning, which are characterized by pain in the abdomen, diarrhea, and nausea. Additionally, these endospores can also produce certain virulence factors that can cause mild but very uncomfortable conditions in the gastrointestinal tracts. The food poisoning related to this bacterium is often under-reported mainly because of its mild, self-limiting, and short duration symptoms (Stenfors et al., 2008).

Outbreak in a recent study was implicated from the contamination of rice dishes with *B. cereus* where fried rice was caused highest contamination (68%) (Bennett et al., 2013). Meat and poultry also contributed to a great degree of disease outbreak caused by this pathogen. Outbreak of severe vomiting was reported after a birthday party in Italy caused by the emetic toxin (Martinelli et al., 2013). Similarly, after a college event in Thailand, close to half of the students reported illnesses, such as vomiting, diarrhea, and abdominal cramping. The cause was later traced back to the consumption of cream-filled eclairs with contaminated cream. Several other reports have shown the outbreak of *B. cereus* infections leading to emetic and diarrheal symptoms (Wijnands, 2008).

3.2.1.2 ESCHERICHIA COLI

E. coli, a Gram-negative, non-motile (NM), non-spore-forming rod-shaped bacteria is present in the most diverse environmental conditions. This facultative anaerobe has the ability to ferment sugars such as lactic acid, formic acid, and acetic acid. While most *E. coli* strains are relatively harmless, some of the strains have acquired the ability to produce toxins and therefore pathogenic to humans (Mitscherlich and Marth, 2012). Pathogenic types of *E. coli* invite greater health concerns causing a high amount of morbidity and mortality globally.

Prospective entry-points for the transmission of *E. coli* include food and water sources that are contaminated with the human and animal fecal matter (Croxen et al., 2010). Even the use of contaminated manure (feces of animals) contributes by large for the contamination of agricultural crops. The pathogenic model of *E. coli* has led to classifying them into six pathotypes (Croxen et al., 2010; Garcia-Angulo et al., 2010) namely:

- Enteropathogenic *E. coli* (EPEC)
- Enterohemorrhagic *E. coli* (EHEC, also known as Shiga toxin-producing *E. coli* [STEC] and formerly referred to as verotoxin-producing *E. coli* [VTEC])
- Enterotoxigenic *E. coli* (ETEC)
- Enteroaggregative *E. coli* (EAggEC)
- Enteroinvasive *E. coli* (EIEC), and
- Attaching and effacing *E. coli* (A/EEC)

Among the various toxins, STEC leads to hemorrhagic colitis (HC) and post diarrheal hemolytic uremic syndrome (HUS), which causes increased bleeding, kidney disorders and death. About 63,000 illnesses and 2,100 hospitalization have been reported because of STEC alone leading to close to 20 deaths annually. It primarily resides in the intestinal tract of cattle but can also exist within the intestine of various other animals. STEC strain O157:H7 is responsible for the production of this toxin (Scallan et al., 2011), which was first observed in an outbreak of bloody diarrhea after the consumption of food from a fast-food chain demonstrated the presence of *E. coli* that expressed O-antigen 157 and H-antigen 7 (Wells et al., 1983). It was named STEC because of the production of toxin similar to the Shiga toxin produced by *Shigella dysenteriae* that was reported for the first time in this outbreak. Further studies on this toxin demonstrated that this strain of *E. coli* was responsible for 85–95% of HUS (Garcia-Angulo et al., 2010) whereas the other toxin-producing strains caused only 5–15% of disease outbreaks globally (Armstrong et al., 1996). The greatest outbreak of STEC-led diseases was observed in 2011 causing over 4,000 illnesses in 16 countries of the European Union just in a span of 2 months (Rasko et al., 2011). This led to 800 HUS cases and 50 deaths in these countries. The characteristics of this outbreak were that several adults suffered this disease that was also associated with neurological disorders. The culmination of typical features of enteroaggregative *E. coli* (EAEC) as well as the ability to produce Shiga toxin was the primary reason for the clinical characteristics observed in this study. Eventually, the studies on other serogroups such as O26, O103, O91, O145, O146, and O128 were also undertaken that lead to the awareness of the diseases caused by these pathogens (Frank et al., 2011).

One such outbreak, leading to over 183 illnesses and one death, was associated with the consumption of contaminated spinach. This led FDA to implement policies to prevent the consumption of fresh and uncooked spinach (CDC, 2006). Studies on the epidemiology of this disease traced it back to a shift in the processing plant of the Natural Selections Foods Company at San Juan Batista, California which produced packed, ready-to-eat (RTE) baby spinach that was contaminated (Jay et al., 2007). Tracing it further down, it was identified that the water source near particular field growing organic spinach was contaminated with the cattle and pig feces (Jay et al., 2007; Berger et al., 2010). In Germany, sprout seed mixes dispatched to various parts of the world were contaminated with STEC

O157, but this consignment could not be retrieved from the market owing to a widespread distribution of this product to the whole of the European Union (EFSA, 2011). The product was eliminated from the store shelves only after reaching its expiry date and this caused a lot of illness because of the production of this toxin (EFSA, 2011). Similarly, another outbreak that was implicated due to the contamination of STEC O157 was in the flour and flour-based batters and doughs that were used for the preparation of pizza bases (Zhang et al., 2007; CDC, 2016). General Mills, which was responsible for the distribution of these flours, recalled the packets in 2016 and suggested that consumption of raw dough, batter, or flour even for tasting purposes was not safe and can lead to illness (Zhang et al., 2016).

3.2.1.3 CAMPYLOBACTER JEJUNI

They are from the *Campylobacteriaceae* causing diarrheal illness. They are Gram-negative, spiral forming bacteria 0.2–0.9 μm wide and 0.2–5.0 μm long (Humphrey et al., 2007). They are microaerophilic bacteria (growing in approximately 10% CO_2 and 5% O_2) and grow in a narrow temperature range between 30–46°C (Humphrey et al., 2007). This alone is responsible for 85,000 morbidities and close to 8,500 hospitalizations annually (Scallan et al., 2011). As per a report given by WHO, approximately 1% of the European population are infected with this bacterium annually that affect the intestines of birds and mammals including humans (Humphrey et al., 2007). The bacteria penetrate inside the epithelial cells by first adhering to it and then damaging the cells. This damage causes conditions such as diarrhea and other systemic infections (NCBI, 2017).

The bacteria is a part of the normal intestinal flora of healthy animals such as sheep, cattle, goats, pigs, ducks, chicken, birds, dogs, cats, and several other marine mammals. Even the natural and groundwater can be positive for the presence of this bacterium, whereas RTE foods and processed foods are at a greater risk of contamination from this bacterium. Under stressed conditions, campylobacter attains "viable but non-culturable state," which has been characterized by elevated uptake of amino acids by the intact membrane but fail to grow on selective medium (Stern et al., 1994; Altekruse et al., 1999; Schaffter et al., 2004). In one of the outbreaks in 2012, 44 people from a wedding reception in Sweden fell ill after the consumption of chicken liver pate that was deliberately undercooked (Lahti et al., 2017). This was also backed by several such outbreaks caused by the consumption of chicken liver pate worldwide.

In addition, consumption of raw milk was also associated with the symptoms of gastrointestinal illness. Similarly, in a jogging rally conducted at Switzerland *C. jejuni* enteritis was observed after the consumption of a drink from raw milk that attacked over 75% of the participants (Hui, 2018). A similar gastroenteritis was also observed in 56 of 235 students and guests of a university residential college after the consumption of food during a function in Australia (Moffatt et al., 2016). However, the overall reports of infections caused by *Campylobacter* species are underestimated owing to the deficit for optimal isolation and detection methods (Lastovica, 2006).

3.2.1.4 CLOSTRIDIUM BOTULINUM

They are the Gram-positive, spore-forming bacteria belonging to the *Bacillaceae* family that include obligatory anaerobic or aero tolerant rods. The vegetative cells appear rod-shaped (either straight or curved), from short coccoid to long filamentous structures that are round, tapered, or blunt existing in single, paired or chain-like appearance. This shape is primarily attributed to the endospores produced by the bacteria under stress, which further produce sporulation under favorable conditions such as anaerobic state or elevated nutrient supply. They are highly prevalent in the soil and intestines of various animals. It is widely distributed in freshwaters, marine sediments, soil, fish, meat, canned foods such as corn, peppers, soups, asparagus, olives, spinach, mushrooms, etc. (Bacon and Sofos, 2003). The mobility of *Clostridium botulinum* is facilitated by means of peritrichous flagella (FDA, 2012). They produce highly lethal neurotoxins that are classified into seven types from A through G on the basis of their antigenic specificity. Ingestion of these toxins, especially A, B, E, and F cause botulism in humans whereas those having C and D antigenic specificity leads to botulism in birds and mammals (Bacon and Sofos, 2003; FDA, 2012).

This bacterium is considered to be the most ancient one and evolutionarily led to the formation of the present-day Mollicutes/Eubacteria (NCBI, 2017). Some of the *Clostridium* species have been identified to produce solvents such as ethanol and acetone, whereas some others produce molecular hydrogen and other useful secondary metabolites. However, ingestion of toxins from *Clostridium* spp. have been reported to cause illnesses (NCBI, 2017). Traditionally botulism was known to be acquired through the consumption of undercooked sausages and canned foods, but recent years witness a large variety of food sources such as cheese, yogurt, salads, sautéed and stored onions, garlic sauce, etc.

Botulism exhibits symptoms of nausea and vomiting after the ingestion of the neurotoxin (FDA, 2012). This was further augmented by more devastating neurological symptoms such as visual impairment, acute flaccid paralysis of the head, face, and pharynx which descends further to the muscles of the thorax. These abnormalities finally lead to the death succumbing to the respiratory failure. Although this toxin is synthesized during the growth of the cells, they are released into the environment upon cell lysis and activated by the proteolytic cleavage (NCBI, 2017). Botulism can be of four categories namely foodborne-caused by the ingestion of toxin containing food, wound botulism-caused by the growth of organism within an infected wound, infant botulism-caused within the intestinal tracts of infants and botulism resulting from the colonization of the bacteria inside the intestine of older children and healthy adults (FDA, 2012; NCBI, 2017). Outbreak of botulism infections was observed in several cases. For instance, in Thailand, 209 cases of consumption of home-canned bamboo shoots led to a large outbreak of this disease (Carter and Peck, 2015). In 2007, eight botulism cases were reported in the USA caused by the consumption of canned hotdog chili sauce and the entire consignment of 39 million cans were recalled from the market (Juliao et al., 2013). Further research on these cases indicated that undercooked canned foods are the most potent sources for the growth of *Clostridium botulinum* causing botulism.

3.2.1.5 LISTERIA MONOCYTOGENES

This bacterium is one of the leading organisms causing deaths from foodborne pathogen. It has been known to cause devastating infections in newborns, the elderly, pregnant women, and immuno-compromised population. Especially in pregnant women, this infection potentially causes stillbirths, miscarriages, and birth defects. This pathogen is mainly found in dead and decaying matters such as vegetable remains, sewage water and soil, etc. This pathogen has a very wide temperature requirement and can thrive in extreme conditions such as 1–45°C. *Listeria* is regarded as an extremely dangerous foodborne pathogen found in a variety of raw and uncooked vegetables and meat. It infects the intestine and spreads through the nervous system and the fetal placental unit (Buchanan et al., 2017). There are about 13 serotypes identified in this species causing meningitis, gastroenteritis, and septicemia (FDA, 2012). Unlike most of the pathogens discussed so far, listeriosis is a relative deadly disease with high morbidity and mortality rates (Jemmi and Stephan, 2006). They even have the potential to multiply under refrigeration and therefore enhance the risk of more contamination (Gandhi and Chikindas, 2007). The spread of this pathogen gets exponential in damp and wet environment and studies have suggested that once it enters the production unit of any food industry, it can remain in their system for years causing food contamination and elimination of this pathogen is one of the greatest challenges (Jemmi and Stephan, 2006; Gandhi and Chikindas, 2007).

The outbreak of *Listeria moncytogenes* alone caused over 2,500 cases and close to 270 deaths with 17.7% fatality (EFSA, 2016). This pathogen caused 2,455 hospitalization and 255 deaths in the USA and has witnessed increasing infections worldwide (Scallan et al., 2011). In other food-related environments (such as farms and retail shops) also the presence of *Listeria* has been reported.

Until 2015, apples were not implicated for causing foodborne illnesses, but in a study, it was proved that caramel apples led to a nationwide *Listeria* outbreak with 97% hospitalization and 20% death cases, while the others recovered (Angelo et al., 2017). This case study demonstrated the intensity of *Listeria*-caused illness and the need to avoid its contamination in food sources. A multi-state outbreak of *Listeria* pathogen was witnessed in the USA in the year 2011. The whole cantaloupe was associated with *Listeria* for the first time suggesting that fresh food products could also house the growth of this pathogen. During this outbreak, over 146 people were ill with 29 death cases and 1 miscarriage. The produce was then recalled from the market at a rate of 1–5 million melons within at least 28 countries (FDA, 2011). Studies showed that this pathogen could be eliminated by the cooking and pasteurization processes. Therefore, RTE foods that require storage in the refrigerators are at greater risk of contamination because they do not require any further cooking before consumption.

3.2.1.6 SALMONELLA SPP.

This pathogenic group belongs to the *Enterobacteriaceae* family and responsible for causing the most common enteric infections globally. This organism was first isolated by the Scientist Dr. Daniel Salmon and therefore it is named after him (Bacon and Sofos, 2003). Among the various species from this genus, *S. enterica* and *S. bongori* are highly prospective of causing diseases, and they are further divided into serotypes on the basis of Kaufmann-White typing scheme that differentiates the species based on their surface and flagellar antigen. *S. enterica* is involved in causing serious diseases such as typhoid. The pathogen contains several sections on its genome known as pathogenicity islands (PAIs) that code for various virulence-related proteins facilitating it to infect and colonize within the host system (FDA, 2012).

It was estimated by the CDC that infections caused by *Salmonella* are among the leading foodborne diseases reported in the US with over 1 million reported, 19,000 hospitalization and 380 deaths annually (Scallan et al., 2013). Food sources contaminated with fecal matter from the human or animal origin can cause *Salmonella* infection. They are primarily contracted from eggs, poultry, and meat sources, but at times even fruits and vegetables also are contaminated with this bacterium (CDC, 2012). Recently, in the year 2015 alone 94,625 cases were reported in the European Union (EFSA, 2016). In a previous report, about 138,000 gallons of ice cream from a single batch were contaminated by *Salmonella* causing illness to over 225,000 individuals across 15 states in Europe (Hennessy et al., 1996). Contamination of peanuts and its products by *Salmonella* spp. led to its recall owing to 714 individuals affected in 46 States of US because of the disease outbreak. This was markedly the largest product recall because the products were used by more than 200 companies for the preparation of brownies, cakes, pie, and a variety of other foods (Cavallaro et al., 2011). Approximately 3,900 products were thus recalled owing to salmonellosis illness. This pathogen could be passively internalized to the foods and a similar case was witnessed in the US after the import of mangoes from Brazil (Hennessy et al., 1996; Cavallaro et al., 2011). As a pre-treatment to eliminate fruit flies, the fruits were dipped in hot water and immediately chilled in a cold water bath. This cold water in the water bath was contaminated with *Salmonella* spp., which was further internalized by the mangoes and led to illness (Sivapalasingam et al., 2003; Penteado et al., 2004).

3.2.1.7 SHIGELLA SPP.

The entire genus of *Shigella* belong to *the Enterobacteriaceae* family having four serogroups namely serogroup A as *Shigella dysenteriae*, serogroup B as *Shigella flexneri*, serogroup C as *Shigella boydii* and serogroup D as *Shigella sonnei* (Bacon and Sofos, 2003). They are the class of Gram-negative, rod shaped, non-spore-forming, and NM facultative anaerobes. They grow at a range of temperatures from 6–48°C and pH from 4.8–9.3 (ICMSF, 1996). They share DNA homology to *E. coli* and share similar biochemical and antibody characteristics. Despite the similarities, the disease symptoms caused by both the bacteria are different. They are predominantly found in the environmental samples with compromised sanitation and lack of hygienic conditions. Shigellosis, the disease caused by the food contamination with *Shigella* spp., spreads by transmission from person to person but also occurs by the contamination of food and water with feces. Outbreaks have been reported by the consumption of contaminated milk, chicken, salads, and shellfish. More often the store-bought and restaurant served foods are prospective of such contamination (Bacon and Sofos, 2003). International travels were among the top sources for shigellosis in most countries including the US. Yet, in developing countries, *S. flexneri* and *S. dysenteriae* type 1 constitute the commonly found serotypes witnessed in the outbreaks from Southern Africa, Asia, and Central America (NCBI, 2017).

Shigella spp. primarily causes gastrointestinal infections, post-incubation period anywhere between 12–50 hrs symptoms such as fever, malaise, fatigue, and abdominal cramps occur. All the serogroups of *Shigella* are capable of causing symptoms diarrhea, but *S. dysenteriae* type 1 is the most frequently screened for severe diarrhea and its associated complications including HUS (Vargas et al., 1999). One of the causes of disease outbreak was observed in Columbia where 32% of the football players were affected by Shigellosis caused after the consumption of cold sandwich prepared by the airline kitchen (Hedberg et al., 1992). Similarly, close to 240 passengers were ill from 24 states and four countries following the same reason in just a span of one month. Even the fresh product was found to be contaminated with *Shigella* spp. and in such case study, three of family members were get ill after the consumption of food at a food fair. Another 32 people were affected similarly, and all the cases showed food contamination with *S. sonnei* as the causative agent of this condition (Crowe et al., 1999).

3.2.1.8 STAPHYLOCOCCUS AUREUS

They are the Gram-positive, NM cocci that exist in single, pairs, tetrads, short fragments, or clusters. The range of temperature requirement is between 7 and 47.8°C but produce enterotoxins between 10 and 46°C (Bacon and Sofos, 2003). Likewise, it thrives in a broad pH range of 4.5 to 9.3 but is optimally growing between 7 and 7.5 (ICMSF, 1996; Bacon and Sofos, 2012; FDA, 2012). These facultative anaerobic bacteria are found everywhere including skin, mouth, intestine, mammary glands, respiratory tracts, and urogenital tracts of infected organisms. The pathogen also has the ability to survive outside the host system for a long duration under dry conditions and was successfully isolated from various sources such as dust, sewage water, air, and many other surfaces. Contamination in food includes pork, beef, salmon, steaks, shrimp, turkey, sausage, and delicatessen salads (Bacon and Sofos, 2003).

Symptoms of infection caused by *S. aureus* include boils on the skin surface, cellulitis, postoperative infections, and impetigo. In addition, they are associated with serious infections such as pneumonia, cerebritis, meningitis, bacteremia, abscesses of the muscles, CNS, and urogenital tracts (Bacon and Sofos, 2003). *S. aureus* infection is also associated with a condition known as toxic shock syndrome that is similar to the septic shock caused by the production of shock syndrome toxin 1. The pathogen entry into the food sources can occur due to the improperly cooked foods, inadequate refrigeration, and lack of personal hygiene during the cooking period. Ingestion of the enterotoxin or the pathogen itself can cause illness that causes symptoms such as headache, vomiting, nausea, cramps of the abdomen, dizziness, excessive perspiration, diarrhea, and muscle cramping (Bacon and Sofos, 2003). Annually, *S. aureus* infection alone claims 241,188 illnesses, 1,064 hospitalizations, and 6 deaths (FDA, 2012). A total of 9 enterotoxins have been identified from this bacterium and are designated as A, B, C1, C2, C3, D, E, F, and G. These enterotoxins have the characteristic property of acting as superantigen thus stimulating T cell production at a great percentage (Mossel et al., 1995). Among all the toxins, A type is the most heat-sensitive whereas B and C require about 80–100°C to lose its serological properties (Bacon and Sofos, 2003).

3.2.1.9 CLOSTRIDIUM PERFRINGENS

This pathogen from the Bacillaceae family is among the important foodborne pathogens. These rod-shaped, NM, spore-forming bacteria produce protein toxins under stress conditions such as radiation, heat, and salt variations (Bacon and Sofos, 2003). They grow at an optimum temperature between 43 and 47°C and pH between 6 and 7.2 (ICMSF, 1996; Bacon and Sofos, 2003). They are the most prevalent species with widespread existence and implicated in simple wounds, Clostridia cellulitis, post-abortion infections, intra-abdominal sepsis, bacteremia, intravascular, pneumonia, and brain abscesses (Rajkowski and Smith, 2001; Bacon and Sofos, 2003). They are the second most frequently encountered foodborne illness in the world, affecting over 2,000 people annually (Grass et al., 2013). Between 1998 and 2010, 15,208 cases were reported with 83 hospitalization and 8 deaths caused due to *Clostridium perfringens* infection. The pathogen entry from outside food consumption was the most common mode of infection. Restaurants contributed about 43% of infections whereas other settings included catering facility caused 19%, private home 16%, prison or jail 11%, and other 10%. Around 144 (50%) outbreaks attributed to a single food commodity, beef was the most common commodity (66 outbreaks, 46%), followed by poultry (43 outbreaks, 30%), and pork (23 outbreaks, 16%) (Grass et al., 2013). There have been 5 toxins identified from this species designated from A through E producing alpha-toxin (phospholipase) involved in myonecrosis. In the US, almost all the cases of gastroenteritis were caused by *Clostridium perfringens* or the ingestion of its toxins (Bacon and Sofos, 2003). Outbreaks of *Clostridium* infections have been reported from the consumption of mince, vegetables, and gravy that were poorly cooked. In a particular mince food were cooked, refrigerated, and then reheated before serving. This caused an outbreak of the disease owing to the production of the enterotoxin by the pathogen. The person responsible was charged under the General Food Regulations

(2004) and the Food Hygiene (England) Regulations (2006) and was convicted (Acheson et al., 2016).

3.2.1.10 VIBRIO SPP.

This genus belongs to the Vibrionaccae family with more than 35 identified species and is highly pathogenic to humans. This non-spore-forming bacterium is Gram-negative, motile, rod-shaped facultative anaerobe. They produce a single polar flagellum that facilitates motility. *Vibrio cholerae*, *Vibrio parahaemolyticus* and *Vibrio vulnificus* are the most pathogenic among all the species from this genus (Bacon and Sofos, 2003). They grow optimally at a temperature between 30 and 37°C and pH of 7.6 and sodium chloride concentration of 0.5% (ICMSF, 1996). This pathogen enters into human system upon ingestion of contaminated foods such as molluscs, undercooked crustaceans, or exposure to contaminated water (Bacon and Sofos, 2003). Symptoms can often go asymptomatic for a long period after which it caused mild to moderate diseases. Close to 5 days after the exposure, the pathogen establishes itself in the host system and causes muscle cramping, fever, vomiting, loose stools, and mucus flecked diarrhea which are the characteristic feature of cholera (Wu et al., 2014). These symptoms last for about 2–3 days but under severe conditions result in primary septicemia, dysentery, and possibility of death. Pre-existing conditions like alcoholism, liver cirrhosis, diabetes, peptic ulcers, and immune disorder augment to the condition can intensify the symptoms (Bacon and Sofos, 2003).

The last two decades in China has witnessed the most common bacteria disease outbreak to be caused by *V. parahaemolyticus* which was also seen in many other Asian countries including India (Wu et al., 2014). The year 2003–2008 witnessed gastroenteritis outbreaks caused by *V. parahaemolyticus* in 12 provinces of China. The outbreak resulted in 9,041 illnesses and 3,948 hospitalizations but no death cases (Ma et al., 2014). Meat and meat products were implicated to be responsible for this outbreak in more than 50% of the cases. Animal-based (i.e., meats, such as poultry, internal organs, beef, deli meat and cured meat, and aquatic products, such as crustacean, molluscs, and fish) foods were the most common commodity reported (61%), followed by mixed foods (19%), and other foods (17%) (Wu et al., 2014).

3.2.2 FUNGAL PATHOGENS

Unlike bacteria and other microbial pathogens, the presence of fungi in food does not directly cause diseases. In fact, several fungi are useful for adding flavor to the foods and therefore considered useful. However, most fungi have an innate ability to produce toxins known as mycotoxins that are harmful for life (Bennett and Klitch, 2003). For several centuries, mycotoxins have been implicated in many diseases and death in both animals and humans. Around 1960, the death of more than lakh poultry animals owing to the consumption of contaminated peanuts led to the identification of mycotoxins for the first time (Alshannaq and Yu, 2017). This mycotoxin was identified as aflatoxin produced by *Aspergillus flavus*. After this discovery, several other studies were focused on the identification of molds producing an array of mycotoxins, and presently over 60 mycotoxins have been identified. Ergot was the first mycotoxin known to cause illness in humans (Bennett and Klitch, 2003). Way back in 1857, ergot poisoning was observed in the Rhine Valley and the most recent outbreak of the same was witnessed in 1952 in France.

Direct evidence on the correlation of aflatoxins and liver cancer in humans, FDA has often raised concerns over the consumption of these toxins even in an extremely negligible levels could prospectively be carcinogenic. The level of aflatoxins has drastically been reduced owing to the advancement in harvesting and storage methods. FDA proposed in the Federal Register of December 6, 1974, a regulation establishing a tolerance of 15 ppb for total aflatoxins in shelled peanuts and peanut products used as human food. Among the various mold-related disease outbreaks, reports of aspergillosis are caused by non-*fumigatus* species of *Aspergillus*. Even the infections caused by yeast and *Trichosporon* species have been reported (Arora, 2003; Bennet and Klitch, 2003).

3.2.3 VIRAL PATHOGENS

3.2.3.1 HEPATITIS A

Hepatitis A virus (HAV) belongs to the Picornaviridae family of the genus Hepatovirus (Cuthbert, 2001), with six genotypes. Among the six genotypes, I, II, and III comprising strains associated with human infections

(FDA, 2012). It consists of 7.5 kb single stranded, polyadenylated RNA genome enclosed within hydrophobic, non-enveloped, icosahedral capsids of approximately 30 nm diameter (Mattison et al., 2009). Hepatitis A is transmitted by fecal to oral route via contaminated food, water, utensils, and either direct contact with infected person or through indirect contact when contaminated food or water is consumed (FDA, 2012; Koopmans and Duzier, 2004). Once ingested, it enters the gastrointestinal tract and multiplies in the liver. Though not able to grow in the environment, it is said to withstand extreme conditions like heat, chemicals, desiccation, and freezing (FDA, 2012).

The incidence and severity of Hepatitis A infection may vary in different countries based on hygienic standards. The infection is said to be endemic in developing countries, infecting at early stages of life and adults being immune to it (Koopmans and Duzier, 2004; Mattison et al., 2009). The endemicity is low in developed countries, affecting adults as they are susceptible. Thus, it constitutes an alarming public health concern. Viral infection is acute, but its resolution provides life-long protection against future infections. It is characterized by symptoms that include fever, abdominal cramps, jaundice, dark colored urine with light colored stools and occasional diarrhea. It spreads extensively without showing any symptoms for 10–14 days during which virus shedding occurs (Koopmans and Duzier, 2004; FDA, 2012). Therefore, the accurate detection of foodborne sources becomes difficult. Foodborne infections of Hepatitis A are associated with different food types, primarily with shellfish due to growing contamination of waters.

The first recorded outbreak of Hepatitis A resulted in over 600 cases, caused by oysters stored in fecally contaminated water. A clam-associated outbreak in Shanghai may have been the largest outbreak to date with 292,301 cases. In addition to this, sporadic viral illness related with shellfish has also been demonstrated. As it majorly deals with shellfish available in coastal regions, it is difficult to avoid the bias in these studies (Koff et al., 1967; Halliday et al., 1991). Shellfish growing waters are typically contaminated with fecal coliform group of *Escherichia coli*. Due to the presence of these bacteria, it becomes difficult to predict the presence of Hepatitis A (Wait and Sobsey, 1983). Therefore, shellfishes are cooked by an approved method to avoid contamination. In 2003, over 1,000 cases were reported from a series of Hepatitis A outbreaks. The outbreaks were linked to green onions imported from Mexico, and soon FDA banned imports from those farms (CDC, 2003). During 2013 and 2014 multinational outbreaks occurred across Europe with over 1,400 cases linked to fresh and preserved strawberries initiated the sequencing of viral genome (Chiapponi et al., 2014). The interpretation of sequencing was difficult due to various methodologies followed across Europe. However, results from whole-genome sequencing (WGS) efficiently replaced numerous different methodologies that were in use in the human and veterinary laboratories for investigation of outbreaks. These methods possess the potential for early detection of foodborne pathogens and resultant data can be used for assessment of risk management (Blackwell et al., 1985). The epidemic potential of a virus genotype and its inheritable characters along with their capability to survive, genetic stability and mutation can be studied.

3.2.3.2 NOROVIRUS

Noroviruses from family Caliciviridae are diverse in nature and classified into six genogroups. In these genogroups, only three, i.e., I, II, and IV are capable of infecting human beings; among them, only 30 genotypes have been described to date. These are non-encapsulated, icosahedral viruses with 30–35 nm diameters with a single stranded RNA of approximately 7.5 kb length (Iturriza and O'Brien, 2016). They are usually transmitted through fecal-oral pathway and are capable of surviving in harsh conditions of environment ranging from freezing to 60°C. Its ability to persist on food commodities and environmental surfaces enables it to disseminate rapidly through secondary spread, usually via food handlers to consumers. The food commodities can be contaminated either by source (food and water) or during preparation. As the virus can spread through environmental surfaces, prevention becomes difficult. Worldwide genome analysis over a period of 10 years of Norovirus reveals that pandemic strains are spread through international food trades (Glass et al., 2009). Though, animal reservoirs pose a potential threat and direct zoonotic transmission is rare (Koopmans, 2008; Verhoef et al., 2011).

Cruise ships traveling in international waters with passengers from different geographical conditions provide an ideal condition for rapid and extensive infection that can spread across the globe. With people disembarking at several ports can result in simultaneous intake and spreading of the virus. 10% of all reported outbreaks are attributed to cruise ships in the

US. In 2010, a cruise ship named "Mercury" reported 1,058 cases within a month resulted from more than 10–22% of passengers and 2–4% of crew infected (McCarter, 2009). Norovirus are the chief reason behind acute gastroenteritis (AGE) outbreaks around the world, causing an estimated 5.4 million cases of foodborne illnesses in the US alone annually (Scallan et al., 2013). According to WHO, most of the children less than a year are suffering from AGE caused by Norovirus (WHO, 2015).

A study reported the presence of 10 human enteric viruses using antigenic or genetic detection methods from 286 fecal specimens of 88 oyster-linked gastroenteritis outbreaks to determine the prevalence of these infections. Results revealed that 96.6% of common enteric virus and 68.9% of fecal specimens contain Norovirus, indicating high prevalence of the disease with the consumption of raw and undercooked oysters (Iritani et al., 2014). Europe encounters frequent outbreaks of Norovirus without reference to oysters or shellfishes. Six outbreaks reported within three months were associated with frozen raspberries. Later it was found that bags of raspberries were contaminated with Geno group I and II Noroviruses, including one typable – GI.6 (Müller et al., 2015). This study revealed that independent outbreaks were originated from a single source of contamination with different batches in Serbia. This led to a series of talks between food handlers and government, resulting in modification of rules to supply heat-treated raspberries in professional catering establishments (Tulandhar et al., 2015). The role of asymptomatic food handlers at different stages of food production contributing to outbreaks has become an alarming concern, with ¼th of the disease spread attributed to them. As the virus persists on environmental surfaces, it is advised to use soaps or sanitizers to wash the hands to remove Norovirus (Ionidis et al., 2016).

3.2.3.3 INFLUENZA VIRUSES

Influenza viruses belong to the family of Orthomyxoviridae, and are divided into types A, B, and C based on the variation in the nucleoprotein antigen. Type C is said to be genetically stable whereas types A and B have hemagglutinin (HA) and neuraminidase (NA) antigens undergoing mutation (Lamb and Krug, 1996). Sequencing has confirmed that these viruses have genetically diverged and exchanged viral RNA within each genus. These viruses are reported to have a common ancestry. Influenza viruses are possessing standard nomenclature procedure that gives them two subtypes: HA and NA (Couch, 1996). Only 3 HA (H1, H2, and H3) and 2 NA (N1 and N2) are reported to cause human epidemics. The virus is enveloped, filamentous, and 300 nm in length and 80–100 nm in diameter. Type A virion is studded with HA and NA (4:1) glycoprotein spikes projecting from host-cell derived lipid membrane (Lamb and Krug, 1996). Helically formed nucleoprotein capsid comprises of segmented, single stranded antisense RNA as genome. The virus binds to the cytoplasm through HA. Transcription and nucleocapsid assembly take place in the nucleus followed by an assemblage of virions in cytoplasm and bud from the cell membrane, which kills the host cell (Couch, 1996; Bouvier and Palese, 2008).

Influenza viruses are transmitted into the aerosols from infected persons to healthy individuals. It multiplies in the respiratory mucosa, resulting in cellular inflammation and destruction. It affects people irrespective of age groups especially in winters every year. It is a febrile illness of sudden onset, characterized by marked myalgias and tracheitis. Symptoms include sudden occurrence of fever, chills, headache, malaise, anorexia, and sore throat. A non-productive cough is characteristic along with sneezing, rhinorrhea, and nasal obstruction (Couch, 1996). Some patients reported to have nausea, vomiting, diarrhea, and abdominal cramps. The replication takes about 6 hours and symptoms last for 1–5 days. It is worthwhile to note that the virus may also spread from chickens and fowls. Open bird markets are of alarming concern as birds may get exposed to infected flocks, humans, or other animals. Thus, avian flu outbreaks can also be expected from influenza viruses.

In the past 100 years, several outbreaks resulted in the deaths of thousands of people. The worst was the 1918 pandemic, causing 20 million deaths across the globe and 500,000 deaths in US alone (Taubenberger, 2005). The recent appearance of H5N1 is a major concern, owing to its high mortality rates once it is contracted from chickens. Though capable of infecting human being, it has not shown a high propensity to spread among humans (Kingsley, 2018). As the virus possesses the ability to adapt, there must be some interference in persistence of virus in humans and human-to-human transmission. Several other influenza avian flu poses threat to mankind including H7N9 (Lu et al., 2013), H9N2, and H7N3 (Belser et al., 2013; Lopez-Martinez et al., 2013). But H7N9 is said to be less pathogenic for chickens with only 29% reported fatality rate for poultry workers (Pascua and Choi, 2014). Although influenza is not typically recognized as a foodborne pathogen, there are also incidents of H571 occurrence among people drinking duck blood (Shao et al., 2011).

Influenza can be diagnosed with enzyme immunoassays like ELISA or any immunofluorescence tests using viral antigens present in cells obtained from the nasopharynx of infected person. The virus is usually isolated from respiratory tracts and grown in tissue cultures followed by detection of hemadsorption, where RBC are added to the culture to adhere with the virus. In case of positive results, serological tests with specific antisera are performed. Regarding treatment, two specific antiviral drugs namely amantadine and rimantadine are available. They are effective only against type A in case of prophylaxis. There is also possibility of occurrence of viral resistance during the treatment. Inactivated influenza viral vaccines have been used for 40 years to prevent the outbreaks. These vaccines were prepared using chick embryos, which led to allergic reactions in some patients. The vaccine is administered parenterally in the fall according to the dose requirement of the patient. As side effects like the development of Guillain-Barre syndrome occurred in the USA (1976) after vaccination, research was initiated to develop live attenuated vaccines as alternatives to conventional vaccines (Couch, 1996).

3.2.4 PARASITIC PATHOGENS

3.2.4.1 CYCLOSPORA CAYETANENSIS

Cyclospora cayetanensis are the single celled, coccidian protozoan parasites of the Eimeriidae family that persist in the small intestine of the host (Dorny et al., 2009). They complete their intermediary life cycle in the cytoplasm of intestinal enterocytes and immediately produce four sporozoites encapsulated by two sporocysts embedded in oocysts (Bacon and Sofos, 2003). It takes around 7–15 days to sporulate after subsequent shedding of the oocytes. Oocytes are spherical in shape, ranging between 7.5–10 μm in diameter and possess 50 nm thick walls with an outer filamentous coat known as wrinkle. Development is believed to be robust in case of shedding on feces (Ghimire, 2010). According to the data retrieved from NCBI, analysis of two of the genomes has been completed, and the total median length of the genome is 44.2991 Mb (NCBI, 2017).

Cyclospora cayetanensis is capable of infecting both immunocompetent and immunocompromised hosts and can cause prolonged illness (up to 6 weeks or longer). The characteristic symptoms include nausea, fatigue-less, bloating, non-bloody diarrhea, vomiting, abdominal cramps, fever, and malaise (Bacon and Sofos, 2003).

The first documented outbreak in the US was reported in Chicago, where 23 cases were found to be infected by hospital water supply. In 1994, British soldiers and their dependents in Nepal were stationed at a small detachment where 12 of them (92%) were found to be infected (Dawson, 2005). From 1996 to 1998, *Cyclospora cayetanensis* was found in several outbreaks as an etiological agent in the US and Canada. It was identified in baby lettuce, raspberries, and basil (Mead et al., 1999). Currently, there are about 15,000 cases of *Cyclospora cayetanensis* infection reported every year in the US (Ortega and Sanchez, 2010). In 1996, it was reported that a total of 1,465 cases in 20 States, the District of Columbia and 2 Canadian Provinces were infected after consuming fresh raspberries from Guatemala contaminated with *Cyclospora cayetanensis* (Tauxe, 2002). As the parasite is endemic in developing tropical and subtropical areas, infections, and their seasonality is poorly understood. North American outbreaks occur most frequently during summer due to the importing of food commodities from tropical countries (Ortega and Sanchez, 2010).

The detection and diagnosis of *Cyclospora cayetanens*is becomes difficult due to its small size, inability to take up stains and intracellular habitat. Once the oocysts were recovered from stool samples, they are subjected to PCR amplification of their genome. The microscopic identification is done with a phase-contrast microscope to check for the oocysts followed by acid-fast staining and autofluorescence with UV rays. A recent study reported using sucrose solution in centrifuging stool samples helped to get remarkable results with 84% positive results (Riner et al., 2007). Most people with mature and healthy immune system are reported to recover without any treatment. But in case of people with poor health or weak immune system may suffer from prolonged or severe illness. However, a drug known as trimethoprim-sulfamethoxazole (TMP-SMX) is being used to treat the illness as it reduces the oocyst excretion and diarrhea within 2–3 days. As there is no vaccine available against this parasite, safe, and hygienic practices like washing hands, avoiding visiting places with seasonal endemic and underdeveloped countries are advised to both travelers and people (CDC, 2020).

3.2.4.2 *TOXOPLASMA GONDII*

Toxoplasma gondii is an intracellular, coccidian, obligate protozoan parasite belonging to phylum Apicomplexa, which inhabit liver and muscle tissue of warm-blooded animals to cause a disease known as toxoplasmosis (Bacon and Sofos, 2003; Weiss and Kim, 2011). It is generally found in cats which are used as its primary reservoir and can also be found in other warm-blooded animals used as intermediate hosts. The parasite can be found residing in body tissues in its different life cycles as tachyzoites, bradyzoites, or sporozoites, which are the three stages of its life cycle (Weiss and Kim, 2011). Tachyzoites said to proliferate and destroy the host cells, whereas bradyzoites multiply within tissue cysts. Oocysts present in cat feces contain sporozoites which sporulate after 1–5 days within oocysts. They survive in oocysts for several days as they are resistant to disinfectants, drying, and freezing (Bacon and Sofos, 2003). Up to date, a total of 17 genomes of *Toxoplasma gondii* have been analyzed according to NCBI, and the median total length of the genome was found to be 64.1936 Mb (CDC, 2020). It became evident that out of 3 genotypes (I, II, and III) genotype II is associated with human toxoplasmosis (Smith and Evans, 2009). Immunocompetent hosts infected with *Toxoplasma gondii* generally show non-clinical symptoms like malaise, fever, rashes, headache, pyrexia, myalgia, and lymphadenopathy (Smith and Evans, 2009; FDA, 2012). Toxoplasmosis is transmitted to humans by the consumption of raw or under-cooked meat contaminated with tissue cysts. In addition to meat, cysts can also be transferred from vegetables, milk, and dairy products (Dorny et al., 2009; Smith and Evans, 2009).

Being one of the world's most common parasites, it has a disease burden like Salmonellosis and Campylobacteriosis. *Toxoplasma gondii* infections in the US alone are estimated to cause about 87,000 illnesses, 4,400 hospitalizations, and 330 deaths annually. This makes it the second major leading cause of foodborne mortality in US and third leading cause of foodborne hospitalizations. Around 22.5% population of the US over the age of 12 has been infected. It has reached 95% for some countries (Mack et al., 2012). The alarming concern is that the parasite can affect the growing fetus as it can easily penetrate into the placenta from infected mother (congenital toxoplasmosis). This condition can lead to a wide range of consequences, including mental retardation, blindness, seizures, and can even result in death (Singh, 2016). Healthy people with robust immune system would not require any treatment for toxoplasmosis. As the parasite mostly appears in cat litter, it is advised to maintain hygienic practices in homes where cats are used as pets. Meat should be cooked at high temperatures (60–70°C) to avoid accumulation of pathogens. It is also advised to wash hands with soaps or sanitizers before having food (Dorny et al., 2009).

3.2.4.3 TRICHINELLA SPIRALIS

Trichinella spiralis belongs to phylum Nematoda and is a viviparous parasite that affects rodents and mammals including humans. The disease caused by *Trichinella spiralis* is known as Trichinosis (Hill and Dubey, 2010). Known to be the smallest nematode parasite affecting human beings, it possesses an unusual lifecycle exists between primary host and secondary host. Males measure between 1.4–1.6 mm long, appear flattened. Anus is on the posterior end, with large copulatory pseudo bursa on each side. Females are larger than males, almost twice their size. They bear vulva near esophagus carry developing eggs in the posterior uterus (Schmidt et al., 1977). Currently, 2 genomes of *Trichinella spiralis* have been analyzed and the resultant median length of the genome is 56.77 Mb, according to NCBI (NCBI, 2017).

It begins its life cycle in the intestinal wall of pigs, followed by muscle tissue cells (Dorny et al., 2009). When humans consume pork, it is transferred to the human gut. Mainly it inhabits human intestine cells and spends most of its time there (Schmidt et al., 1977). Besides pigs, it also affects most of the carnivorous animals. Larvae are passed from female into the bloodstream from where they reach muscle fiber to encyst. These encysted forms are viable for a long time. The migration and encystment processes bring up the symptoms like fever, chills, pain, night sweating, nausea, vomiting, edema, neurological disorders and even death (Dorny et al., 2009; Murrell and Krompton, 2009). Adult nematodes can live up to 8 weeks in duodenal and jejunal mucosal epithelium. During this transient period, female releases around 1,500 larvae into the bloodstream to reach muscle tissues, where they reside for several years before being expelled. These larvae develop, mature, and are encapsulated by a calcified cell wall in skeletal muscles after 6–18 months. The same host facilitates the development of both larvae and adult stages. The encysted larvae can survive in host body for up to 10 years and freed by digestive enzymes of second host after ingestion (Bacon and Sofos, 2003). The nematodes can be inactivated by thermal treatment. Therefore, it is recommended to cook pork and other meat products at 76.7°C. According to a study, the parasite has caused about 52 cases of foodborne illnesses annually in the US, with a case fatality rate of 0.003 (Mead et al., 1999).

3.3 FOOD-RELATED MICROORGANISMS

Most foods have an innate microflora of bacteria that offer benefits in terms of preservation or fermentation whereas others can lead to illness and spoilage of food. In the previous section, the microorganisms that cause illness have been extensively discussed, whereas this section is stipulated for the microbial flora of food that is important for the production of useful products. Apart from these benefits, several microorganisms, including lactobacillus have been recognized as probiotics that are essential for maintaining health and wellness by improving the ability to take up macro and micro-nutrients better within the systems. Bacteria, viruses, and molds are useful for the production of some of the other foods and therefore will be discussed in subsections.

3.3.1 BACTERIA

They constitute the most widespread and important organisms of the food processing industry. Many of them live as a native of the normal microflora of the living system within most animals including humans. Lactic acid bacteria (LAB) are the primary contributors of food fermentation and significantly affect the flavor and other characteristics of fermented foods. Diary starter cultures always have Propionibacteria that are responsible for the production of the characteristic eye and flavors in cheese (Cogan and Accolas, 1996). Five species of dairy Propionibacteria are currently recognized: *P. freudenreichii* subsp. *freudenreichii*, *P. freudenreichii* subsp. *shermanii*, *P. thoenii*, *P. acidipropionici* and *P. jensenii*. Vinegar industry requires the action of *Acetobacter* that are Gram-negative rods and cocci (*Acetobacter aceti*, *A. pasteurianus*, and *A. peroxydans*). LAB and associated bacteria such as *Propionibacterium* and *Bifidobacterium* are often involved in the production of lactic, acetic, propionic, formic, phenyllactic, diacetyl, acetoin, reuterin, reutericyclin, peroxides, and are often represented as biopreservatives of food (Leroy and De Vuyst, 2010).

3.3.2 YEASTS AND MOLDS

Vegetable foods often prepared from raw materials such as soyabean, wheat nuts, cassava are termed as mold-ripened (Beuchat, 1987). They are eaten immediately such as tempeh, sufu, etc., or seasoned to form miso, shoyu, etc. Yeasts are the oval shaped fungi that reproduce by budding. They efficiently grow on damp surfaces and are often found on fruits, breads, and any other moist surfaces. The structures are known as hyphae and they appear like a matte on the surface of the hosts (Hamlyn et al., 1987). While yeasts are involved in the fermentation process for the preparation of several kinds of alcoholic beverages, molds are essentially useful for the production of several other food products such as Shoyu, miso, katsuobushi, tempeh, sufu, oncom, and ankak mostly produced as a part of the Asian foods. European market uses *P. camemberti* for the production of cheese (Seitz, 1990). Yeast is the oldest known microorganism used in the brewing industry and therefore known as the oldest domesticated microorganism. Even propionic acid and citric acid are produced as the secondary metabolites produced from yeasts along with acetates, ethyl esters, monoterpenic alcohols, aldehydes, and fatty acids. These organic acids add to the characteristic flavor and aroma of the food product. The action of various exoenzymes adds significant flavors to the food substance and is an important part of the delicacies in each continent (Kinsella et al., 1976). Mold-fermented foods require a relatively higher pH owing to its metabolic activity. Yeasts and molds efficiently grow on moist surfaces including damp regions of building, equipment, etc.

3.4 CONCLUSION

Food plays an important role for the determination of health and wellness of any organism. Consumption of fresh and healthy food is the key to living a healthy lifestyle. The technological advances have led to the use of microorganisms for the production of various food substances. Several bacteria and fungi are used to produce savories and beverages and they render a characteristic aroma and flavor to the food substances. Beverages produced from the fermentation of yeast are widely acclaimed and are in great demand all over the world. Similarly, products obtained from the fermentation of milk using LAB possess high nutritious value. Many other fermentation products include food additives such as citric acid and propionic acids that are used as preservatives, whereas some other fermentation process produces drugs such as antibiotics.

Similarly, food also hosts a wide array of harmful microorganisms that can cause mild to severe diseases. Diseases caused by such pathogens are collectively known as foodborne illness. The global health issues are critically impaired by the presence of such pathogens in food. Time and again newer disease outbreaks are witnessed owing to improper handling of food substances. Research, surveillance, and vigilance are therefore essential to understand the emergence and spread of foodborne pathogens. Risk assessments should be undertaken to assess the intensity of food hazards that can occur. The primary mode of entry of the pathogen can be at various stages such as soil, harvesting, storage, and others. Domestic and international government agencies should collaborate with healthcare and pharmaceutical firms in order to ensure the safety of food substances. Collectively, although the presence of microorganisms in food substances can be useful for various aspects, its undesirable presence can often cause illness and sometimes can be fatal.

KEYWORDS

- **bacteria**
- **foodborne illnesses**
- **foodborne pathogens**
- **fungal parasitic species**
- **gastrointestinal diseases**
- **neurological diseases**
- **pathogenesis**
- **toxins**
- **viruses**

REFERENCES

Acheson, P., Bell, V., Gibson, J., Gorton, R., & Inns, T., (2016). Enforcement of science - using a *Clostridium perfringens* outbreak investigation to take legal action. *Journal of Public Health, 38*(3), 511–515.

Alshannaq, A., & Yu, J. H., (2017). Occurrence, toxicity, and analysis of major mycotoxins in food. *International Journal of Environmental Research and Public Health, 14*(6), 632.

Altekruse, S. F., Stern, N. J., Fields, P. I., & Swerdlow, D. L., (1999). *Campylobacter jejuni* - an emerging foodborne pathogen. *Emerging Infectious Diseases, 5*(1), 28.

Andersson, A., Ronner, U., & Granum, P. E., (1995). What problems does the food industry have with the spore-forming pathogens *Bacillus cereus* and *Clostridium perfringens*? *International Journal of Food Microbiology, 28*(2), 145–155.

Angelo, K. M., Conrad, A. R., Saupe, A., et al., (2017). Multistate outbreak of *Listeria monocytogenes* infections linked to whole apples used in commercially produced, prepackaged caramel apples: United States, 2014–2015. *Epidemiology and Infection, 145*(5), 848–856.

Armstrong, G. L., Hollingsworth, J., & Morris, Jr. J. G., (1996). Emerging foodborne pathogens: *Escherichia coli* O157: H7 as a model of entry of a new pathogen into the food supply of the developed world. *Epidemiologic Reviews, 18*(1), 29–51.

Arora, D. K., (2003). *Fungal Biotechnology in Agricultural, Food, and Environmental Applications*. Boca Raton: CRC Press.

Bacon, R. T., & Sofos, J. N., (2003). Characteristics of biological hazards in foods. In: Shnidt, R. H., & Rodrick, G. E., (eds.), *Food Safety Handbook* (pp. 157–195). New Jersey: John Wiley and Sons, Inc.

Belser, J. A., Davis, C. T., Balish, A., et al., (2013). Pathogenesis, transmissibility, and ocular tropism of a highly pathogenic avian influenza A (H7N3) virus associated with human conjunctivitis. *Journal of Virology, 87*(10), 5746–5754.

Bennett, J. W., & Klich, M., (2003). Mycotoxins. *Clinical Microbiology Reviews, 16*(3), 497–516.

Bennett, S. D., Walsh, K. A., & Gould, L. H., (2013). Foodborne disease outbreaks caused by *Bacillus cereus*, *Clostridium perfringens*, and *Staphylococcus aureus* - United States, 1998–2008. *Clinical Infectious Diseases, 57*(3), 425–433.

Berger, C. N., Sodha, S. V., Shaw, R. K., et al., (2010). Fresh fruit and vegetables as vehicles for the transmission of human pathogens. *Environmental Microbiology, 12*(9), 2385–2397.

Beuchat, L. R., (1987). *Food and Beverage Mycology*. New York: Springer Science and Business Media.

Bhalla, T. C., (2017). Yeasts and traditional fermented foods and beverages. In: Satyanarayana, T., & Kunze, G., (eds.), *Yeast Diversity in Human Welfare* (pp. 53–82). Singapore: Springer.

Bisen, P. S., Debnath, M., & Prasad, G. B., (2012). *Microbes: Concepts and Applications*. New York: John Wiley and Sons Inc.

Blackwell, J. H., Cliver, D. O., Callis, J. J., et al., (1985). Foodborne viruses: Their importance and need for research. *Journal of Food Protection, 48*(8), 717–723.

Blaser, M. J., (2011). Deconstructing a lethal foodborne epidemic. *New England Journal of Medicine, 365*, 1835–1836.

Bouvier, N. M., & Palese, P., (2008). The biology of influenza viruses. *Vaccine, 26*(S4), 49–53.

Buchanan, R. L., Gorris, L. G. M., Hayman, M. M., Jackson, T. C., & Whiting, R. C., (2017). A review of *Listeria monocytogenes*: An update on outbreaks, virulence, dose-response, ecology, and risk assessments. *Food Control, 75*, 1–13.

Carter, A. T., & Peck, M. W., (2015). Genomes, neurotoxins and biology of *Clostridium botulinum* group I and group II. *Research in Microbiology, 166*(4), 303–317.

Cavallaro, E., Date, K., Medus, C., et al., (2011). *Salmonella typhimurium* outbreak investigation team. *Salmonella typhimurium* infections associated with peanut products. *New England Journal of Medicine, 365*(7), 601–610.

CDC, (2003). Hepatitis A outbreak associated with green onions at a restaurant - Monaca, Pennsylvania, (2003). *Morbidity and Mortality Weekly Report, 52*(47), 1155.

CDC, (2006). Ongoing multistate outbreak of *Escherichia coli* serotype O157: H7 infections associated with consumption of fresh spinach - United States, (2006). *Morbidity and Mortality Weekly Report, 55*, 1–2.

CDC, (2012). Multistate outbreak of *Salmonella* Bareilly and *Salmonella* Nchanga infections associated with a raw scraped ground tuna product. https://www.cdc.gov/salmonella/bareilly-04-12/index.html (accessed on 20 December 2021).

CDC, (2020). Parasites - Cyclosporiasis (*Cyclospora* infection). https://www.cdc.gov/parasites/cyclosporiasis/prevent.html (accessed on 20 December 2021).

CDC, (2020). *Parasites - Cyclosporiasis* (*Cyclospora* infection). https://www.cdc.gov/parasites/cyclosporiasis/treatment.html (accessed on 20 December 2021).

Chiapponi, C., Pavoni, E., Bertasi, B., et al., (2014). Isolation and genomic sequence of hepatitis A virus from mixed frozen berries in Italy. *Food and Environmental Virology, 6*(3), 202–206.

Cogan, T. M., & Accolas, J. P., (1996). *Dairy Starter Cultures* (pp. 101–127). New York: VCH Publisher.

Couch, R. B., (1996). Orthomyxoviruses. In: Baron, S., (ed.), *Medical Microbiology*. University of Texas Medical Branch at Galveston.

Crowe, L., Lau, W., & Mcleod, L., (1999). Outbreaks of *Shigella sonnei* infection associated with eating fresh parsley - United States and Canada. *Morbidity and Mortality Weekly Report, 48*(14), 285–289.

Croxen, M. A., Law, R. J., Scholz, R., Keeney, K. M., Wlodarska, M., & Finlay, B. B., (2013). Recent advances in understanding enteric pathogenic *Escherichia coli*. *Clinical Microbiology Reviews, 26*(4), 822–880.

Cuthbert, J. A., (2001). Erratum: Hepatitis A: Old and new. *Clinical Microbiology Reviews, 14*(3), 642.

Dawson, D., (2005). Foodborne protozoan parasites. *International Journal of Food Microbiology, 103*(2), 207–227.

Desai, R., Yen, C., Wikswo, M., et al., (2011). Transmission of norovirus among NBA players and staff, winter 2010–2011. *Clinical Infectious Diseases, 53*(11), 1115–1117.

Dorny, P., Praet, N., Deckers, N., & Gabriël, S., (2009). Emerging foodborne parasites. *Veterinary Parasitology, 163*(3), 196–206.

EFSA, (2011). Tracing seeds, in particular fenugreek (*Trigonella foenum-graecum*) seeds, in relation to the Shiga toxin-producing *E. coli* (STEC) O104: H4 2011 outbreaks in Germany and France. *EFSA, 8*(7), 176E.

Erkmen, O., & Bozoglu, T. F., (2016). *Food Microbiology, Principles into Practice*. New York: John Wiley and Sons Inc.

Estes, M. K., Prasad, B. V., & Atmar, R. L., (2006). Noroviruses everywhere: Has something changed? *Current Opinion in Infectious Diseases, 19*(5), 467–474.

FDA, (2011). *Factors Potentially Contributing to the Contamination of Fresh Whole Cantaloupe Implicated in a Multi-State Outbreak of Listeriosis*. US Food and Drug Administration. https://www.qualityassurancemag.com/article/qa-102011-fda-reports-factors-potentially-contributing-to-cantaloupe-listeria-outbreak/ (accessed on 20 December 2021).

FDA, (2012). *Large and Ongoing Outbreak of Haemolytic Uraemic Syndrome*. Germany. Bad bug book. https://www.fda.gov/files/food/published/Bad-Bug-Book-2nd-Edition-%28PDF%29.pdf (accessed on 20 December 2021).

Florou-Paneri, P., Christaki, E., & Bonos, E., (2013). Lactic acid bacteria as source of functional ingredients. In: Kongo, M., (ed.), *Lactic Acid Bacteria - R and D for Food, Health and Livestock Purposes*. London: IntechOpen.

Franchini, S., Natter, B., Augschiller, M., et al., (2014). Hepatitis A outbreak in Italy, 2013: A matched case-control study. *Eurosurveillance, 19*(37), 20906.

Frank, C., Werber, D., Cramer, J. P., et al., (2011). Epidemic profile of Shiga-toxin–producing *Escherichia coli* O104: H4 outbreak in Germany. *New England Journal of Medicine, 365*(19), 1771–1780.

Gandhi, M., & Chikindas, M. L., (2007). *Listeria*: A foodborne pathogen that knows how to survive. *International Journal of Food Microbiology, 113*(1), 1–15.

García, A., Fox, J. G., & Besser, T. E., (2010). Zoonotic enterohemorrhagic *Escherichia coli*: A One Health perspective. *ILAR Journal, 51*(3), 221–232.

Garcia-Angulo, V. A., Kalita, A., & Torres, A. G., (2013). Advances in the development of enterohemorrhagic *Escherichia coli* vaccines using murine models of infection. *Vaccine, 31*(32), 3229–3235.

Ghimire, T. R., (2010). Redescription of genera of family Eimeriidae Minchin, 1903. *International Journal of Life Sciences, 4*, 26–47.

Glass, R. I., Parashar, U. D., & Estes, M. K., (2009). Norovirus gastroenteritis. *New England Journal of Medicine, 361*(18), 1776–1785.

Grass, J. E., Gould, L. H., & Mahon, B. E., (2013). Epidemiology of foodborne disease outbreaks caused by *Clostridium perfringens*, United States, 1998–2010. *Foodborne Pathogens and Disease, 10*(2), 131–136.

Halliday, M. L., Kang, L. Y., Zhou, T. K., et al., (1991). An epidemic of hepatitis A attributable to the ingestion of raw clams in Shanghai, China. *Journal of Infectious Diseases, 164*(5), 852–859.

Hamlyn, P. F., Wales, D. S., & Sagar, B. F., (1987). Extracellular enzymes of *Penicillium*. In: Pebberdy, J. F., (ed.), *Penicillium and Acremonium* (pp. 245–284). Boston, MA: Springer.

Hedberg, C. W., Levine, W. C., White, K. E., et al., (1992). An international foodborne outbreak of shigellosis associated with a commercial airline. *The Journal of the American Medical Association, 268*(22), 3208–3212.

Hennessy, T. W., Hedberg, C. W., Slutsker, L., et al., (1996). A national outbreak of *Salmonella enteritidis* infections from ice cream. *New England Journal of Medicine, 334*(20), 1281–1286.

Hill, D. E., & Dubey, J. P., (2010). Foodborne parasites. In: Juneja, V. K., & Sofos, J. N., (eds.), *Pathogens and Toxins in Foods* (pp. 195–217). American Society of Microbiology (ASM) Press.

Hofstra, H., Van, D. V. J. M. B. M., & Van, D. P. J., (1994). Microbes in food processing technology. *FEMS Microbiology Reviews, 15*(2, 3), 175 183.

Hui, Y. H., (2018). *Foodborne Disease Handbook: Bacterial Pathogens* (Vol. 1). Florida: CRC Press.

Humphrey, T., O'Brien, S., & Madsen, M., (2007). *Campylobacter*s as zoonotic pathogens: A food production perspective. *International Journal of Food Microbiology, 117*(3), 237–257.

Hutt, P. B., & Hutt, P. B. I., (1984). A history of government regulation of adulteration and misbranding of food. *The Food and Drug Law Journal, 39*, 2.

ICMSF, (1996). *Microorganisms in Foods 5: Characteristics of Microbial Pathogens* (Vol. 5). Springer Science and Business Media.

Ionidis, G., Hübscher, J., Jack, T., et al., (2016). Development and virucidal activity of a novel alcohol-based hand disinfectant supplemented with urea and citric acid. *BMC Infectious Diseases, 16*(1), 77.

Iritani, N., Kaida, A., Abe, N., et al., (2014). Detection and genetic characterization of human enteric viruses in oyster-associated gastroenteritis outbreaks between 2001 and 2012 in Osaka City, Japan. *Journal of Medical Virology, 86*(12), 2019–2025.

Iturriza-Gomara, M., & O'Brien, S. J., (2016). Foodborne viral infections. *Current Opinion in Infectious Diseases, 29*(5), 495–501.

Jay, M. T., Cooley, M., Carychao, D., et al., (2007). *Escherichia coli* O157: H7 in feral swine near spinach fields and cattle, central California coast. *Emerging Infectious Diseases, 13*(12), 1908.

Jemmi, T., & Stephan, R., (2006). *Listeria monocytogenes*: Foodborne pathogen and hygiene indicator. *Revue Scientifique et Technique, 25*(2), 571–580.

Julia, C., Haston, M. D., & Larry, K. P., (2018). Foodborne disease outbreaks: An ongoing public health problem. *Morbidity and Mortality Weekly Report Surveillance Summaries, 67*(10), 1–11.

Juliao, P. C., Maslanka, S., Dykes, J., et al., (2013). National outbreak of type A foodborne botulism associated with a widely distributed commercially canned hot dog chili sauce. *Clinical Infectious Diseases, 56*(3), 376–382.

Kingsley, D. H., (2016). Emerging foodborne and agriculture-related viruses. *Microbiology Spectrum, 4*(4). doi: 10.1128/microbiolspec.PFS-0007-2014.

Kinsella, J. E., Hwang, D. H., & Dwivedi, B., (1976). Enzymes of *Penicillium roqueforti* involved in the biosynthesis of cheese flavor. *Critical Reviews in Food Science and Nutrition, 8*(2), 191–228.

Koff, R. S., Grady, G. F., Chalmers, T. C., Mosley, J. W., Swartz, B. L., & Boston Inter-Hospital Liver Group, (1967). Viral hepatitis in a group of Boston hospitals: Importance of exposure to shellfish in a nonepidemic period. *New England Journal of Medicine, 276*(13), 703–710.

Koopmans, M., & Duizer, E., (2004). Foodborne viruses: An emerging problem. *International Journal of Food Microbiology, 90*(1), 23–41.

Koopmans, M., (2008). Progress in understanding norovirus epidemiology. *Current Opinion in Infectious Diseases, 21*(5), 544–552.

Koutsoumanis, K. P., Lianou, A., & Sofos, J. N., (2014). Food safety: Emerging pathogens. *Encyclopedia of Agriculture and Food Systems*, 250–272.

Lahti, E., Löfdahl, M., Ågren, J., Hansson, I., & Olsson, E. E., (2017). Confirmation of a campylobacteriosis outbreak associated with chicken liver Pâté using PFGE and WGS. *Zoonoses and Public Health, 64*(1), 14–20.

Lamb, R. A., & Krug, R. M., (1996). Orthomyxoviridae: The viruses and their replication. In: Knipe, D. M., Howley, P. M., & Fields, B. N., (eds.), *Fields Virology.* Philadelphia: Lippincott-Raven Press.

Lastovica, A. J., (2006). Emerging *Campylobacter* spp.: The tip of the iceberg. *Clinical Microbiology Newsletter, 28*(7), 49–56.

Leroy, F., & De Vuyst, L., (2010). Bacteriocins of lactic acid bacteria to combat undesirable bacteria in dairy products. *Australian Journal of Dairy Technology, 65*(3), 143–149.

Lopez-Martinez, I., Balish, A., Barrera-Badillo, G., Jones, J., Nuñez-García, T. E., Jang, Y., et al., (2013). Highly pathogenic avian influenza A (H7N3) virus in poultry workers, Mexico- 2012. *Emerging Infectious Diseases, 19*(9), 1531.

Lu, S., Zheng, Y., Li, T., et al., (2013). Clinical findings for early human cases of influenza A(H7N9) virus infection, Shanghai, China. *Emerging Infectious Diseases, 19*(7), 1142–1146.

Ma, C., Deng, X., Ke, C., et al., (2014). Epidemiology and etiology characteristics of foodborne outbreaks caused by *Vibrio parahaemolyticus* during 2008–2010 in Guangdong Province, China. *Foodborne Pathogens and Disease, 11*(1), 21–29.

Mack, A., Hutton, R., Olsen, L., Relman, D. A., & Choffnes, E. R., (2012). *Improving Food Safety Through a One Health Approach: Workshop Summary*. Washington DC: National Academies Press.

Marshall, K. E., Nguyen, T. A., Ablan, M., et al., (2020). Investigations of possible multistate outbreaks of *Salmonella*, Shiga toxin-producing *Escherichia coli*, and *Listeria monocytogenes* infections-United States- 2016. *Morbidity and Mortality Weekly Report Surveillance Summaries, 69*(6), 1–14.

Martinelli, D., Fortunato, F., Tafuri, S., et al., (2013). Lessons learnt from a birthday party: A *Bacillus cereus* outbreak, Bari, Italy. *Annali Dellistituto Superiore di Sanita, 49*(4), 391–394.

Mattison, K., Bidawid, S., & Farber, J., (2009). Hepatitis viruses and emerging viruses. *Foodborne Pathogens*, 891–929.

Maula, M. S., (2010). *Agroindustry of Fermented Products*. http://agroindustriindonesia.blogspot.com/2010/10/agroindustry-of-fermented-products.html (accessed on 20 December 2021).

McCarter, Y. S., (2009). Infectious disease outbreaks on cruise ships. *Clinical Microbiology Newsletter, 31*(21), 161–168.

Mead, P. S., Slutsker, L., Dietz, V., et al., (1999). Food-related illness and death in the United States. *Emerging Infectious Diseases, 5*(5), 607.

Mishra, S. S., Ray, R. C., Panda, S. K., & Montet, D., (2017). Technological innovations in processing of fermented foods an overview. In: Ray, R. C., & Montet, D., (eds.), *Fermented Foods* (Part II, pp. 21–45). London: CRC Press.

Mitscherlich, E., & Marth, E. H., (2012). *Microbial Survival in the Environment: Bacteria and Rickettsiae Important in Human and Animal Health*. Springer.

Moffatt, C. R. M., Greig, A., Valcanis, M., et al., (2016). A large outbreak of *Campylobacter jejuni* infection in a university college caused by chicken liver pâté, Australia, 2013. *Epidemiology and Infection, 144*(14), 2971–2978.

Mossel, D. A. A., Corry, J. E., Struijk, C. B., & Baird, R. M., (1995). *Essentials of the Microbiology of Foods: A Textbook for Advanced Studies*. New York: John Wiley and Sons Inc.

Müller, L., Schultz, A. C., Fonager, J., et al., (2015). Separate norovirus outbreaks linked to one source of imported frozen raspberries by molecular analysis, Denmark, 2010–2011. *Epidemiology and Infection, 143*(11), 2299–2307.

Murrell, D., & Crompton, D. W. T., (2009). Foodborne helminth infections. In: Blackburn C. W., & McClure, P. J., (eds.), *Foodborne Pathogens: Hazards, Risk Analysis and Control*, 1009–1041. London: CRC Press.

NCBI, (2017). *National Centre for Biotechnology Information*. https://www.ncbi.nlm.nih.gov/genome (accessed on 20 December 2021).

Ortega, Y. R., & Sanchez, R., (2010). Update on *Cyclospora cayetanensis*, a foodborne and waterborne parasite. *Clinical Microbiology Reviews, 23*(1), 218–234.

Pascua, P. N. Q., & Choi, Y. K., (2014). Zoonotic infections with avian influenza A viruses and vaccine preparedness: A game of. *Clinical and Experimental Vaccine Research, 3*(2), 140–148.

Penteado, A. L., Eblen, B. S., & Miller, A. J., (2004). Evidence of *Salmonella* internalization into fresh mangos during simulated postharvest insect disinfestation procedures. *Journal of Food Protection, 67*(1), 181–184.

Rajkowski, K. T., & Smith, J. L., (2001). Update: Food poisoning and other diseases induced by *Bacillus cereus*. In: Hui, Y. H., Pierson, M. D., & Gorham, J. R., (eds.), *Foodborne Disease Handbook*, 61–76.

Rasko, D. A., Webster, D. R., Sahl, J. W., et al., (2011). Origins of the *E. coli* strain causing an outbreak of hemolytic–uremic syndrome in Germany. *New England Journal of Medicine, 365*(8), 709–717.

Rezac, S., Kok, C. R., Heermann, M., & Hutkins, R., (2018). Fermented foods as a dietary source of live organisms. *Frontiers in Microbiology, 9*, 1785.

Riner, D. K., Mullin, A. S., Lucas, S. Y., Cross, J. H., & Lindquist, H. A., (2007). Enhanced concentration and isolation of *Cyclospora cayetanensis* oocysts from human fecal samples. *Journal of Microbiological Methods, 71*(1), 75–77.

Sahu, L., & Panda, S. K., (2018). Innovative technologies and implications in fermented food and beverage industries: An overview. In: Panda, S. K., & Halady, P. K. S., (eds.), *Innovations in Technologies for Fermented Food and Beverage Industries* (pp. 1–23). New York: Springer.

Scallan, E., Hoekstra, R. M., Angulo, F. J., et al., (2011). Foodborne illness acquired in the United States - major pathogens. *Emerging Infectious Diseases, 17*(1), 7.

Schaffter, N., Zumstein, J., & Parriaux, A., (2004). Factors influencing the bacteriological water quality in mountainous surface and groundwaters. *Acta Hydrochimica et Hydrobiologica, 32*(3), 225–234.

Schmidt, G. D., Roberts, L. S., & Janovy, J., (1977). *Foundations of Parasitology*. New York: McGraw Hill Publishers.

Seitz, E. W., (1990). Microbial and enzyme-induced flavors in dairy foods. *Journal of Dairy Science, 73*(12), 3664–3691.

Shao, D., Shi, Z., Wei, J., & Ma, Z., (2011). A brief review of foodborne zoonoses in China. *Epidemiology and Infection, 139*(10), 1497–1504.

Singh, S., (2016). Congenital toxoplasmosis: Clinical features, outcomes, treatment, and prevention. *Tropical Parasitology, 6*(2), 113.

Sivapalasingam, S., Barrett, E., Kimura, A., et al., (2003). A multistate outbreak of *Salmonella enterica* serotype Newport infection linked to mango consumption: Impact of water-dip disinfestation technology. *Clinical Infectious Diseases, 37*(12), 1585–1590.

Smith, H., & Evans, R., (2009). Parasites: *Cryptosporidium*, *Giardia*, *Cyclospora*, *Entamoeba histolytica*, *Toxoplasma gondii* and pathogenic free-living amoebae (*Acanthamoeba* spp. and *Naegleria fowleri*) as foodborne pathogens. In: Blackburn, C. W., & McClure, P. J., (eds.), *Foodborne Pathogens* (pp. 930–1008). New York: Woodhead Publishing.

Smith, J. L., & Fratamico, P. M., (2018). Emerging and re-emerging foodborne pathogens. *Foodborne Pathogens and Disease, 15*(12), 737–757.

Steinkraus, K., (2004). *Industrialization of Indigenous Fermented Foods*. Baca Roton: CRC Press.

Stenfors, A. L. P., Fagerlund, A., & Granum, P. E., (2008). From soil to gut: *Bacillus cereus* and its food poisoning toxins. *FEMS Microbiology Reviews, 32*(4), 579–606.

Stern, N. J., Jones, D. M., Wesley, I. V., & Rollins, D. M., (1994). Colonization of chicks by non-culturable *Campylobacter* spp. *Letters in Applied Microbiology, 18*(6), 333–336.

Tamang, J. P., Cotter, P. D., Endo, A., et al., (2020). Fermented foods in a global age: East meets West. *Comprehensive Reviews in Food Science and Food Safety, 19*(1), 184–217.

Taubenberger, J. K., Reid, A. H., Lourens, R. M., Wang, R., Jin, G., & Fanning, T. G., (2005). Characterization of the 1918 influenza virus polymerase genes. *Nature, 437*(7060), 889–893.

Tauxe, R. V., (2002). Emerging foodborne pathogens. *International Journal of Food Microbiology, 78*(1, 2), 31–41.

Tuladhar, E., Hazeleger, W. C., Koopmans, M., Zwietering, M. H., Duizer, E., & Beumer, R. R., (2015). Reducing viral contamination from finger pads: Handwashing is more effective than alcohol-based hand disinfectants. *Journal of Hospital Infection, 90*(3), 226–234.

Vargas, M., Gascon, J., De Anta, M. T. J., & Vila, J., (1999). Prevalence of *Shigella* enterotoxins 1 and 2 among *Shigella* strains isolated from patients with traveler's diarrhea. *Journal of Clinical Microbiology, 37*(11), 3608–3611.

Venturini, C. M., (2019). Yeasts and molds in fermented food production: An ancient bioprocess. *Current Opinion in Food Science, 25*, 57–61.

Verhoef, L., Kouyos, R. D., Vennema, H., et al., (2011). An integrated approach to identifying international foodborne norovirus outbreaks. *Emerging Infectious Diseases, 17*(3), 412.

Wait, D. A., & Sobsey, M. D., (1983). Method for recovery of enteric viruses from estuarine sediments with chaotropic agents. *Applied and Environmental Microbiology, 46*(2), 379–385.

Weiss, L. M., & Kim, K., (2011). *Toxoplasma Gondii: The Model Apicomplexan, Perspectives and Methods*. Elsevier.

Wells, J. G., Davis, B. R., Wachsmuth, I. K., et al., (1983). Laboratory investigation of hemorrhagic colitis outbreaks associated with a rare *Escherichia coli* serotype. *Journal of Clinical Microbiology, 18*(3), 512–520.

WHO, (2015). *WHO Estimates of the Global Burden of Foodborne Diseases: Foodborne Disease Burden Epidemiology Reference Group 2007–2015*. WHO, Geneva.

Wijnands, L. M., (2008). *Bacillus cereus associated Foodborne Disease: Quantitative Aspects of Exposure Assessment and Hazard Characterization*. Wageningen University, The Netherlands.

Wu, Y., Wen, J., Ma, Y., Ma, X., & Chen, Y., (2014). Epidemiology of foodborne disease outbreaks caused by *Vibrio parahaemolyticus*, China, 2003–2008. *Food Control, 46*, 197–202.

Zhang, G., Ma, L., Patel, N., Swaminathan, B., Wedel, S., & Doyle, M. P., (2007). Isolation of *Salmonella typhimurium* from outbreak-associated cake mix. *Journal of Food Protection, 70*(4), 997–1001.

CHAPTER 4

Foodborne Outbreaks: Sources and Mode of Transmission of Foodborne Pathogenic Microorganisms

SMITHA CHANDRASHEKAR KOLLARAHITHLU,
SUHASINI SATHIYAMOORTHY,
SARANYA THIRUVOTTRIYUR SHANMUGAM, KAROLIEN DE WAEL,
JAYABRATA DAS, and PANDIYARASAN VELUSWAMY

ABSTRACT

The foodborne pathogens and microorganisms have been playing a prevalent role in the ebb and flow of the economy worldwide. The increasing population has strained the food processing industry to produce food in large quantity, which in turn has affected the quality of food. To curb this issue there is immense pressure to produce and maintain quality food within a short time frame. Hence, high throughput technology is used to determine and timely assess the safety and hygiene of food. Further, the revolution of the food industry has also seen an upsurge of new pathogens and microorganisms, thereby increasing the risk of exposure towards rarest diseases to a larger population. This chapter sheds light on the different types of foodborne pathogens affecting the food industry and its social impact. It further emphasizes on the safety measures to be taken on the prevention of the disease from the farm to the processing industries and in turn to the household.

Global Food Safety: Microbial Interventions and Molecular Advancements. Saher Islam, PhD, Devarajan Thangadurai, PhD, Jeyabalan Sangeetha, PhD, & Natália Cruz-Martins, PhD (Editors)

4.1 INTRODUCTION

The primary need for every living species on this biosphere is "food" which is essential for acquiring energy and breeding. Hence, the food quality has to be preserved to protect humankind and the surrounding species from foodborne pathogens. Although stringent measures are taken to conserve food safety, foodborne pathogens have posed a significant threat to the consumers. These pathogens are bacterial agents such as bacteria, viruses, or parasites that cause illness due to infected food. This illness could occur either by foodborne infection or by foodborne intoxication. When the toxigenic pathogen is ingested with food and makes the human a host, which is known as "foodborne infection." For example, anaerobic bacterium like *Clostridium botulinum* once it gains entry into a host produces the toxin "botulinum," which leads to neuroparalytic disorder (Sugiyama, 1980). The amount of microorganisms required to cause an infection varies with the organism. When the human host ingests the food product infected with a pathogen, it is known as "foodborne intoxication," e.g., *Staphylococcus aureus* produces an enterotoxin, selects a specific target such as the gastrointestinal tract or the nervous system and produces symptoms such as vomiting, diarrhea, or severely disrupts muscle function. However, the revolution of the food industry has also seen an upsurge of drug-resistant organisms in the millennia. This has increased the risk of exposure to new foodborne illness. The understanding of emerging and re-emerging of the foodborne pathogens have been due to the change in lifestyle, economic, and climatic conditions (Schelin et al., 2011). The daunting aspect of the new constantly evolving pathogens has posed significant health problems for past decades. Thus, the necessity for developing new molecules to fight these microorganisms (Ribeiro et al., 2017) is the need of the hour. Given the past decades, a primary concern has to protect the health of the human population and prevent the spread of these pathogens further. Many food laws have been implemented since ancient times to control the emergence and re-emergence of pathogens. However, with the increase in consumption of exotic animals, raw fresh food and ingredients, the world seems to be a witness to the foodborne illness caused by these microorganisms despite all these measures (Andjelković et al., 2017).

4.2 HISTORY OF FOODBORNE OUTBREAKS

The evidence on the association of human diseases and food consumption dates back to 460 BC (Bintsis, 2017). The influence of certain conditions, such as cholera and typhoid fever, was still not known in the mid-1800s (Tulodziecki, 2011). However, the miasma theory was later proposed, which believed that the organic matter decays and decomposes into new toxic compounds that are released into the surroundings (Julia and Valleron, 2011; Tulodziecki, 2011). The cholera outbreak followed this in 1853–1854, wherein John Snow correlated the cholera mortality with a water source (Koch and Denike, 2009).

Later on, in the 1960s, the world saw a rise in foodborne outbreaks due to microorganisms like *Salmonella*, *Shigella*, *C. botulinum*, and *S. aureus*. In subsequent illness from the 1970s–1980s due to Norwalk virus, *C. jejuni*, *Cryptosporidium*, *Vibrio cholerae*, and a new strain of *E. coli* O157:H7 witnessed. Further, in 1985, the consumption of Mexican style soft cheese made from unpasteurized milk led to an infection from *L. monocytogenes*. In 2010, 600 million foodborne illnesses and 4,20,000 deaths were witnessed due to 31 foodborne hazards as estimated by the WHO (Havelaar et al., 2015). Although stringent measures are followed to conserve the quality of food and hygiene, the death toll due to foodborne illness still seems to persist worldwide. One such outbreak of hemorrhagic diarrhea observed due to *E. coli* O104:H4 in 2011 from sprouted fenugreek seeds with more than 4,000 cases and 50 deaths (Garcia et al., 2015). In 2012, the bacterium *Salmonella cubana* was identified in Alfalfa sprouts (Tanguay et al., 2017). From 2014–2017, a total of 14 foodborne outbreaks were observed that were caused by fruits such as mango and papaya linked to *Salmonella* spp. (Carstens et al., 2019). Thus, it can be observed that the spread of foodborne pathogens seemed to increase at an exponential rate.

4.3 SOURCE AND MODE OF TRANSMISSION OF FOODBORNE PATHOGENS/MICROORGANISMS

The ubiquitous nature of foodborne pathogens has increased the risk of exposure to a variety of foodborne outbreaks in recent years. Foodborne illness has not only affected food producers and processors but also had a drastic impact on the economy worldwide. This contamination occurs at different stages, either in the fresh produce, during transportation, or in the processing stage. Foodborne pathogens are categorized based on the type of food consumed, as reported by the US (FDA, 2010). Some of the commonly found pathogens are depicted in Table 4.1.

TABLE 4.1 List of Major Foodborne Microorganisms

Microorganism	Example	Mortality (%)	Occurrence
Bacteria	*Clostridium perfringens*	26	Uncooked meats, vegetables, unpasteurized milk, and soft cheese
	V. parahaemolyticus	4	
	Bacillus spp.	0	
	Staphylococcus aureus	6	
	V. vulnificus	36	
	Listeria monocytogenes	255	
	Yersinia	29	
	Vibrio cholerae	0	
Parasites	*Cryptosporidium parvum*	4	Meat, water
	Cyclospora cayetanensis	0	
	Toxoplasma gondii	327	
	Giardia intestinalis	2	
Viruses	Avian influenza virus	60	Oysters, ready to eat food, raw or undercooked food and water, fruit bats or nonhuman primates
	Norovirus	149	
	Rotavirus	0	
	Astrovirus	0	
	Hepatitis A	7	
	COVID-19	21.9*	

*Still counting.

In general, these foodborne pathogens affect children, the elderly, and people whose immune systems are compromised (Fleury et al., 2008). The burden of foodborne infectious disease has increased with the aging population and increasing global mobility. These pathogens, once they have gained entry to a host, replicate by continuously destroying the first line of defense, i.e., cellular, and humoral barriers and establish themselves in the host and make it susceptible to transmission to a new host.

Based on the mode of transmission from the host to vector, pathogens are classified into three types zoonotic, genetic (soil, water, and decaying plant materials) and human origin (person to person). Most of the foodborne pathogens are zoonotic, wherein the infective agents such as bacteria, fungi, viruses, and parasites are transmitted from animals to humans (Klous et al., 2016). As there are many ways of entry into the food chain, the zoonotic mode of transmission is the most difficult to control.

4.4 TYPES OF TRANSMISSION

The ability of the microorganism to proliferate at an exponential rate has raised an alarming concern worldwide. Based on the amount of population affected, the outbreak is classified as an epidemic, endemic, or pandemic.

The epidemic is defined as a sudden outbreak in a small population, e.g., infection due to *Vibrio cholerae* in Southeast Asia. Whereas endemic is characterized as the occurrence of disease persistently in a particular area, e.g., infection by *Cyclospora cayetanensis* from raspberries in Guatemala. Pandemic, on the other hand, is the spread of the disease worldwide, e.g., the recent COVID-19 virus, which has spread all over the world, infecting millions of people. The transmission occurs either through (a) direct contact; or (b) indirect contact.

1. **Direct Contact:** It is a transmission that takes place from a contaminated site to a susceptible host by surface contact. Examples include infections spread through skin-to-skin contact or mucous membranes, soil, or contaminated vegetation.
2. **Indirect Contact:** It is a transmission that takes place with the help of a vector/vehicle or through air particles (coughing/ sneezing). A vector could be water, biological agents such as blood, and other vehicles that can carry pathogens, e.g., clothing, bedding, and handkerchiefs.

4.5 EMERGING AND RE-EMERGING FOODBORNE PATHOGENS

The significant factors affecting food safety is industrial food production with a high number of animals and larger farms, thus creating a questionable problem of waste disposal and treatment. Furthermore, among the various pathogens which have accounted to date, the human population has been a witness to emerging and re-emerging pathogens. Emerging pathogens are the microorganisms that have infected humans for the first time. For example, the novel Coronavirus (COVID-19), which has emerged from bats and the infections were transmitted directly to humans through market civets and dromedary camels (Huang et al., 2020). The emergence of these new pathogens could be due to unsafe food handling, uncooked burgers, and meat, ice cream contaminated with raw liquid eggs, and increased consumption of chilled stored chicken.

The re-emerging pathogens are the pathogens which were present in the past decade and tend to reoccur in new locations in drug-resistant forms; e.g., H1N1 influenza pandemic virus observed to emerge from pigs in 2009, where they formed complex changes in human, swine, and avian influenza genes (Morens et al., 2009). Later on, another subtype of the influenza virus H1N1 evolved from wild birds, which amplified viral transmission from domestic poultry to humans.

Antibiotics have played a minor role in fighting against myriad diseases and infections in the modern-day. However, the persistent and over usage of these antibiotics has increased the pressure on the organisms and made them resistant to antibiotics (Andersson and Hughes, 2014). Such organisms are called 'Antibiotic-resistant bacteria' (ABR). The ARB enters the food chain either directly by ingestion or by ingestion of food contaminated with ARB or by contact with infected animals (Founou et al., 2016). Thus, these resistant microbes could increase the risk of exposure to a new form of the disease and risking the health of the public (Akbar and Anal, 2015). The ABR bacteria are estimated to cause a death toll of 10 million people by 2050, with a cumulative cost of US$100 trillion (Thapa et al., 2019). Hence, the need for the detection of these pathogens becomes highly mandatory.

4.6 PREVENTATIVE MEASURES

The first step in the prevention of spreading foodborne outbreaks is the identification and recognition of health problems. The prevention of the disease outbreak by maintaining high sanitary measures and mitigating the illness is the need of the hour (Odeyemi et al., 2018). The diverse places of food consumption of the people on a daily basis, such as restaurants, hotels, hospital, and street shops make it the responsibility of the vendors to follow regulatory and control measures in maintaining sanitation and hygiene to prevent the outbreak of diseases (Jianu and Golet, 2014). The revolution of the food industry, economic, and demographic transformation has paved for the possibility of new safety problems. The diversity of the food chain increased food production to meet the needs of the rapidly growing global population requires a crucial assessment of the food quality before it reaches the consumers. The sheds light on new challenges like detection, investigation, control, and prevention of foodborne illness (Dharma et al., 2014).

The spread of the foodborne outbreak can be prevented at the farming level; in developing countries, the farmers who still use old techniques and live near the animals have a higher chance of exposure to zoonotic transmission. Hence, they need to be educated on present-day technology and prevent the contamination of food. The food processing industries need to know good manufacturing practices, hazard analysis and critical control point (HACCP) systems, and quality control (CDC, 2003). Avoiding cross-contamination of cooked food and raw materials and preventing the consumption of uncooked food products is important.

The prevention of zoonotic diseases as put forth by CDC are by taking some measures like recognition of the illness, investigation, collaboration, development of advanced structures for diagnosis and surveillance, international, and disciplinary interventions, applied epidemiological and ecological research, training, updating on the current technology and information/communication (Binder, 1998). Hence, the CDC has considered the following objectives for the prevention of the disease:

1. Surveillance and response:
    - Focus on strengthening infectious disease surveillance and response, globally as well as in the United States;
    - Improving methods for gathering and evaluating surveillance data.
2. Applied research:
    - Improvised tools for identifying and understanding infectious diseases;
    - Determining risk factors for infectious diseases;
    - Development and evaluation of prevention control strategies.

3. Infrastructure and training:
   - Enhancing epidemiologic and laboratory capacity in the US and globally;
   - Improvising CDC's capacity to serve as a reference center for the diagnosis of infectious diseases and drug-resistant testing;
   - Enhancing the nation's capacity to respond to outbreaks, and ensure future generations to be able to respond to emerging threats.
4. Prevention and control:
   - Implementation, support, and evaluation of disease prevention in the US and globally;
   - Demonstration of programs to develop, evaluate, and promote strategies to help health care providers and other individuals change behaviors that facilitate disease transmission.

4.7 SOCIAL IMPACT OF THE OUTBREAK

The foodborne pathogens have been causing a menace worldwide, creating havoc among the population each time there is a foodborne outbreak. These foodborne pathogens affect one-third of the world's population annually and approximately 48 million people in the US (Scallan et al., 2011). Every year *Clostridium jejuni* causes 2.4 million cases in the US and is one of the causative agents for diarrheal diseases (5–14%) worldwide (CDC, 2008). The *Salmonella* infection has observed to increase from 14.53 cases per 100,000 in 2014 to a rate of 15.19 in 2013 with a yearly variation ranged from 0.5% to 16.8%, and the average annual percent change was an increase of 1.3% from 2005 to 2013 (Johnson et al., 2014). In France and Belgium, the *Salmonella enterica* serotype Chester; *Salmonella chester* caused an outbreak of 162 cases between 2014 and 2015, it was more likely that the people have eaten in a restaurant and visited the coast of Morocco. The drastic increase in death toll every year due to the sudden outbreaks of foodborne illness has impeded the socio-economic development worldwide (Fonteneau et al., 2017). The consumption of cream-filled eclairs contaminated with *Bacillus cereus* led to 470 people falling sick with vomiting, nausea, and abdominal pain. It has also reported as a primary causative agent in the Netherlands in 2006 with an outbreak of 5.4% and 32% in Norway in 2000 (Tewari and Abdulla, 2015). *Campylobacter*-induced enteritis caused illness in children aged 0–4 years in Mexico and Thailand. Around 300,000 cases/million population occur yearly in the Netherlands are also affected by the same bacterium.

An estimated 6 million cases with 350,000 hospitalizations and 9,000 deaths associated with foodborne infections each year have been observed in the US. Also, a decline in foodborne pathogens was observed from 1996–1998; also, the outbreak of *Shigella* decreased by 43%, *Yersinia* species by 49%, *Listeria monocytogenes* by 32%, *Campylobacter* species by 30%, EHEC O157: H7 by 29%, and *Salmonella typhimurium* by 42%; however, the incidence of *Salmonella enteritidis* and *S. Heidelberg* increased, each by 25%, and *S. javiana* by 82%.

4.8 CONCLUSION

The outbreak of the foodborne illness worldwide makes it difficult to monitor and control the spread of the disease. Foodborne illness occurs due to the ingestion of the food contaminated with pathogenic bacteria, viruses, or parasites. The foodborne pathogens could be the zoonotic, genetic, or human origin, and consumption of these pathogens leads to illness. These foodborne pathogens gain entry to the human body through different routes, either through intoxication, toxicoinfection or infection. However, few pathogens tend to re-emerge time and again; when they do, they evolve into a more mutated species, which makes it difficult for the researchers to obtain a cure immediately. These pathogens tend to affect infants, older people, and people whose immune system is compromised. These pathogens tend to be drug-resistant and may not respond to the drug which previously applied for the cure. The pathogen, once it gains entry through the food, majorly causes gastrointestinal problems, respiratory problems, and neuroparalytic disorders. These pathogens then multiply and express their virulence, thus damaging the host cell. The psychrophilic and mesophilic nature of the pathogens does not alter the esthetic quality of foods. Their pathogenic nature is acquired through plasmids, bacteriophages, or mutation.

The sudden rise in the illness affects the products, producers, retailers, and in turn leading to a recall of the products. Further, its inadequate supply to fulfill the treatment of the affected individuals in the developing countries poses a serious threat to a mass population. Thus, the consequences of the outbreak were observed to transform the economy and human lives drastically. As seen to date, the significant spread of the foodborne pathogens takes place from the markets and food producers. This takes place due to a lack of training for the farmers and the food handlers. Hence, the need to train these people about the updated technology and to follow safety guidelines is mandatory. Although stringent rules developed to maintain food quality and hygiene, the rising population has increased the demand for food production. Further, new techniques with detection methods will also play a very critical role in the early detection of pathogens.

Recent foodborne statistics state a decline in the foodborne outbreak of several pathogens; however, there is a possible threat of the foodborne pathogen in the near future. The emergence of these pathogens could be mitigated by regular surveys of the food quality from the raw material to the finished product. Educating the global population on the same and scrutinizing the product for pathogens during production and distribution, thus saving the global population from the occurrence and distribution of foodborne pathogens.

KEYWORDS

- **diseases**
- **food impact**
- **microorganisms**
- **pathogens processing**
- **quality**
- **upsurge**

REFERENCES

Akbar, A., & Anal, A. K., (2015). Isolation of *Salmonella* from ready-to-eat poultry meat and evaluation of its survival at low temperature, microwaving and simulated gastric fluids. *Journal of Food Science and Technology, 52*(5), 3051–3057. https://doi.org/10.1007/s13197-014-1354-2.

Andersson, D. I., & Hughes, D., (2014). Microbiological effects of sublethal levels of antibiotics. *Nature Reviews Microbiology, 12*(7), 465–478. https://doi.org/10.1038/nrmicro3270.

Andjelković, U., Šrajer, G. M., Gašo-Sokač, D., Martinović, T., & Josić, D., (2017). Foodomics and food safety: Where we are. *Food Technology and Biotechnology, 55*(3), 290–307. https://doi.org/10.17113/ftb.55.03.17.5044.

Binder, S., Levitt, A. M., & Hughes, J. M., (1999). Preventing emerging infectious diseases as we enter the 21st century: CDC's strategy. *Public Health Reports, 114*(2), 130–134. https://doi.org/10.1093/phr/114.2.130.

Bintsis, T., (2017). Foodborne pathogens. *AIMS Microbiology, 3*(3), 529–563. https://doi.org/10.3934/microbiol.2017.3.529.

Carstens, C. K., Salazar, J. K., & Darkoh, C., (2019). Multistate outbreaks of foodborne illness in the United States associated with fresh produce from 2010 to 2017. *Frontiers in Microbiology, 10*, 1–15. https://doi.org/10.3389/fmicb.2019.02667.

Coker, A. O., Isokpehi, R. D., Thomas, B. N., Amisu, K. O., & Obi, C. L., (2002). Human campylobacteriosis in developing countries. *Emerging Infectious Diseases, 8*(3), 237–243.

Dhama, K., Rajagunala, S., Chakrabort, S., Verma, A. K., Kumar, A., Tiwari, R., & Kapoor, S., (2013). Foodborne pathogens of animal origin-diagnosis, prevention, control and their zoonotic significance: A review. *Pakistan Journal of Biological Sciences, 16*(20), 1076–1085. https://doi.org/10.3923/pjbs.2013.1076.1085.

Fleury, M. D., Stratton, J., Tinga, C., Charron, D. F., & Aramini, J., (2008). A descriptive analysis of hospitalization due to acute gastrointestinal illness in Canada, 1995–2004. *Canadian Journal of Public Health, 99*(6), 489–493. https://doi.org/10.1007/bf03403783.

Fonteneau, L., Jourdan Da, S. N., Fabre, L., et al., (2017). Multinational outbreak of travel-related *Salmonella chester* infections in Europe, summers 2014 and 2015. *Eurosurveillance, 22*(7), 1–11. https://doi.org/10.2807/1560-7917.ES.2017.22.7.30463.

Founou, L. L., Founou, R. C., & Essack, S. Y., (2016). Antibiotic resistance in the food chain: A developing country perspective. *Frontiers in Microbiology, 7*, 1–19. https://doi.org/10.3389/fmicb.2016.01881/

Havelaar, A. H., Kirk, M. D., Torgerson, P. R., et al., (2015). World Health Organization global estimates and regional comparisons of the burden of foodborne disease in 2010. *PLoS Medicine, 12*(12), 1–23. https://doi.org/10.1371/journal.pmed.1001923/

Jianu, C., & Goleţ, I., (2014). Knowledge of food safety and hygiene and personal hygiene practices among meat handlers operating in western Romania. *Food Control, 42*, 214–219. https://doi.org/10.1016/j.foodcont.2014.02.032.

Johnson, E. A., & Schantz, E. J., (2017). *Seafood Toxins, Foodborne Diseases*. Elsevier Inc. https://doi.org/10.1016/B978-0-12–385007-2.00017-6.

Julia, C., & Valleron, A. J., (2011). Louis-René villermé (1782–1863), a pioneer in social epidemiology: Re-analysis of his data on comparative mortality in Paris in the early 19th century. *Journal of Epidemiology and Community Health, 65*(8), 666–670. https://doi.org/10.1136/jech.2009.087957.

Klous, G., Huss, A., Heederik, D. J. J., & Coutinho, R. A., (2016). Human-livestock contacts and their relationship to transmission of zoonotic pathogens, a systematic review of literature. *One Health, 2*, 65–76. https://doi.org/10.1016/j.onehlt.2016.03.001.

Koch, T., & Denike, K., (2009). Crediting his critics' concerns: Remaking John snow's map of Broad Street cholera, 1854. *Social Science and Medicine, 69*(8), 1246–1251. https://doi.org/10.1016/j.socscimed.2009.07.046.

Morens, D. M., Taubenberger, J. K., & Fauci, A. S., (2009). The persistent legacy of the 1918 influenza virus. *New England Journal of Medicine, 361*(3), 225–229. https://doi.org/10.1056/NEJMp0904819.

Navarro-Garcia, F., (2014). *Escherichia coli* O104:H4 pathogenesis: An enteroaggregative *E. coli*/Shiga toxin-producing *E. coli* explosive cocktail of high virulence. *Microbiology Spectrum, 2*(6), 1–19. https://doi.org/10.1128/microbiolspec.ehec-0008-2013.

Odeyemi, O. A., Sani, N. A., Obadina, A. O., et al., (2019). Food safety knowledge, attitudes and practices among consumers in developing countries: An international survey. *Food Research International, 116*, 1386–1390. https://doi.org/10.1016/j.foodres.2018.10.030.

Poulain, B., (2010). La neurotoxine botulinique. *Revue Neurologique, 166*(1), 7–20. https://doi.org/10.1016/j.neurol.2009.09.004.

Rewar, S., & Mirdha, D., (2014). Transmission of Ebola virus disease: An overview. *Annals of Global Health, 80*(6), 444–451. https://doi.org/10.1016/j.aogh.2015.02.005.

Ribeiro-Santos, R., Andrade, M., Madella, D., Martinazzo, A. P., De Aquino, G. M. L., De Melo, N. R., & Sanches-Silva, A., (2017). Revisiting an ancient spice with medicinal purposes: Cinnamon. *Trends in Food Science and Technology, 62*, 154–169. https://doi.org/10.1016/j.tifs.2017.02.011.

Scallan, E., Hoekstra, R. M., Angulo, F. J., et al., (2011). Foodborne illness acquired in the United States - major pathogens. *Emerging Infectious Diseases, 17*(1), 7–15. https://doi.org/10.3201/eid1701.P11101.

Schelin, J., Wallin-Carlquist, N., Cohn, M. T., Lindqvist, R., Barker, G. C., & Rådström, P., (2011). The formation of *Staphylococcus aureus* enterotoxin in food environments and advances in risk assessment. *Virulence, 2*(6), 580–592. https://doi.org/10.4161/viru.2.6.18122.

Tewari, A., & Abdullah, S., (2015). *Bacillus cereus* food poisoning: International and Indian perspective. *Journal of Food Science and Technology, 52*(5), 2500–2511. https://doi.org/10.1007/s13197-014-1344-4.

Thapa, S. P., Shrestha, S., & Anal, A. K., (2020). Addressing the antibiotic resistance and improving the food safety in the food supply chain (farm-to-fork) in Southeast Asia. *Food Control, 108*, 106809. https://doi.org/10.1016/j.foodcont.2019.106809.

Tulodziecki, D., (2011). A case study in explanatory power: John Snow's conclusions about the pathology and transmission of cholera. *Studies in History and Philosophy of Science Part C: Studies in History and Philosophy of Biological and Biomedical Sciences, 42*(3), 306–316. https://doi.org/10.1016/j.shpsc.2011.02.001.

Part II

Characterizing Foodborne Pathogens Using Molecular Approaches

CHAPTER 5

Epidemiology of Foodborne Bacterial Diseases and Molecular Techniques to Assure Food Safety

AJITH ATHIRA, BENSSAN K. VARGHESE, and
RADHAKRISHNAN PREETHA

ABSTRACT

The epidemiological study investigates the causes and development of the disease, treatment, and control methods. The infections due to some foodborne microorganisms can cause severe illness and even death. Controlling these diseases is crucial because the majority of them are contagious. It is mandatory to identify the organism and collect all the possible information, including ancestry, behavior, ecology, virulence, and resistance to control foodborne diseases. Epidemiologists worldwide rely on modern molecular techniques for the identification and characterization of pathogens. This chapter describes the epidemiological analysis of pathogens including *Listeria monocytogenes, Salmonella enterica, Campylobacter jejuni, Yersinia enterocolitica, Yersinia pseudotuberculosis, Clostridium botulinum, Staphylococcus aureus, Shigella sonnei, Shigella dysenteriae, Shigella boydii, Shigella flexneri,* and *Escherichia coli* using techniques such as PCR, PFGE, RFLP, MLST, RAPD, MLVA, WGS, CRISPR-HRMA, MALDI-TOF, and DNA microarray.

Global Food Safety: Microbial Interventions and Molecular Advancements. Saher Islam, PhD, Devarajan Thangadurai, PhD, Jeyabalan Sangeetha, PhD, & Natália Cruz-Martins, PhD (Editors)

5.1 INTRODUCTION

Several diseases transmit through food (Faour-Klingbeil and Todd, 2020). Reports suggest a large number of deaths throughout human history due to the naturally occurring foodborne pathogens. Among these, only a smaller number of foodborne illnesses could be identified and cured (Awasti and Anand, 2020). When considering the foodborne outbreaks that happened from 1995, it is evident that bacterial diseases were more predominant (Doren et al., 2013). Epidemiology refers to the area of science, which deals with the occurrence, spread, and control of diseases (Hartge, 2014). Molecular epidemiology deals with utilizing molecular genetic techniques for the epidemiological investigation of disease conditions (Foxman, 2012). The epidemiological data considered as keystone in foodborne outbreak research. Many types of molecular procedures are for foodborne biological pathogen identification (Table 5.1). Typing techniques had employed for differentiating foodborne pathogens at the species and subspecies level to identify the pathogen. Molecular typing methods have become an inevitable technique in the epidemiological study of foodborne pathogens (Ferrari et al., 2017). A wide range of molecular techniques had used for detecting, classifying, and characterizing foodborne pathogens. The commonly used molecular techniques include PFGE (pulsed-field gel electrophoresis), MLST (multilocus sequence typing), RAPD (random amplified polymorphic DNA), deoxyribonucleic acid sequencing, multiplex polymerase chain reaction (PCR), and many more (Adhikary et al., 2019).

Nowadays, molecular genetic methods are used to identify, isolate, and study the genetic diversity of the biological pathogen. While selecting a technique, some essential basic facts need to be considered; for example, a PCR is well suited for identifying the virulence genes of *C. perfringens* but not suitable for *Salmonella* strains (Lukinmaa et al., 2004). However, the species genetic diversity had explained by typing methods such as serotyping or phage typing. The selected typing method should have high reliability and characterizing power and should be cheap, usable, and easily understand the results.

Along with the molecular typing results, epidemiological information is also necessary to conclude an expected infection (Lukinmaa et al., 2004). Different molecular techniques have to be used to identify, analyze, compare, and diagnose various foodborne biological pathogens (Figure 5.1).

TABLE 5.1 Foodborne Pathogens Outbreaks and Molecular Techniques Used for the Epidemiological Study

Biological Pathogen	Source	Diseases/Infections	Countries in which Reported	Molecular Techniques Used	References
Listeria monocytogenes	Dairy products, cheese products, fish, sausage, milk, ready-to-eat foods	Septicemia, meningitis, maternal illness, neonatal illness	Chile, Brazil, New Zealand, Italy	Real-time PCR, multiplex PCR, PFGE	Montero et al. (2015)
Salmonella enterica	Meat, raw fruits, and vegetables, unpasteurized milk, other dairy products, eggs	Enteric fever, gastroenteric disease, septicemia, typhoidal fever, diarrhea	USA, Spain, Switzerland, Italy, Hungary, Japan, Russia	Microarray techniques, whole-genome sequencing, genome assembly, bioinformatics	Chiu et al. (2004)
Campylobacter jejuni	Poultry meat, sheep, pigs, pet animals, cattle	Mild, watery to bloody diarrhea, Guillan-Barre syndrome	European Countries	PFGE analysis, computer analysis, multiplex PCR, RAPD, CRISPR-HRMA, MLST, PFGE	Guirado et al. (2020)
Yersinia enterocolitica and *Yersinia pseudotuberculosis*	Meat, dairy products, salad vegetables	Swelling of the lymph nodes, diarrhea, and septicemia	England, Japan, Nigeria, New Zealand, Iran	Whole-genome sequencing, PFGE, and MLST	Hunter et al. (2019)
Clostridium botulinum	Poultry processing units, wildlife	Physical disability, lethality	Canada, Europe, USA	Real-time PCR, mouse bioassay	Gratiet et al. (2020)
Staphylococcus aureus	Meat, milk, dairy products	Pimples, scalded skin syndrome, pneumonia, sepsis, meningitis, toxic shock syndrome, osteomyelitis	Africa, India, Germany	MALDI-TOF, DNA microarrays	Pal et al. (2020)

TABLE 5.1 *(Continued)*

Biological Pathogen	Source	Diseases/Infections	Countries in which Reported	Molecular Techniques Used	References
Shigella spp.	Food and water contaminated with feces	Diarrhea, dysentery, fever, stomach cramps	Israel, England, Pakistan	qPCR, whole-genome sequencing	Lindsay et al. (2013)
Escherichia coli	Raw milk, raw or uncooked food, vegetables contaminated with feces	Meningitis, septicemia, cystitis, nephritis, urinary tract infections	Nigeria, Europe	PCR, next-generation sequencing methods, MLST, RFLP, PFGE	Matsukawa et al. (2019)

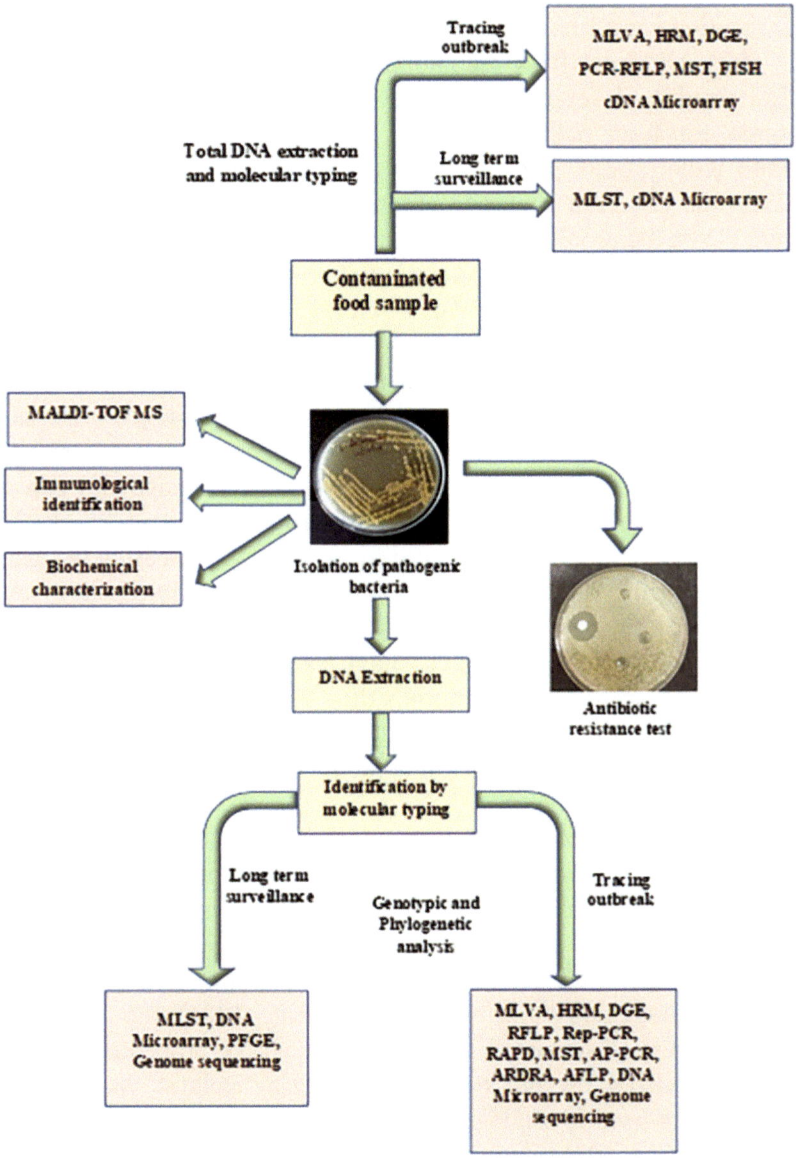

FIGURE 5.1 Identification methods for foodborne pathogens.

PCR has several advantages over other detection methods like typing, classification, and characterization of foodborne pathogens (Lazcka et al., 2007). Different PCR techniques such as real-time PCR, multiplex PCR, nested PCR, reverse-transcription PCR, and many more had widely reported for food pathogen studies (Lukinmaa et al., 2004). Another method widely used is PFGE, in which restriction patterns had used to analyze and compare bacterial genomes. In this method, genomic DNA isolated from the organism gets digested using enzymes, and DNA separation occurs by using a pulsed electric field. However, this method is time-consuming, but a faster version of PFGE is also available. It has accuracy and can be used in short-term epidemiological examinations. However, it is a labor-intensive process (Verweij and Stensvold, 2014).

In ribotyping, genomic DNA is digested and resolved by electrophoresis using a specific probe for ribosomal RNA genes. Automated ribotyping methods are beneficial in epidemiological surveys because they can determine many bacterial strains in a short duration (Lukinmaa et al., 2004). Plasmid analysis is used for plasmid fingerprinting, detecting the plasmid genes, and identifying the plasmid profile. For epidemiological investigations, plasmid fingerprinting and plasmid genes detection are very useful (Liebana et al., 2001). Pathogens have extrachromosomal factors, and plasmids are related to the bacteria's drug resistance or virulence factors (Brown et al., 1991). Pathogens can lose or attain plasmids, limiting the application of plasmid analysis in epidemiological surveys (Horby et al., 2003).

Typing by sequencing involves a variety of methods such as MLST, whole-genome sequencing (WGS), variable number of tandem repeats (VNTRs), multi-locus variable-number tandem repeat analysis (MLVA) and DNA microarray (de Benito et al., 2004). Sequencing helps to make a systematic list of all genetic elements. It also provides information on bacterial pathogenicity and helps to determine the source and extent of bacterial illness (Revazishvili et al., 2004). MLST procedure was developed and utilized for the typing of bacterial isolates. Whole-genome sequence data used to analyze and study the variable number of tandem repeats (VNTRs) (Bertelli and Greub, 2013).

Another method used in the field of epidemiological investigations is biosensor technology. This technology assures consistently good results in less time. The biosensor is a rapidly emerging pathogen identification technology. It is a device consisting of biological materials (tissue, organelles, microorganisms, cell receptors, enzymes, antibodies, natural products, and nucleic acids) or biologically derived materials (recombinant antibodies, engineered proteins) or biomimetics (synthetic catalysts and imprinted polymers) as receptors. Along with the bioreceptors mentioned above, different transducing microsystems had used in the biosensors (Chang et al., 2017).

This chapter discusses the epidemiology of foodborne bacterial diseases caused by *Listeria monocytogenes*, *Salmonella enterica*, *Campylobacter jejuni*, *Yersinia enterocolitica*, *Yersinia pseudotuberculosis*, *Clostridium botulinum*, *Staphylococcus aureus*, *Shigella sonnei*, *Shigella dysenteriae*, *Shigella boydii*, *Shigella flexneri*, and *Escherichia coli* (Table 5.1). In addition to that, identification, typing, classification, and characterization of foodborne bacterial pathogens using different molecular techniques are explained in detail.

5.2 MOLECULAR EPIDEMIOLOGICAL METHODS FOR FOODBORNE PATHOGENS

5.2.1 *LISTERIA MONOCYTOGENES*

Listeria monocytogenes is a biological pathogen that causes food-related severe illness in humans. The infection caused by this pathogen is known as listeriosis. It is one of the most severe zoonotic diseases (Pombinho et al., 2020). This organism's vital characteristic is its existence in a distinct environment such as stale areas, salty areas, cold areas, and areas with a broad pH range. The disease caused by the organism shows symptoms of inflammatory diseases that affect those with low immunity. A significant outbreak caused by *L. monocytogenes* was reported in 2009 in Chile by consuming soft cheese. Vital sources of this pathogen are meat, sausages, and dairy products (Montero et al., 2015). The outbreak caused in Chile has proved the importance of food in public health and the need for an immediate epidemiological examination. Two methods, PCR, and PFGE, were opted for identifying its occurrence, behavior, and control of the disease. The first and foremost step in any epidemiological study is the isolation and identification of the disease-causing biological pathogen. In the above case, PCR had suggested detecting the serotypes and identify the virulence-related genes (Paduro et al., 2020).

Listeria species can survive in water, soil, or any materials. In an epidemiological study, the primary concern is that one should understand the pathogen's ecology. So, the existence of *L. monocytogenes* in humans and animals, and its persistence in food and environment, should be monitored. Epidemiology paves the way for recognizing the microorganism's capacity, virulence, and resistance. The genus of *Listeria* consists of 21 identified species and six subspecies.

Listeriosis in humans commonly arises from ready-to-eat (RTE) foods, and the main strains involved are *L. innocua* and *L. monocytogenes*. Frequently *L. monocytogenes* are reported as a significant causative organism causing foodborne outbreaks (Luth et al., 2019). *Listeria* strains, about 120, were isolated from cheese products (Wagner and Stessl, 2014). Multiplex PCR technique was commonly used for *Listeria* differentiation (Bubert et al., 1999). Biochemical profiling had also used to select and identify them correctly (Kaszoni-Ruckerl et al., 2020).

Figure 5.2 represents the invasion of *L. monocytogenes* in the host cell through phagocytosis, which leads to cell lysis and proliferation of the microorganism. This cycle repeats as the pathogen moves into the next cell, and thus the spread happens.

L. monocytogenes had majorly found in food items such as fish, sausage, milk, and cheese. The disease commonly occurs through animal sources. Some *L. monocytogenes* can survive in the food environments even after applying some inactivation methods (Carpentier and Cerf, 2011). In this case, the problem arises not only due to food contamination but also due to some new resistant strains. A detailed analysis should carry out to discover new strains (Langsrud et al., 2003). Cross-contamination can be prevented only by regular inspections of *L. monocytogenes*, affecting food sources. As per another report from Brazil, *Listeria monocytogenes* contamination occurred mainly in beef-based products. Brazil is the largest producer of beef and carries out export activities also. Because of the continuous contamination in this area, epidemiological investigations had done (Teixeira et al., 2019).

Traditional systems such as the culturing method had used for primary detection processes. To confirm the detection of the causative organism, real-time PCR had opted. Another main concern was the assessment of the serotype and ancestry of the pathogen. These could assess by using a multiplex PCR. Another essential criterion to be found out is the genetic resemblance, and as per the study, the technique suggested was PFGE (Teixeira et al., 2019).

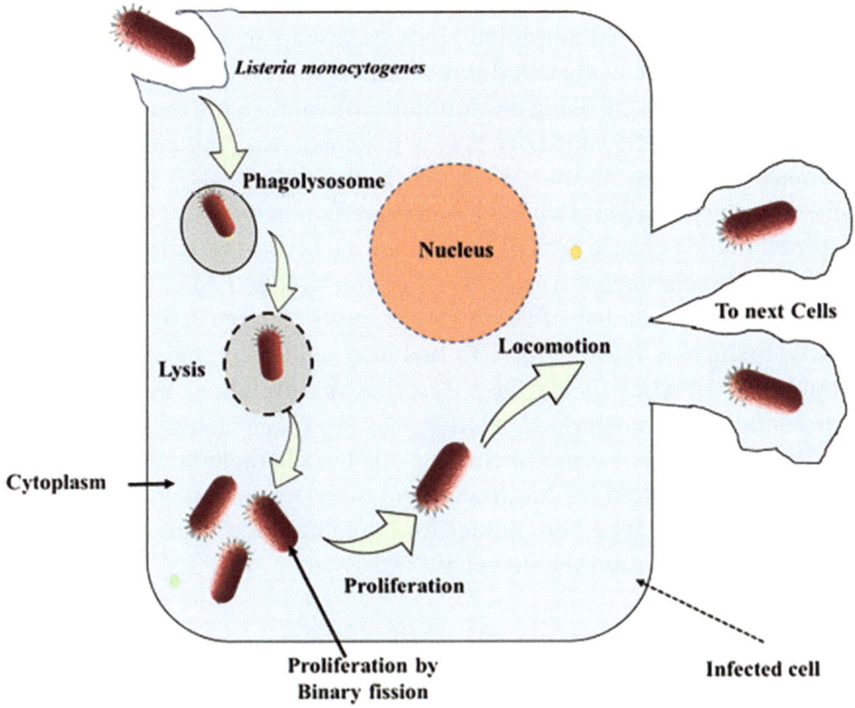

FIGURE 5.2 Inversion of *Listeria monocytogenes* in the host cells.

Another report from Brazil explained the role of *L. monocytogenes* in exported beef; the analysis of the types and genetic variations of the pathogen was also available. Molecular techniques such as real-time PCR and fluorimetric analysis were used in this study to confirm and quantify pathogens (Traunsek et al., 2011). Optimization of PCR-based detection of *L. monocytogenes in processed* chicken meat is also available in the literature (Moura et al., 2019). Multiplex PCR is another molecular technique used to examine the ancestry and the serotype of the confirmed *Listeria monocytogenes*. It can discriminate between different serotypes of *L. monocytogenes*. Multiplex PCR enables rapid detection of *Listeria* species and ensures food safety. Compared to conventional detection techniques, multiplex PCR is more specific and stable (Feng et al., 2020).

Another concern is to identify genetic similarities. The technique suggested as the better option for the separation of pathogens based on their genetic similarity is PFGE. The separation process meant the PulseNet or CDC protocol for *Listeria monocytogenes* (Graves and Swaminathan, 2001). They separated amplified DNA fragments in agarose gel (1%) and then stained them using ethidium bromide for visualization. One could analyze the results by using appropriate software. Using the PFGE profile, they can test genetic similarity between the isolates (Teixeira et al., 2019). Another technique similar to PFGE is multiple-locus variable-number tandem-repeat analysis (MLVA). A comparison between PFGE and MLVA is problematic because both of them examine several genetic events. PFGE relies on the changes in restriction enzyme site, and MLVA relies on the change in copy number of the tandem repeats (Sperry et al., 2008). Multi Locus Sequence Typing (MLST) had also helped to study the bacteria's population genetics. It can also describe the mutation, inheritance, and horizontal gene transfer in *L. monocytogenes* (Meinersmann et al., 2004). WGS is also a convenient technique for the characterization of *Listeria* isolates, and it provides the highest possible resolution in strain typing. It had referred to as a new model for contamination source tracking and outbreak investigation (Hyden et al., 2016).

5.2.2 SALMONELLA ENTERICA

Salmonella enterica belongs to the family Enterobacteriaceae, which affects mainly animals and humans. A global outbreak of this pathogen in poultry has been reported from 2008 to 2015 (Cohen et al., 2020). A significant outbreak of the disease reported in 2000 (Feasy et al., 2012). The transfer of bacterial pathogens mainly occurs through water and food polluted with human excreta. It happened in countries that had poor practices of cleanliness. The causative organism is acid susceptible, so that it will have to overcome the gastric acid barrier, which acts as an obstacle for initiating the infection (Crump et al., 2015).

The strains of *Salmonella* such as *S. enterica*, *S. typhimurium*, and *S. infantis* are specific in the host environment, and they infect mammals, reptiles, and birds. The pathogens mentioned above mainly target persons having less immunity. The illness is associated with gastrointestinal diseases. Reports say that approximately 80 million infections and 60,000 deaths happen worldwide every year (Havelaar et al., 2015). *S. infantis* is another well-known biological pathogen that causes severe illness next to *S. enterica.* It occurs from several food sectors, especially from poultry production units (Cohen et al., 2020). This organism is mainly responsible for the diseases and infections caused to majorly infants and then young children. They adhere to the host cells, even more, better than *S. typhimurium*. When compared to *S. typhimurium*, *S. infantis* rarely invade the host cell. The outbreaks caused by *S. infantis* have been reported in many countries (Cohen et al., 2020).

The non-typhoidal *Salmonella* species are majorly responsible for the death reported in the USA. One of the most dangerous *Salmonella* species is *S. choleraesuis*, which caused several infections in Thailand and USA (Crim et al., 2015). After identifying the causative organism, the next step in the epidemiological study is finding a defense mechanism against them. *S. choleraesuis* infection was studied further using the pig as a model of the experiment (Gray and Cray, 2001). The pathogen strains were administered to pigs orally. The isolates formed colonies and occupied the intestinal epithelium, resulting in the host organism's septicemia condition, and this resulted in the establishment of the carrier state in pigs. After seven days, the pig came back to good health. Similarly, they tested *S. typhimurium* in mice (Aviv et al., 2019).

Multilocus variable-number tandem-repeat analysis (MLVA) discriminates *Salmonella enterica* serovar *Enteritidis* and provides more timely information through databases (Hopkins et al., 2011). Whole-genome sequencing (WGS) is a fast and rapid technique. The sequencing of *Salmonella* species using WGS had done within 48 hours. So, WGS has considered an alternative to conventional subtyping methods. This sequence-based data had used to examine further outbreaks caused by *Salmonella* (Deng et al., 2012). Genome assembly had done after whole-genome sequencing, where they developed the hybrid assembly of the strain. The data obtained can be deposited in the databases such as NCBI (National Center for Biotechnology Information) (Cohen et al., 2020). This data helps to assess the risk and hazards related to the use of drugs associated with Salmonellosis and the resistance patterns of *Salmonella* (McDermott et al., 2016). The microarray technique has been reported recently to identify *S. enterica* genomes (Cohen et al., 2020). Genome assembly is another technique wherein some of the short DNA sequences can be taken and replaced in the same position to make a model of the original chromosome. It helps in the identification of similar *Salmonella* outbreaks and the identification of related organisms (Cohen et al., 2020).

Biosensors had also used in the detection of *Salmonella* species. A biosensor is a powerful tool that could operate for screening purposes. Hence, it will help to implement the HACCP (hazard analysis and critical control points) food production sector (Cinti et al., 2017). Different biosensors used in *Salmonella*'s epidemiological study include surface plasmon resonance biosensor, fiber optic, light scattering sensors, and microfluidic SERS biosensor. Also reported surface plasmon resonance biosensor (BIOCORE) for *Salmonella* detection via antibodies reacting with *Salmonella* group A, B, D, and E (Bokken et al., 2006). Fiber optic and light scattering sensors help in the screening of *S. enterica* in food items during hygienic farm practices, pre-harvest, and post-harvest operations, food manufacture, and product distribution (Abdelhaseib et al., 2016).

5.2.3 CAMPYLOBACTER JEJUNI

Campylobacter jejuni is a bacterial pathogen, usually causing gastroenteritis in some parts of the world. It is a commonly occurring bacterial foodborne illness in European countries (Blaser and Engberg, 2008). It causes diarrhea in humans and leads to death (Frazao et al., 2019). The infection has commonly transferred due to the uptake of unclean poultry meats. A significant vector that carries this disease is the wild birds (French et al., 2009). It leads to a condition of acute neuropathy, called Guillan-Barre syndrome (GBS). The disorder starts from the legs and immediately reaches up to the arms (Koga et al., 2006). Genetic studies of the pathogen indicated the existence of some of the hypervariable sequences in the genes. These sequences have coded with the proteins which participated in developing the structures at the surface. These variable sequences also contained short homopolymeric nucleotide tracts (Guirado et al., 2020).

The *Campylobacter* group typically consists of two organisms that cause primary foodborne illnesses. They are *C. jejuni* and *C. coli* (Scallan et al., 2011). Among these, *C. jejuni* is the most common. It commonly occurs in poultry and affects sheep, pigs, pet animals, and cattle (Wiekzorek et al., 2012). The contamination occurs mainly due to the use of half-cooked meat and contamination of prepared foods during processing and storage. The positive part of this disease is that the treatment doesn't need any antibiotics. Antibiotics should take when the symptoms extend to septicemia or other illness (Wiekzorek et al., 2012). The infection can transmit to approximately six people out of 10,000 per year (Tauxe et al., 1988).

The main concern is that this disease can affect any age group of persons. The pathogen could pass to soil and even through water, and the transmission occurs from human to human. After the pathogen's ingestion, the pathogen's incubation duration is 24 to 72 hours, and indications frequently involve cold headache, and fever (Tauxe and Blake, 1992). This disease commonly affects children rather than adults. The condition is more prevalent in developing countries than in developed countries. In developing countries, the illness is mostly involved with young children. The disease also affects immunocompetent people (Tauxe et al., 1988). The main reason for more occurrences of the pathogen in developing countries is overexposure to the organism (Blaser et al., 1983). The disease can occur due to the consumption of improperly cooked foods of animal origin and untreated water. Personnel hygiene maintenance is essential for avoiding the disease. The direct examination of the organism has conducted using a phase-contrast microscope (Blaser, 1997). The further recognition of the pathogen has done by using a multiplex PCR. An epidemiological investigation commonly relies on PFGE analysis (Broman et al., 2002).

The analysis of potential pathogens can be done by analyzing the genetic components. Molecular identification and screening of the virulence genes of *C. jejuni* had made by PCR-based techniques (Melo et al., 2019). The lifecycle of *Campylobacter* species had shown in Figure 5.3. CRISPR-HRMA (clustered regularly interspaced short palindromic repeats locus by high-resolution melting analysis), MLST (Multi Locus Sequence Typing), and PFGE (Pulsed Field Gel Electrophoresis) have used for examining the genetic variations and the epidemiological aspects related to *C. jejuni* (Frazao et al., 2019).

The strains of *C. jejuni* are found in wild birds as well as in broiler chicken. One of the conventional methods used to identify and purify the *C. jejuni* strain is PCR amplification (Melo et al., 2019). In a multiplex PCR, molecular identification to the genus and species level happens. The isolates identified in the test had taken to cultivate on the blood agar plates for a time range of 24 hours (Kim et al., 2017). By analyzing the genotypes using PFGE, genetic similarities between the strains of *Campylobacter* can be evaluated (Kim et al., 2013).

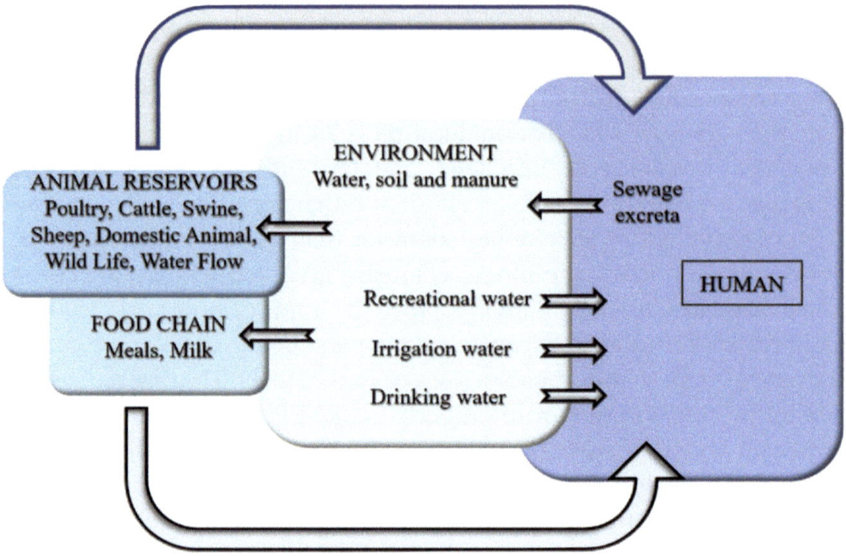

FIGURE 5.3 Life cycle of *Campylobacter* species.

RAPD as a PCR-based technique can be used to examine the microbial isolate's genetic variations. PCR amplification had performed using primers, and the PCR mixture involves template DNA, KCl, Tris-HCl, dNTP, $MgCl_2$, Taq DNA polymerase, and primer. PCR amplification and electrophoresis had conducted, and then DNA fingerprint gel images should be examined using appropriate software (Melo et al., 2019). RAPD has also applied for checking the diversity of *Campylobacter jejuni* present in a sample. They reported RAPD for typing the isolates of *C. jejuni*, for examining isolated patterns, and finding the similarity between them. It can also help to identify the source of the organism (Han et al., 2020). It will help to examine the genetic diversity of the bacteria and help to understand the epidemiology of diseases to a great extent (Dingle et al., 2002).

CRISPR-HRMA is a genome-editing technique through which a specific change had made in the DNA of the microorganism can be identified (Guha and Edgell, 2017). Primarily a reaction should be conducted using a reaction mixture that contains genomic DNA, primer, RNase, and

DNase. Then PCR is performed. The HRMA should be done immediately after the PCR cycle (Frazao et al., 2019). This is a specific and sensitive method used for detecting the genetic variation of *C. jejuni* isolates. In the epidemiological investigation studies, CRISPR-HRMA had implemented for identification, genotyping, cross-examination of single nucleotide polymorphisms (SNPs), and antibiotic resistance identification (Tamburro and Ripabelli, 2017).

5.2.4 *YERSINIA ENTEROCOLITICA* AND *YERSINIA PSEUDOTUBERCULOSIS*

Yersiniosis is one of the most important zoonotic diseases caused by the *Yersinia* sp., *Yersinia enterocolitica,* and *Yersinia pseudotuberculosis*. This disease transfers from one person to another through direct contact with the affected animals and their contaminated surroundings (Oda et al., 2015). It can cause severe gastroenteric diseases to the host animal. Many reports regarding the outbreaks of yersiniosis say that it occurred from dairy products and the meat exposed to the strains of *Yersinia* spp. (MacDonald et al., 2014). According to a new report, the microbial strains are active in vegetable salads through which humans can be affected (Williamson et al., 2016). After consuming the contaminated food items, gastroenteric symptoms can occur within three weeks. Primary symptoms had treated without antibiotics, but severe infection enhances immunity-related diseases such as arthritis, glomerulonephritis, erythema nodosum, and Reiter's disease (Mair and Fox, 1986).

Identification of *Y. enterocolitica* and *Y. pseudotuberculosis* has done using traditional culture methods and conventional biochemical testing. The pathogenicity also varies between different species among *Yersinia* (Hunter et al., 2019). *Yersinia enterocolitica* can cause septicemia and gastroenteritis. *Yersinia pseudotuberculosis* shows several indications such as diarrhea, lymph node swelling, and septicemia in human beings. Disease-causing *Yersinia* species widely occurs in wild animals (Takahashi et al., 2020).

People living in the rural areas of Africa used Quarry (Hunted Meat) in their diet. Some outbreaks occurred in Africa, where Yersinia spread vigorously through the Quarry directly or indirectly (Ojo et al., 2019). They had reported Yersiniosis outbreaks from the pig farms located in

Europe. The outbreak has also been reported in Australia, and the source was sheep. The causative organism had identified as *Yersinia pseudotuberculosis*. Simultaneously, the Yersiniosis outbreak said from New Zealand occurred in farmed deer because of the same causative organism (Stanger et al., 2018).

The *Y. enterocolitica* strains identified in the pigs were resistant to some antibiotics, indicating an immediate requirement of prevention methods that could prevent *Y. enterocolitica* (Drummond et al., 2012). Typing of the bacterial strains carried out by the use of short-read sequence typing (SRST), metric oriented sequence typing (MOST), PFGE, and MLST (Xue et al., 2020). *Yersinia enterocolitica* and *Yersinia pseudotuberculosis* strains, and their sequencing are an essential step in the epidemiological study. WGS has performed well in this case. It allows a complete comparison among multiple samples of *Yersinia* species. This comparison will help the epidemiologists to examine the organism genetically (Thomson et al., 2006). PFGE and MLST are two different serotyping methods adopted for the *Yersinia* species. These methods cannot perform before whole-genome sequence analysis. One can obtain serotypes from genomic data (Hunter et al., 2019), and PFGE has been used to differentiate and discriminate *Yersinia* isolates. Compared to PFGE, MLST is not suitable for minute differentiation of the species (Revazishvili et al., 2008).

5.2.5 CLOSTRIDIUM BOTULINUM

Clostridium botulinum causes severe infections in humans. Usually, the disease caused by the bacteria is non-communicable, as well as neuroparalytic. *Clostridium botulinum* produces toxins called botulinum toxins (Silva et al., 2020). Botulinum toxins are a type of neurotoxin. The neurotoxin production had reported by different *Clostridium* species such as *C. butyricum*, *C. parabotulinum*, *C. baratti*, *C. sporogenes*, *C. argentinense*, and *C. botulinum*. Non-toxigenic members are also present in the species mentioned above (Smith et al., 2020).

Clostridium botulinum can enter into the food at the time of harvesting, processing, and storage. During this period, the toxin had produced by the microbe grown in the food item. These bacteria can withstand the destructive treatments of radiation treatment, high-pressure treatment, and heat treatments. The bacteria can produce endospores; hence they have an opposing tendency against the harmful treatments (Kim et al., 2003). According to the World Health Organization (WHO), approximately 500 foodborne botulism cases have been reported per year (Kirk et al., 2015). Foodborne botulism symptoms involve abdominal cramps, vomiting, nausea, double vision, paralysis, difficulty speaking and swallowing, facial weakness, and dry mouth (Harris et al., 2020).

When considering foodborne botulism's worldwide occurrence, the primary sources of outbreaks are poultry processing units and wildlife (Sobel, 2005). Researchers have isolated toxin-producing strains to study the behavior and genetic variations. The methods used for the isolation and detection of naturally occurring strains include PCR and the Mouse Bioassay. In both neuro-toxigenic and non-toxigenic strains, the identification has been done using real-time PCR (Gratiet et al., 2020). Also, real-time PCR was used to detect the BoNT gene in *Clostridium botulinum* (Gratiet et al., 2020). In the case of foodborne botulism outbreaks, it is essential to detect many samples for investigating the contamination source (Anniballi et al., 2012). MLST technique had used for detecting the presence of genetic linkages in *Clostridium botulinum*. Along with the genetic diversity, the phylogenetic analysis for *C. botulinum* can also examine (Jacobson et al., 2008).

5.2.6 STAPHYLOCOCCUS AUREUS

Staphylococcus aureus belongs to the *Staphylococcaceae* family. It is one of the dangerous species in this family because of its severe toxicity towards humans and animals. They can also persist in water, soil, and air. The infections usually spread through the foods contaminated by the pathogen (Butaye et al., 2007). This bacterium has a unique ability to prevent phagocytosis by coagulating the plasma, thereby converting the plasma fibrinogen into fibrin. Another critical feature of *S. aureus* is that they are resistant to antibiotics. It is said to be multi-drug resistant because *S. aureus* opposes various types of medicines. Methicillin-resistant *S. aureus* (MRSA) stain is the commonly reported antibiotic-resistant (ABR) strain. MRSA-based *S. aureus* is classified into two divisions. They are CA-MRSA (community-associated MRSA) and HA-MRSA (health-associated MRSA). These strains have similar biological characters, but they are different in antibiotic resistance, pathogenicity, and genetic characteristics (Pal et al., 2020).

Matrix-assisted laser desorption ionization-time of flight (MALDI-TOF) is also mainly used for identifying *S. aureus* strains. It is one of the sensitive, most straightforward, and fastest methods. The isolates of *S. aureus* were cultured after attaining optimum growth; the whole culture has homogenized. Then centrifuged it, and the pellet has taken for MALDI-TOF analysis (Song et al., 2017).

Genetic variations study has also reported using DNA Microarrays (Blomfeldt et al., 2016). They used DNA microarrays for identifying the genetic diversity of an organism. The DNA microarrays meant for *S. aureus* are having a chip attached to every well of a microtiter strip. The chip is having various target sequences for *S. aureus*. They used data obtained from this study to examine the genes, and a comparison between them had made. DNA microarrays can perform thousands of reactions at a time. The examination of *S. aureus* gene expression patterns helps in epidemiological surveillance (Brennan et al., 2011).

5.2.7 SHIGELLOSIS

Shigellosis is a disease that is prevalent across the world and is more common in developing countries. The primary symptom of Shigellosis is diarrhea, but dysentery has also grown in severe cases. Shigellosis in children starting with these symptoms gradually leads to death in extreme cases. The major four *Shigella* strains responsible for the disease are *S. sonnei*, *S. dysenteriae*, *S. boydii*, and *S. flexneri*. Shigellosis is a highly contagious disease in which the disease transmission had associated with the food and water contaminated with the faces. The condition can transfer from one person to another by direct contact. Reports showed that the disease could carry by vectors such as flies (Cohen et al., 2014).

Another mode of transfer of the disease is sexual transmission. Reports say that disease outbreaks due to *S. flexneri* and *S. sonnei* occur through sexual relationships in Europe, Asia, and North America. Shigella outbreaks, commonly regarded as travel-associated diseases (Bardsley et al., 2018). This diarrheal infection has a mortality rate ranging from 5 to 15% (Cash et al., 2014). Shigellosis is associated with the food and water contaminated with feces, so the diagnosis was performed by taking stool samples: traditional culture methods and modern diagnostic technologies adopted to identify the species. In addition to these methods, several technicians and researchers have developed several molecular methods for testing the pathogens (Cohen et al., 2014). Researchers have claimed that there are some similarities between the *Shigella* species and *E. coli*. This similarity causes some troubles in the testing and diagnosis of *Shigella*. The *Shigella* has to oppose the stomach's acid environment to become capable of infecting the host (Lima et al., 2016).

DNA amplification techniques are the most convenient methods used for the detection of the *Shigella* species. When compared to the conventional methods, this technique has shown accuracy and regularity. Some reports were initially tested negative by traditional culture methods and were turned positive in quantitative PCR (qPCR) (Lindsay et al., 2013). *Shigella*'s study, especially about the occurrence and spread caused by the species, can be studied using WGS. It is also evident from the recent analysis that the strains are resistant to specific antibiotics such as ciprofloxacin, azithromycin, and trimethoprim-sulfamethoxazole (TMP-SMX) (Bardsley et al., 2020).

qPCR had used to detect the *Shigella* species, and it is one of the fastest techniques for screening *Shigella* species (Lindsay et al., 2013). Compared to the other PCR techniques, qPCR has high specificity and sensitivity. This assay decreases the risk of cross-contamination (Siala et al., 2017). The whole-genome sequencing of the samples had also reported for the detection of the organism. This technique had performed with the help of organism-specific algorithms and bioinformatics. A comparison had made between the sample and identified strains in bioinformatics software data. In this way, the similarity had calculated and reported (Chattaway et al., 2019).

5.2.8 *ESCHERICHIA COLI*

Escherichia coli belong to the Enterobacteriaceae family. It is the bacteria that can generally live in the human intestine. It also resides in the gut of some animals (Saeedi et al., 2017). Enterohemorrhagic *E. coli* (EHEC) serotype O157:H7 causes severe diarrhea in humans and leads to the hemolytic uremic (HU) syndrome. In some cases, it can also cause kidney failure in humans (Stones et al., 2017). Pathogenic strains of *E. coli* affect the gastrointestinal tracts of both humans and animals. This type of pathogenic bacteria can affect the poultry sector and causes an extensive production loss (Barnes et al., 2003). Its pathogenic activity becomes severe when transmitted to other humans and animals. It has some resistance against the existing antibiotics also. This problem is critical not only for *E. coli*, but many other pathogenic bacteria are also acquiring resistance against certain known antibiotics. If this problem persists for a long time, this could lead to severe outbreaks and a lack of medicine. These ABR strains can enter into a human host, mostly during traveling (Sharma et al., 2016).

Antibiotics can usually treat these infections. But some strains of *E. coli* can develop enzymes such as beta-lactamase. This group shows resistance against antibiotics such as cephalosporins and penicillins. These antibiotics have a beta-lactam ring, which acts as an essential part of their structure, and it prevents the cell wall synthesis of the bacteria. When the bacteria produce beta-lactam enzymes, the beta-lactam ring of antibiotic gets destroyed, and the antibiotic becomes inactive (Lewis, 2016).

E. coli involves bacteremia and urinary tract infection (Rodhe et al., 2008). The strains of *E. coli* had isolated by using the standard methods (Manges et al., 2019). Different PCR techniques were used for the detection and characterization of *E. coli*. Researchers performed multiplex PCR assays for detecting and characterizing Shiga toxigenic *Escherichia coli* (STEC). STEC are toxin-producing *Escherichia coli,* which cause severe gastrointestinal illness (Paton et al., 1998).

The next-generation sequencing method has become excellently sufficient for the rapid determination of the genomic characteristics. Next-generation sequencing helps to determine the emergence of resistance to antibiotics such as sulfonamides, trimethoprim, beta-lactamase, aminoglycosides tetracycline, and fluoroquinolones (Sharma et al., 2019). MLST is the genotyping method used in the epidemiological study of *E. coli*. The detection and classification of the strains of *E. coli* had done by using this technique. This technique had used to determine the virulence and antimicrobial resistance (AMR) genes. Molecular techniques such as RFLP and PFGE were also reported for the epidemiological study of *E. coli*, especially to find out the similarity of strains. They determined the organism's similarity index, and the closely related species had been identified (Matsukawa et al., 2019). They also used the typing method such as MLVA to analyze the colonization of various isolates of *E. coli*. It is having a discriminatory power greater than PFGE (Keys et al., 2005).

5.3 CONCLUSION

Epidemiological studies are needed to prevent the spread of pathogenic foodborne diseases. Molecular methods are having a specific role in epidemiological investigations. The main problem that the world is going to face in the future is the lack of antibiotics. In addition to that, the pathogens are acquiring resistance against the existing antibiotics. A solution is needed before the situation goes out of control. In this context, the information regarding the microorganisms, such as the serotype, genetic diversity, and variations, is crucial. This chapter provides an insight into the molecular epidemiological study of *Listeria monocytogenes*, *Salmonella enterica*, *Campylobacter jejuni*, *Yersinia enterocolitica*, *Yersinia pseudotuberculosis*, *Clostridium botulinum*, *Staphylococcus aureus*, *Shigellosis*, and *Escherichia coli*.

KEYWORDS

- ***Campylobacter jejuni***
- **disease outbreak**
- **epidemiology**
- **foodborne diseases**
- **molecular methods**

REFERENCES

Abdelhaseib, M. U., Singh, A. K., Bailey, M., Singh, M., El-Khateib, T., & Bhunia, A. K., (2016). Fiber optic and light scattering sensors: Complimentary approaches to rapid detection of *Salmonella enterica* in food samples. *Food Control, 61*, 135–145.

Adhikary, S., Bisgaard, M., Boot, R., Benga, L., Nicklas, W., & Christensen, H., (2019). Development of multilocus sequence typing (MLST) of *Rodentibacter pneumotropicus*. *Vet. Microbiol, 231*, 11–17.

Anniballi, F., Auricchio, B., Woudstra, C., et al., (2013). Multiplex real-time PCR for detecting and typing *Clostridium botulinum* group III organisms and their mosaic variants. *Biosecur. Bioterror., 11*(S1), S207–214.

Aviv, G., Cornelius, A., Davidovich, M., et al., (2019). Differences in the expression of SPI-1 genes pathogenicity and epidemiology between the emerging *Salmonella enterica* serovar *infantis* and the model *Salmonella enterica* serovar *typhimurium J. Infect. Dis, 220(6), 1071–1081.*

Awasti, N., & Anand, S., (2020). The role of yeast and molds in dairy industry: An update. In: Minj, J., Sudhakaran, V. A., & Kumari, A., (eds.), *Dairy Processing: Advanced Research to Applications* (pp. 243–262). Springer, Singapore.

Bardsley, M., Jenkins, C., Mitchell, H. D., et al., (2020). Persistent transmission of shigellosis in England is associated with a recently emerged multi-drug resistant strain of *Shigella sonnei*. *J Clin Microbiol, 58*(4), e01692-19.

Barnes, H. J., Vaillancourt, J., & Gross, W. B., (2003). Colibacillosis. In: Saif, Y. M., (eds.), *Diseases of Poultry* (pp. 631–652). Ames, IA: Iowa State University Press.

Bertelli, C., & Greub, G., (2013). Rapid bacterial genome sequencing: Methods and applications in clinical microbiology. *Clinical Microbiology and Infection, 19*(9), 803–813.

Blaser, M. J., & Engberg, J., (2008). Clinical aspects of *Campylobacter jejuni* and *Campylobacter coli* infections. In: Nachamkin, I., & Blaser, J. M., (eds.), *Campylobacter* (pp. 99–121). American Society for Microbiology (ASM) Press: Washington DC, USA.

Blaser, M. J., (1997). Epidemiologic and clinical features of *Campylobacter jejuni* infections. *J. Infect. Dis., 176(S2), 103–105.*

Blaser, M. J., Wells, J. G., Feldman, R. A., Pollard, R. A., & Allen, J. R., (1983). *Campylobacter enteritis* in the United States. *Ann. Intern. Med., 98*(3), 360–365.

Blomfeldt, A., Eskesen, A. N., Aamot, H. V., Leegaard, T. M., & Bjornholt, J. V., (2016). Population-based epidemiology of *Staphylococcus aureus* bloodstream infection: Clonal complex 30 genotype is associated with mortality. *Eur. J. Clin. Microbiol. Infect. Dis., 35*(5), 803–813.

Bokken, G. C. A. M., Corbee, R. J., Knapen, F., & Bergwer, A. A., (2003). Immunochemical detection of *Salmonella* group B, D and E using an optical surface plasmon resonance biosensor. *FEMS Microbiology Letters, 222*(1), 75–82.

Brennan, G. I., Shore, A. C., Corcoran, S., Tecklenborg, S., Coleman, D. C., & Connell, B., (2011). Emergence of hospital- and community-associated panton-valentine leukocidin-positive methicillin-resistant *Staphylococcus aureus* genotype ST772-MRSA-V in Ireland and detailed investigation of an ST772-MRSA-V cluster in a neonatal intensive care unit. *J. Clin. Microbiol., 50*(30), 841–847.

Broman, T., Palmgren, H., Bergstrom, S., et al., (2002). *Campylobacter jejuni* in black-headed gulls (*Larus ridibundus*): Prevalence, genotypes, and influence on *C. jejuni* epidemiology. *J. Clin. Microbiol., 40*(12), 4594–4602.

Brown, D. J., Threlfall, E. J., & Rowe, B., (1991). Instability of multiple drug resistance plasmids in *Salmonella typhimurium* isolated from poultry. *Epidemiol. Infect., 106*(2), 247–257.

Bubert, A., Hein, I., Rauch, M., et al., (1999). Detection and differentiation of *Listeria* spp. by a single reaction based on multiplex PCR. *Appl. Environ. Microbiol., 65*(10), 4688–4692.

Butaye, P., Struelens, M., & Uyehera, L., (2007). MRSA, report on zoonotic agent in Belgium. Trend and sources by working group on foodborne infections and intoxication. *Drug Research, 20*, 59–61.

Carpentier, B., & Cerf, O., (2011). Persistence of *Listeria monocytogenes* in food industry equipment and premises. *Int. J. Food Microbiol., 145*(1), 1–8.

Cash, B. A., Rodo, X., Emch, M., et al., (2014). Cholera and Shigellosis: Different epidemiology but similar responses to climate variability. *PLoS One, 9*(9), e107223.

Chang, H., Voyvodic, P. L., Zuniga, A., & Bonnet, J., (2017). Microbially derived biosensors for diagnosis, monitoring and epidemiology. *Microb. Biotechnol., 10*(5), 1031–1035.

Chattaway, M. A., Dallman, T. J., Larkin, L., et al., (2019). The transformation of reference microbiology methods and surveillance for *Salmonella* with the use of whole-genome sequencing in England and Wales. *Front. Public Health, 7*, 317.

Cinti, S., Volpe, G., Piermarini, S., Delibato, E., & Palleschi, G., (2017). Electrochemical biosensors for rapid detection of foodborne *Salmonella*: A critical overview. *Sensors, 17*(8), 1910.

Cohen, D., Bassal, R., Goren, S., & Rouach, T., (2014). Recent trends in the epidemiology of shigellosis in Israel. *Epidemiol. Infect., 142*(12), 2583–2594.

Cohen, E., Rahav, G., & Gal-Mor, O., (2020). Genome sequence of an emerging *Salmonella enterica* serovar *infantis* and genomic comparison with other *S. infantis* strains. *Genome Biol. Evol., 12(3), 151–159.*

Crim, S. M., Griffin, P. M., Tauxe, R., et al., (2015). Preliminary incidence and trends of infection with pathogens transmitted commonly through foodborne diseases active surveillance network, 10 U.S. sites, 2006–2014. *Morbidity and Mortality Weekly Report, 64*(18), 495–499.

Crump, J. A., Sjolund-Karlsson, M., Gordon, M. A., & Parry, C. M., (2015). Epidemiology, clinical presentation, laboratory diagnosis, antimicrobial resistance, and antimicrobial management of invasive *Salmonella* infections. *Clin. Microbiol. Rev., 28*(4), 901–937.

De Benito, I., Cano, M. E., Aguero, J., & Garcia, L. J. M., (2004). A polymorphic tandem repeat is potentially useful for typing the chromosome of *Yersinia enterocolitica. Microbiology, 150*(1), 199–204.

Deng, X., Ran, L., Wu, S., et al., (2012). Laboratory-based surveillance of non-typhoidal *Salmonella* infections in Guangdong province, China. *Foodborne Pathog. Dis., 9*(4), 305–312.

Dingle, K. E., Colles, F. M., Ure, R., et al., (2002). Molecular characterization of *Campylobacter jejuni* clones: A basis for epidemiologic investigation. *Emerg. Infect. Dis., 8*(9), 949–955.

Doren, J. M. V., Neil, K. P., Parish, M., Gieraltowski, L., Gould, L. H., & Gombas, K. L., (2013). Foodborne illness outbreaks from microbial contaminants in spices, 1973–2010. *Food Microbiol., 36*(2), 456–464.

Drummond, N., Murphy, B. P., Ringwood, T., Prentice, M. B., Buckley, J. F., & Fanning, S., (2012). *Yersinia enterocolitica*: A brief review of the issues relating to the zoonotic pathogen, public health challenges, and the pork production chain. *Foodborne Pathog. Dis., 9*(3), 179–189.

Faour-Klingbeil, D., & Todd, E. C. D., (2020). Prevention and control of foodborne diseases in middle-east, North African countries: Review of national control systems. *Environmental Research and Public Health, 17*(1), 70.

Feasey, N. A., Dougan, G., Kingsley, R. A., Heyderman, R. S., & Gordon, M. A., (2012). Invasive non-typhoidal *Salmonella* disease: An emerging and neglected tropical disease in Africa. *Lancet, 379*(9835), 2489–2499.

Feng, Y., Yao, H., Chen, S., Sun, X., Yin, Y., & Jiao, X., (2020). Rapid detection of hypervirulent serovar 4h *Listeria monocytogenes* by multiplex PCR. *Front. Microbiol., 11*, 1309.

Ferrari, R. G., Panzenhagen, P. H. N., & Conte-Junior, C. A., (2017). Phenotypic and genotypic eligible methods for *Salmonella typhimurium* source tracking. *Front Microbiol, 8*, 2587.

Foxman, B., (2012). Applications of molecular tools to infectious disease epidemiology. *Molecular Tools and Infectious Disease Epidemiology*, 23–39.

Frazao, M. R., De Souza, R. A., Medeiros, M. I. C., et al., (2019). Molecular typing of *Campylobacter jejuni* strains: Comparison among four different techniques. *Bacterial, Fungal and Virus Molecular Biology, 51*, 519–525.

French, N. P., Midwinter, A., Holland, B., et al., (2009). Molecular epidemiology of *Campylobacter jejuni* isolates from wild-bird fecal material in children's playgrounds. *Appl. Environ. Microbiol., 75*(3), 779–783.

Gratiet, T. L., Poezevara, T., Rouxel, S., et al., (2020). Development of an innovative and quick method for the isolation of *Clostridium botulinum* strains involved in avian botulism outbreaks. *Toxins, 12*(1), 42.

Graves, L. M., & Swaminathan, B., (2001). PulseNet standardized protocol for subtyping *Listeria monocytogenes* by macrorestriction and pulsed-field gel electrophoresis. *Int. J. Food Microbiol, 65*(1, 2), 55–62.

Gray, J. T., & Fedorka-Cray, P. J., (2001). Survival and infectivity of *Salmonella choleraesuis* in swine feces. *J. Food Prot., 64*(7), 945–949.

Guha, T. K., & Edgell, D. R., (2017). Applications of alternative nucleases in the age of CRISPR/Cas 9. *International Journal of Molecular Sciences, 18*(12), 2565.

Guirado, P., Paytubi, S., Miro, E., et al., (2020). Differential distribution of the wlaN and cgtB genes, associated with Guillain-Barré Syndrome, in *Campylobacter jejuni* isolates from humans, broiler chickens, and wild birds. *Microorganisms, 8*(3), 325.

Han, K., Jang, S. S., Choo, E., Heu, S., & Ryu, S., (2007). Prevalence, genetic diversity, and antibiotic resistance patterns of *Campylobacter jejuni* from retail raw chickens in Korea. *International Journal of Food Microbiology, 114*(1), 50–59.

Harris, R. A., Anniballi, F., & Austin, J. W., (2020). Adult intestinal toxemia botulism. *Toxins, 12*(2), 81.

Hartge, P., (2014). *A Dictionary of Epidemiology. American Journal of Epidemiology, 181*(8).

Havelaar, A. H., Kirk, M. D., Torgerson, P. R., et al., (2015). World Health Organization global estimates and regional comparisons of the burden of foodborne disease in 2010. *PLoS Med., 12*(12), e1001923.

Hopkins, K. L., Peters, T. M., De Pinna, E., & Wain, J., (2011). Standardization of multilocus variable-number tandem-repeat analysis (MLVA) for subtyping of *Salmonella enterica* serovar enteritidis. *Euro Surveill., 16*(32), 19942.

Horby, P. W., O'Brien, S. J., Adak, G. K., et al., (2003). A national outbreak of multi-resistant *Salmonella enterica* serovar *Typhimurium* definitive phage type (DT) 104 associated with consumption of lettuce. *Epidemiol. Infect., 130*(2), 169–178.

Hunter, E., Greig, D. R., Schaefer, U., et al., (2019). Identification and typing of *Yersinia enterocolitica* and *Yersinia pseudotuberculosis* isolated from human clinical specimens in England between 2004 and 2018. *Journal of Medical Microbiology, 68*(4), 538–548.

Hyden, P., Pietzka, A., Lennkh, A., et al., (2016). Whole-genome sequence-based serogrouping of *Listeria monocytogenes* isolates. *Journal of Biotechnology, 235*, 181–186.

Jacobson, M. J., Lin, G., Whittam, T. S., & Johnson, E. A., (2008). Phylogenetic analysis of *Clostridium botulinum* type A by Multi-locus sequence typing. *Microbiology, 154*(8), 2408–2415.

Kaszoni-Ruckerl, I., Mustedanagic, A., Muri–Klinger, S., et al., (2020). Predominance of distinct *Listeria innocua* and *Listeria monocytogenes* in recurrent contamination events at dairy processing facilities. *Microorganisms, 8*(2), 234.

Keys, C., Kemper, S., & Keim, P., (2005). Highly diverse variable number tandem repeat loci in the *E. coli* O157:H7 and O55:H7 genomes for high-resolution molecular typing. *J. Appl. Microbiol., 98*(4), 928–940.

Kim, H. C., Chong, S. T., Nunn, P. V., Jang, W. J., Klein, T. A., & Robbins, R. G., (2013). Seasonal abundance of ticks collected from live-captured small mammals in Gyeonggi Province, Republic of Korea, during 2009. *Syst. Appl. Acarol, 18*(3), 201–211.

Kim, H. S., Hwang, J. H., TakJeong, M. D. S., Lee, Y. T., Suh, Y. L., & Shim, J. S., (2003). Effect of muscle activity and botulinum toxin dilution volume on muscle paralysis. *Developmental Medicine and Child Neurology, 45*(3), 200–206.

Kim, T., Hwang, H. J., & Kim, J. H., (2017). Development of a novel, rapid multiplex polymerase chain reaction assay for the detection and differentiation of *Salmonella enterica* serovars Enteritidis and Typhimurium using ultra-fast convection polymerase chain reaction. *Foodborne Pathogens and Disease, 14*(10), 580–586.

Kirk, M. D., Pires, S. M., Black, R. E., et al., (2015). World Health Organization estimates of the global and regional disease burden of 22 foodborne bacterial, protozoal, and viral diseases, 2010: A data synthesis. *PLoS Med., 12*(12), e1001921.

Koga, M., Gilbert, M., Takahashi, M., Li, J., Koike, S., Hirata, K., & Yuki, N., (2006). Comprehensive analysis of bacterial risk factors for the development of Guillain-Barré syndrome after *Campylobacter jejuni* enteritis. *J. Infect. Dis, 193*(4), 547–555.

Langsrud, S., Sidhu, M. S., Heir, E., & Holck, A. L., (2003). Bacterial disinfectant resistance - A challenge for the food industry. *Int. Biodeterior. Biodegrad., 51*(4), 283–290.

Lazcka, O., Campo, F. J. D., & Munoz, F. X., (2007). Pathogen detection: A perspective of traditional methods and biosensors. *Biosensors and Bioelectronics, 22*(7), 1205–1217.

Liebana, E., Guns, D., Garcia-Migura, L., Woodward, M. J., Clifton-Hadley, F. A., & Davies, R. H., (2001). Molecular typing of *Salmonella* serotypes prevalent in animals in England: Assessment of methodology. *J. Clin. Microbiol., 39*(10), 3609–3616.

Lima, I. F. N., Havt, A., & Lima, A. A. M., (2015). Update on molecular epidemiology of Shigella infection. *Curr. Opin. Gastroenterol., 31*(1), 30–37.

Lindsay, B., Ochieng, J. B., Ikumapayi, U. N., et al., (2013). Quantitative PCR for detection of Shigella improves ascertainment of Shigella burden in children with moderate-to-severe diarrhea in low-income countries. *Journal of Clinical Microbiology, 51*(6), 1740–1746.

Lukinmaa, S., Nakari, U., Eklund, M., & Siitonen, A., (2004). Application of molecular genetic methods in diagnostics and epidemiology of foodborne bacterial pathogens. *APMIS, 112*(11, 12), 908–929.

Luth, S., Boone, I., Kleta, S., & Al Dahouk, S., (2019). Analysis of RASFF notifications on food products contaminated with *Listeria monocytogenes* reveals options for improvement in the rapid alert system for food and feed. *Food Control, 96*, 479–487.

MacDonald, E., Einoder-Moreno, M., Borgen, K., et al., (2014). National outbreak of *Yersinia enterocolitica* infections in military and civilian populations associated with consumption of mixed salad, Norway, 2014. *Euro Surveill., 21*(34), 30321.

Mair, N. S., & Fox, E., (1986). *Yersiniosis - Lab Diagnosis, Clinical Features, Epidemiology* (pp. 1–47). Public Health Laboratory Service.

Matsukawa, M., Igarashi, M., Watanabe, H., et al., (2019). Epidemiology and genotypic characterization of dissemination patterns of uropathogenic *Escherichia coli* in a community. *Epidemiology and Infection, 147*, 1–9.

McDermott, P. F., Tyson, G. H., Kabera, C., et al., (2016). Whole-genome sequencing for detecting antimicrobial resistance in nontyphoidal *Salmonella. Antimicrob. Agents Chemother, 60*(9), 5515–5520.

Meinersmann, R. J., Phillips, R. W., Wiedmann, M., & Berrang, M. E., (2004). Multilocus sequence typing of *Listeria monocytogenes* by use of hypervariable genes reveals clonal and recombination histories of three lineages. *Appl. Environ. Microbiol., 70*(4), 2193–2203.

Melo, R. T., Grazziotin, A. L., Junior, E. C. V., et al., (2019). Evolution of *Campylobacter jejuni* of poultry origin in Brazil. *Food Microbiology, 82*, 489–496.

Montero, D., Bodero, M., Riveros, G., et al., (2015). Molecular epidemiology and genetic diversity of *Listeria monocytogenes* isolates from a wide variety of ready-to-eat foods and their relationship to clinical strains from listeriosis outbreaks in Chile. *Front. Microbiol., 6*, 384.

Moura, G. F., Tomborelli, P. M., Carvalho, R. C. T., et al., (2019). *Listeria monocytogenes* and other species as persistent contaminants in the processing of chicken meat. *J. Appl. Poult. Res., 28*(2), 470–478.

Oda, S., Kabeya, H., Sato, S., et al., (2015). Isolation of pathogenic *Yersinia enterocolitica* 1B/O:8 from *Apodemus* mice in Japan. *J. Wildl. Dis., 51*(1), 260–264.

Ojo, O. E., Ogunjobi, O. O., Oyekunle, M. A., Dipeolu, M. A., & Otesile, E. B., (2019). Prevalence and antimicrobial resistance of *Salmonella* and *Yersinia* in the feces of hunted wildlife in Abeokuta, Nigeria. *Rev. Elev. Med. Vet. Pays Trop., 72*(4).

Paduro, C., Monteroa, D. A., Chamorroa, N., Carrenob, L. J., Vidalc, M., & Roberto, V. R., (2020). Ten years of molecular epidemiology surveillance of *Listeria monocytogenes* in Chile 2008–2017. *Food Microbiology, 85*, 103280.

Pal, M., Kerorsa, G. B., Marami, L. M., & Kandi, V., (2020). Epidemiology, pathogenicity, animal infections, antibiotic resistance, public health significance, and economic impact of *Staphylococcus aureus*: A comprehensive review. *American Journal of Public Health Research, 1*(1), 14–21.

Pitkanen, T., & Hanninen, M. L., (2017). In: Rose, J. B., & Jimenez-Cisneros, B., (eds.), *Members of the Family Campylobacteraceae: Campylobacter jejuni, Campylobacter coli*. In Global Water Pathogen Project.

Pombinho, R., Vieira, A., Camejo, A., et al., (2020). Virulence gene repression promotes *Listeria monocytogenes* systemic infection. *Gut. Microbes., 11*(4), 868–881.

Revazishvili, T., Kotetishvili, M., Stine, O. C., Kreger, A. S., Morris, J. G. Jr., & Sulakvelidze, A., (2004). Comparative analysis of multilocus sequence typing and pulsed-field gel electrophoresis for characterizing *Listeria monocytogenes* strains isolated from environmental and clinical sources. *J. Clin. Microbiol., 42*(1), 276–285.

Revazishvili, T., Rajanna, C., Bakanidze, L., et al., (2008). Characterization of *Yersinia pestis* isolates from natural foci of plague in the Republic of Georgia, and their relationship to *Y. pestis* isolates from other countries. *Clinical Microbiology and Infection, 14*(5), 429–436.

Rodhe, N., Lofgren, S., Matussek, A., et al., (2008). Asymptomatic bacteriuria in the elderly: High prevalence and high turnover of strains. *Scand. J. Infect. Dis., 40*(10), 804–810.

Ronholm, J., Nasheri, N., Petronella, N., & Pagotto, F., (2016). Navigating microbiological food safety in the era of whole-genome sequencing. *Clin. Microbiol. Rev., 29*(4), 837–857.

Saeedi, P., Yazdanparast, M., Behzadi, E., et al., (2017). A review on strategies for decreasing *E. coli* O157:H7 risk in animals. *Microbial Pathogenesis, 103*, 186–195.

Scallan, E., Hoekstra, R. M., Angulo, F. J., et al., (2011). Foodborne illness acquired in the United States – major pathogens. *Emerging Infectious Diseases, 17*(1), 7–15.

Sharma, G., Sharma, S., Sharma, P., et al., (2016). *Escherichia coli* biofilm: Development and therapeutic strategies. *J. Appli. Microbiol., 121*, 309–319.

Silva, A. P., Lima, T. M., Candido, et al., (2020). Epidemiological profile of people affected by botulism in Brazil between 2008 and 2018. *Research, Society and Development, 9*(4), 2525–3409.

Smith, T. J., Xie, G., Williamson, C. H. D., et al., (2020). Genomic characterization of newly completed genomes of botulinum neurotoxin-producing species from Argentina, Australia and Africa. *Genome Biology and Evolution, 12*(3), 229–242.

Sobel, J., (2005). Interdependent preferences and reciprocity. *Journal of Economic Literature, 43*(2), 392–436.

Song, K. H., Kim, M., Kim, C. J., et al., (2017). Korea infectious diseases (KIND) study group. Impact of vancomycin MIC on treatment outcomes in invasive *Staphylococcus aureus* infections. *Antimicrob. Agents Chemother, 61*(3), e01845-16.

Sperry, K. V. E., Kathariou, S., Edwards, J. S., & Wolf, L. A., (2008). Multiple-locus variable-number tandem-repeat analysis as a tool for subtyping *Listeria monocytogenes* strains. *Journal of Clinical Microbiology, 46*(4), 1435–1450.

Stanger, K. J., McGregor, H., Marenda, M., Morton, J. M., & Larsen, J. W. A., (2018). Assessment of the efficacy of an autogenous vaccine against *Yersinia pseudotuberculosis* in young merino sheep. *New Zealand Veterinary Journal, 67*(1), 27–35.

Stones, D. H., Fehr, A. G. J., Thompson, L., et al., (2017). Zebrafish (*Danio rerio*) as a vertebrate model host to study colonization, pathogenesis, and transmission of foodborne *Escherichia coli* O157. *mSphere, 2*(5), e00365-17.

Takahashi, T., Kabeya, H., Sato, S., et al., (2020). Prevalence of *Yersinia* among wild sika deer (*Cervus nippon*) and boars (*Sus scrofa*) in Japan. *Journal of Wildlife Diseases, 56*(2), 270–277.

Tamburro, M., & Ripabelli, G., (2017). High resolution melting as a rapid, reliable, accurate and cost-effective emerging tool for genotyping pathogenic bacteria and enhancing molecular epidemiological surveillance: A comprehensive review of the literature. *Ann. Ig., 29*(4), 293–316.

Tauxe, R. V., & Blake, P. A., (1992). Epidemic cholera in Latin America. *JAMA, 267*(10), 1388–1390.

Tauxe, R. V., Hargrett-Bean, N., Patton, C. M., & Wachsmuth, I. K., (1988). "*Campylobacter*" isolates in the United States, 1982–1986. *Morbidity and Mortality Weekly Report: Surveillance Summaries, 37*(SS-2), 1–13.

Teixeira, L. A. C., Carvalho, F. T., Vallim, D. C., et al., (2019). *Listeria monocytogenes* in export-approved beef from Mato Grosso, Brazil: Prevalence, molecular characterization and resistance to antibiotics and disinfectants. *Microorganisms, 8*(1), 18.

Thomson, N. R., Howard, S., Wren, B. W., et al., (2006). The complete genome sequence and comparative genome analysis of the high pathogenicity *Yersinia enterocolitica* strain 8081. *PLoS Genetics, 2*(12), e206.

Traunsek, U., Toplak, N., Jersek, B., Lapanje, A., Majstorovi, T., & Kovac, M., (2011). Novel cost-efficient real-time PCR assays for detection and quantitation of *Listeria monocytogenes*. *J. Microbiol. Methods, 85*(1), 40–46.

Verweij, J. J., & Stensvold, C. R., (2014). Molecular testing for clinical diagnosis and epidemiological investigations of intestinal parasitic infections. *Clinical Microbiology Reviews, 27*(2), 371–418.

Wagner, M., & Stessl, B., (2014). Sampling the food processing environment: Taking up the cudgel for preventive quality management in food processing environments. *Methods Mol. Biol. 1157*, 275–283.

Wieczorek, K., Szewczyk, R., & Osek, J., (2012). Prevalence, antimicrobial resistance, and molecular characterization of *Campylobacter jejuni* and *Campylobacter coli* isolated from retail raw meat in Poland. *Veterinarni Medicina, 57*(6), 293–299.

Williamson, D. A., Baines, S. L., Carter, G. P., et al., (2016). Genomic insights into a sustained national outbreak of *Yersinia pseudotuberculosis*. *Genome Biol. Evol., 8*(12), 3806–3814.

Xue, Y., Zhai, S., Wang, Z., et al., (2020). The Yersinia Phage X1 administered orally efficiently protects a murine chronic enteritis model against *Yersinia enterocolitica* infection. *Front. Microbiol., 11*, 351.

CHAPTER 6

Molecular Characterization of Foodborne Pathogens

CHARLES OLUWASEUN ADETUNJI, WADZANI DAUDA PALNAM, AYODELE EUGENE AYENI, MUHAMMAD AKRAM, OLUGBENGA SAMUEL MICHAEL, AMNA SIDDIQUE, SAHER ISLAM, OLUGBEMI TOPE OLANIYAN, AJAYI KOLAWOLE TEMIDAYO, KAYODE ADEBISI AROWORA, DEVARAJAN THANGADURAI, JULIANA BUNMI ADETUNJI, BENJAMIN EWA UBI, and DANIEL INGO HEFFT

ABSTRACT

Human diseases caused by foodborne microorganisms have been highlighted as important health concerns, especially in most developing countries. There is a higher demand for quality food by human population, particularly in low- and middle-income countries. To prevent the intra and international spread of pathogens associated with food; various technological advances have been developed for appropriate identification of several genera and species of pathogens isolated from food within the complex food chain beginning from its primary production to the final consumers. It has been documented that molecular strategies enable the successful tracking and tracing of microbial isolates that are available in food sample. This chapter provides a comprehensive overview of classical examples of various molecular approaches used in the detection and characterization of pathogens associated with human food and food-related

Global Food Safety: Microbial Interventions and Molecular Advancements. Saher Islam, PhD, Devarajan Thangadurai, PhD, Jeyabalan Sangeetha, PhD, & Natália Cruz-Martins, PhD (Editors)

microorganisms. It is envisaged that the use of molecular identification will be made into simple and cost-effective forms which could enable countries, especially the developing countries, to monitor and guarantee adequate food safety of their food products.

6.1 INTRODUCTION

The presence of foodborne pathogens across the globe may continue for many more years if not properly addressed by relevant stakeholders most especially when attention is drawn towards its current threat to human safety and health. Many deadly strains of pathogenic bacteria including *Salmonella*, *Vibrio vulnificus*, entero-hemorrhagic *Escherichia coli*, *Listeria monocytogenes*, *Clostridium botulinum*, Norovirus, and *Campylobacter jejuni* are the main causes of illnesses worldwide due to contaminated food consumption (Crerar et al., 1996). Recently, there has been rigorous increased attention from all relevant stakeholders in government and private sectors to ensure that food products are adequately protected from any form of pathogenic infections that may pose a risk to public health. Many interventional programs have been initiated to reduce foodborne diseases, contaminations, toxicants due to the emergence of a new threat and increased risk from the consumption of poor hygienic food. Proper monitoring and sanitary surveillance to reduce biohazards should be set up as a matter of urgency across the globe to prevent any occurrence of a global outbreak of foodborne diseases (Maria et al., 2017).

Various naturally occurring microbes find a way to get into the food products especially processed foods and posing a serious health risk to the consumer. Hence, to minimize the biohazards in food, there must be a deliberate reduction in the production and consumption of processed food. Studies have revealed that farm dairy products, meats, and poultry harbor a large number of microbes including *Campylobacter*, *Escherichia coli*, *Salmonella*, and *Listeria*. Also, seafood and fed supplements are known to contain numerous pathogens like Hepatitis A virus (HAV), *Listeria* spp., and *Vibrio* spp. McAuley et al. (2014) revealed that in the food industry, increased attention has been seen in the role of microbial entry into food products which usually serve as a vehicle for the transmission of many pathogenic bacteria. McAuley et al. (2014) performed a survey to analyze many foodborne pathogen occurrences in dairy products across different weather conditions in Australia. Samples were collected and analyzed for the presence of *Bacillus cereus*, *Clostridium perfringens*, *Campylobacter*, *Cronobacter*, *Listeria*, Shiga-toxigenic *Escherichia coli*, *Salmonella*, *Yersinia enterocolitica* and coagulase-positive Staphylococci prevalence. The results showed that *B. cereus* and *C. perfringens* are the most prevalent in the area but low in most of the samples, which revealed increased standards of pathogen safety with a low incidence of pathogens detection in most of the samples investigated.

Many conventional microorganism detection methods have been employed in the past, such as the conventional culture methods, which is very reliable, accurate, and fast utilizing agar plates to isolate and identify pure cultures plus evaluating cultures using metabolic fingerprinting or phenotypic analysis. This method of approach is known to be labor-intensive, and it takes a longer time to confirm results even with the modern tools or biochemical-based approach (Cancino-Padilla et al., 2017). Another method of detection is antibody method which is based on the principle of antibody binding with target antigen to form an antigen-antibody complex, in reaction to a specific invading pathogen. This technique works on successful expression of antigen on the pathogenic organism which is mostly influenced by temperature, acids, preservatives, salts, and other chemical constituents present in the food. There are many detection methods in antigen-antibody complex reactions including the use of enzyme-linked immunosorbent assay, lateral flow device, and antibody-coated magnetic beads using biosensors. Many of the biosensors such as piezoelectric biosensors, cell-based sensors, microfluidic biochips, electrochemical immunosensors, fiber-optic biosensor and surface plasmon resonance sensor are designed to identify Staphylococcal enterotoxin, botulinum toxin, *E. coli*, *Salmonella* sp. and *Listeria* sp. (Vo-Dinh and Cullum, 2000).

There are numerous advanced tools and techniques developed to help in epidemiological and physiological detection of pathogens as well as techniques that could help in the possible control of foodborne pathogens. Genomic sequencing is generally known to offer a wide range of information about the pathogenic organism's genetic makeup that is used to understand characteristic features of foodborne agents. Biological monitoring would similarly provide data and distribution of pathogens which can adapt and change in any environmental conditions with activation of more virulence potential or capacity (Miguel et al., 2015). Bezirtzoglou et al.

(2000) revealed that food provides a good and conducive environment for microbes to thrive and grow originating from post-harvesting, processing, slaughter, distribution, or storage. These microorganisms are known to cause spoilage to the food products or illnesses to the consumers. This is so due to the unique composition of food products including milk, cheese, and some dairy products.

Illness produced by food causes a number of infections, and these infections are caused by microorganisms so, the identification of these harmful microorganisms has increased the interest in the research of new biomolecular techniques. These new biomolecular techniques are developed to enhance the sensor features like sensitivity, simplicity, and economic viability. In the past few years, there are number of revolutionary advancements have been made in the very sensitive, precipitous, robotic, molecular recognition processes related to food or other animal origin foods (Mohania et al., 2008). Moreover, numerous identification approaches use primers to target distinct series like the 16S ribosomal RNA encoding gene, the 16S-23S rRNA intergenic spacer, 23S rRNA encoding, recA, and ldhD genes (Decallone et al., 1991; Giraffa and Neviani, 2000; Arora et al., 2006). Therefore, this chapter discusses extensively on molecular methods used for the identification of several foodborne pathogens present in food samples.

6.2 FOODBORNE MICROBES

There are several diseases that are caused by numerous food causing microorganisms that could lead to several infections. The identification of these harmful microorganisms has led to an increase in the global scientific interest in the research of new biomolecular techniques. There are several studies that have validated the potential and several applications of strains of lactic acid bacteria (LAB), most especially when they serve as probiotics. The genome methods have been utilized for the identification of bacteria through unconventional methods. There are a number of microorganisms in the food and food products that can cause serious diseases, so this area is of great interest. For the detection and identification of these microorganisms, researchers have developed some procedures like microarrays, biosensor, nanomaterials-based method, pulse-field gel electrophoresis, ARDRA, phenotypic characterization, genotypic characterization, and real-time PCR. The biosensors are the analytical devices that change biological response into electric signals with sensor devices that affect substrates concentration and other parameters of biological significance even where they do not use biological system directly. The use of biosensors can increase accuracy, easy to handle, more sensitive, response time is less, and more specific also. These are mostly used to calculate the biological effects, for example, genotoxicity, biotoxins, endocrine effects, and number of constraints like glucose, lactate, urea, and creatinine in blood or urine samples. Real time PCR is the DNA established procedure which reduces risk to false positives as it is much sensitive in a sense that it can detect one copy of target DNA because of high dynamic range.

Thomas (2017) revealed that foodborne pathogens have wreck serious havoc on the global economy and health system due to the increased outbreaks witnessed in recent years by a number of pathogenic organisms such as *Bacillus cereus*, *Campylobacter jejuni*, *Clostridium botulinum*, *Clostridium perfringens*, *Escherichia coli*, *Cronobacter sakazakii*, *Shigella* spp., *Salmonella* spp., *Staphylococcus aureus*, *Yersinia enterocolitica*, viruses such as Noroviruses, Hepatitis A and parasites such as *Toxoplasma gondii*, *Trichinella spiralis*, and *Cyclospora cayetanensis*. It is generally known that the current classical food risk and safety management approach system is grossly ineffective to reduce the risk in food, hence the need to look for alternative means to address some of these issues. Nawal et al. (2013) demonstrated that pathogens in food are the main cause of sickness and death in any population, particularly the developing countries with an estimated death rate of 1.9 million individuals annually. Many of the pathogenic bacteria causing diseases are referred to as zoonotic commonly found in improperly cooked and unhygienic dairy products. To control an outbreak due to pathogenic infection, many laboratory investigations and increased surveillance must be carried out to detect early signs and symptoms along the food production chain.

Susan et al. (2015) stated that food legislation generally involves appropriate biohazard and risk reduction strategies deployed to reduce foodborne pathogens, chemicals, allergens, and ensuring population safety. Suzzi and Corsetti (2020) revealed that there is clear evidence for the involvement of microbes in food safety and hygiene, food technology, food poisoning, food omics, food genomics, functional foods, and probiotics. Also, the role of microorganisms in probiotics research is seeking more attention, particularly for the safety status of probiotic strains, microbial adhesive ability, antimicrobial compounds production, inhibition of pathogenic growth, and gut microbiome physiology. Many microbial probiotic strains such as *Clostridium*, *Bifidobacterium*, *Faecalibacterium*, and *Bacteroides* could be screened for immunomodulatory capacity, bioactive substances, and anticancer molecules utilizing molecular biology techniques like polymerase chain reaction (PCR), whole-genome sequencing, and metagenomics.

6.3 PHENOTYPIC CHARACTERIZATION

Usually, phenotypical properties can be used for classification of food-borne microorganisms such as LAB for morphology, method of glucose fermentation, growth at changed temperatures, lactic acid configuration, fermentation of numerous carbohydrates or proteins shape inside cell wall (Gatti et al., 1997). Phenotypic methods are limited inherently like its underprivileged reproducibility. One more drawback of the phenotypic analysis is that complete knowledge of the genome is not articulated, because gene expressions are immediately associated with the environmental conditions. So, all these above-mentioned disadvantages badly influence the trustworthiness on phenotypic characterization processes (Tsakalidou et al., 1994).

6.4 GENOTYPIC CHARACTERIZATION

Number of dissimilar genotype techniques applied as tool for two purposes: for identification of species and for the strains differentiation of LAB to its clonal level. This is the DNA-dependent technique, and its main benefit is its bigoted power and its widespread applicability. Strongly associated strains along with the same phenotypical features now become easy to differentiate by DNA-based procedures.

6.5 MOLECULAR RIBOTYPING

This technique involves the utilization of nucleic acid probes for the identification of ribosomal genes in which DNA of bacteria is insulated and constraint patterns are created with the help of hybridization with a 23S and 16S rRNA gene probe (Roy and Sirois, 2001). Then, consequent agarose gel electrophoresis is tracked by Southern blotting, where DNA is moved to a membrane and hybridized with 23S or 16S rRNA probes (William and Sandler, 1971). Normally, the fingerprint patterns are much more constant than those of restriction enzyme analysis (REA). Some drawbacks include the usage of common probes for all species due to resemblance of ribosomal genes in which we could check the effectiveness of the ribotyping with some microorganisms like *L. acidophilus*, *L. plantarum* and *L. fermentum* but ribotyping technique shows great inequitable strength at level of genera than that of strain level (Kimura et al., 1997; Klein et al., 1998; Morelli, 2001; Walter et al., 2001).

6.6 MOLECULAR TECHNIQUES USED FOR THE CHARACTERIZATION OF FOODBORNE PATHOGENS

Molecular techniques are different from the lengthy and monotonous culturing techniques. Many of the molecular techniques for the characterization of microbes are based on simple DNA amplification-based analysis or DNA sequencing-based analysis which includes simple PCR, real-time PCR, multiplex PCR and RAPD-PCR. More complex molecular techniques include methods based on fragment analysis (RFLP and AFLP), targeted gene (PFGE and ribotyping), whole-genome sequencing (WGS) and mass spectrometry (Franco-Duarte et al., 2019).

Babalola (2003) reported the availability of numerous DNA sequences currently used as molecular markers in the detection of complex foodborne microbial pathogenic outbreaks. These techniques are known to be remarkable due to their sensitivity in detecting foodborne pathogens and other food-associated microorganisms. Also, the conventional approaches used in identifying pathogens from food generally take time and are labor-intensive, hence the need for novel methods in the quick detection of pathogens in food. Molecular techniques such as gene cloning and the recombinant DNA techniques are already revolutionizing methods used in detecting pathogens in foods. The following are examples of molecular techniques utilized for characterizing foodborne pathogens and food-related microorganisms.

6.6.1 POLYMERASE CHAIN REACTION (PCR)

A PCR assay was used by Ghaffar et al. (2014) for the rapid detection of pathogens in food, fruits, and vegetables in Egypt using specific primers and comparing with standard culture technique. The species-specific primers of *E. coli* O157:H7, *L. monocytogenes* and *S. typhimurium* were used for the PCR assay and they yielded the expected amplicon size. When validated in artificially infected lettuce plants after enrichment, these primers revealed their specificity by yielding the expected PCR product. The process involved in the detection of pathogens in this manner was confirmed to be quicker, more time-saving, and more sensitive when compared to the conventional plating methods. Katzav et al. (2008) also used a normal enrichment culture and PCR method to test for *Campylobacter* spp. in 194 chicken and turkey meat products in Finland. The C412F-16SrRNA-campR2 gene-based PCR assay with pre-enrichment showed to be faster, more sensitive with a detection limit of 70 CFU/mL and more specific for the detection of *C. jejuni*, *C. coli* and *C. hyoilei*.

6.6.2 MULTIPLEX PCR

Mazza et al. (2015) developed a multiplex PCR for concurrent detection and identification of six species of *Listeria* in a single reaction out of an array of 108 strains of *Listeria* spp., which were isolated from sausages. The *Listeria* species aimed for detection include *L. grayi*, *L. innocua*, *L. ivanovii*, *L. monocytogenes*, *L. seelingeri* and *L. welshimeri*. Multiplex PCR allowed a unique identification of strains that initially exhibited a doubtful identity with the analytical profile index (API) technique. The result showed a higher sensitivity and specificity than API biochemical determination even without the need for pre-purification of colony culture. The study of Jamali et al. (2015) achieved the detection of nine virulence-associated genes in all 43 isolates of *Listeria monocytogenes* obtained from raw fish meat and open-air fish market using two multiplex-PCR assays. The occurrence of nine virulence-determinant genes in the *L. monocytogenes* isolated from the seafood depicted a great risk of human listeriosis in Northern Iran.

Also, in the study by Razei et al. (2017) three important toxin-producing foodborne pathogens: *Listeria monocytogenes*, *Bacillus cereus*, and

Campylobacter jejuni were detected using multiplex PCR technique in which specific primers were designed for the nucleotide sequences of the *NHEB/NHEC* gene, the *hly* gene and *C* gene of *B cereus*, *L. monocytogenes* and *C. jejuni* respectively. Low detection limits of 3–5 pg were observed for the three bacteria using multiplex PCR. This PCR method confirmed specificity and was recommended for quick and precise detection of toxin-producing food bacteria during epidemiological investigations.

Nguyen et al. (2016) established high specific and sensitive multiplex PCR to simultaneously identify *E. coli* O157: H7, *Salmonella* spp., and *L. monocytogenes* at a low value of 10 CFU/mL each after 12 hours pre-enrichment. The multiplex PCR assay was evaluated for its capacity to detect multiple pathogens concurrently in a single reaction without the interference of non-target bacteria. All the species-specific primer pairs produced a single expected PCR product size. The target genes *invA*, *stx*, and *hlyA* of *Salmonella* spp., *L. monocytogenes* and *E. coli* O157:H7 were employed while 16S rRNA gene primer sets were used as an internal amplification control.

A combined approach of the multiplex real-time PCR assay was also developed by Li et al. (2017) to detect *E. coli* O157:H7 and discriminate for non-O157 Shiga toxin-producing *E. coli* using target genes Z3276 and Shiga toxin genes – *stx 1* and *stx 2* respectively. Their work confirmed the high sensitivity and specificity of the multiplex real-time PCR assay as well as its efficiency in detection of *E. coli* O157: H7 when assessed in beef and spinach.

6.6.3 NESTED PCR

This has a modification for enhanced sensitivity and specificity than conventional PCR. The technique involves the use of two primer sets to be used during the two sequential PCR reactions. The result of the initial reaction is used as a template for a final set of primers and a second amplification step. This approach significantly enhanced the sensitivity and precision of DNA amplification (Carr et al., 2010). Emerenini et al. (2013) employed the nested PCR approach using universal 16S rRNA gene primers and LAB specific primers in two separate rounds to identify and characterize the LAB isolates from fruit and vegetable sources. The first-round PCR product of 16S rRNA gene primers yielded 1,500 bp and was used as a template for the nested PCR, which generated an expected amplicon of 900 bp for LAB group. The use of nested PCR method was declared holistic and faster for the characterization of LAB genera.

6.6.4 REAL-TIME PCR

Several authors have established the real-time PCR for monitoring the amplified products by assessing fluorescent signal to achieve higher specificity, sensitivity, and direct quantification of target DNA with evasion of amplicon contamination relative to conventional PCR for efficient detection of foodborne microbes (Chena et al., 2017). Rudi et al. (2005) employed the ethidium monoazide (EMA) with real time-PCR for LEB medium to achieve the detection of viable *L. monocytogenes* on gouda-like cheeses below the limit of detection at 100 CFU/g. In the assay, *L. monocytogenes* primer pairs with the species-specific probe were employed to detect the target DNA from purified EMA treated samples of *L. monocytogenes* strains. It was expected that EMA bound DNA could not be PCR amplified which helped a sorting of dead and viable cells.

Mafu et al. (2009) studied the detection efficiency and limits of the culture-dependent technique and real-time PCR in three foodborne pathogens using a single enrichment medium. For *hylA*, *invA*, and stx_1-*eae* genes of *L. monocytogenes*, *S. enterica,* and *E. coli* O157:H7, respectively, specific primers and TaqMan probe was designed for these pathogens. Microbiological culture results and real-time PCR results of analyzed samples revealed that both methods show the same positive results when $\geq 10^3$ CFU/mL are present in samples but at lower counts, real-time PCR proves to be more sensitive and rapid with enrichment medium.

Recently, a wider array of 12 pathogens found in food and drinking water were detected and quantified simultaneously in a reaction using TaqMan real-time PCR assay by Liu et al. (2019). Species-specific primers of virulence determinant genes in 12 pathogens, were fluorescently labeled and optimized for consistent reaction conditions. The assay showed 100% positive results for the 12 pathogens. When evaluated by spiking with minced meat, the detection limit of the assay was 10^3 CFU/g for *Vibrio parahaemolyticus* and 10^4 CFU/g for the remaining 11 organisms.

The *pagC* gene-based immunomagnetic beads (IMBs)-qPCR assay was developed by Wang et al. (2018) to identify *Salmonella* found in milk and pork samples. Full specificity was observed in the whole 34 *Salmonella* strains examined for target *pagC* gene with no amplicon yield observed in non-target bacterial strains. The IMBs-qPCR was confirmed to be a promising method for quick, efficient, and precise detection of *Salmonella* strains.

6.6.5 REVERSE-TRANSCRIPTION PCR

In reverse-transcription PCR, ribonucleic acid (RNA) is used as the starting nucleic acid template instead of DNA. The process involves the initial reverse transcription of the RNA to complementary DNA (cDNA) and subsequent amplification of DNA in the PCR process. Though not always true, RNA is believed to be a viability indicator since it is thought that RNA degrades more quickly than DNA and therefore would be found only in metabolically active cells. Due to this reason, reverse-transcription PCR is useful in detecting only viable cells of pathogens (Adzitey et al., 2013). More so, this technique has been used for detecting common foodborne RNA viruses including human Noroviruses, which are non-culturable in the laboratory. Moreover, the technique is being utilized in monitoring the growth and metabolic activity of food bacteria, controlling processes such as fermentation, and food spoilage. It is also useful for performing risk analysis during processing and storage in studying the pathogen's responses and virulence in food (Ceuppens et al., 2014). The drawback of the reverse-transcriptase PCR is that RNA is unstable and requires more skill during handling.

6.7 DNA MICROARRAY

DNA microarray with subsequent bioinformatics analysis provides an avenue for the simultaneous identification of thousands of genes and targeted DNA sequences, thus offering great usefulness for studying foodborne microorganisms. The technology consists of solid support, usually silicon, or glass, upon which DNA fragments or oligonucleotides are attached. The nucleic acid to be investigated is then attached to the microarray and could be a cDNA, genomic DNA, mRNA, or a product from PCR. More specifically, in experimental usage, fluorescently labeled DNA from a biological sample may be hybridized to the microarray, or the extracted nucleic acid may be PCR amplified, subsequently labeled by fluorescence, and hybridized to the array. The procedure has several benefits: it is useful for the specific characterization of microorganisms as it provides information for specific identification of isolates based on genetic information, understanding the virulence attribute of microbes based on presence or absence of certain virulence genes, and providing information on how the pathogenic strains emerged. DNA microarray technology is not just restricted to the characterization of bacteria. They are also useful for studying viruses as well as parasites. There are two types of DNA microarrays mainly produced: oligonucleotide arrays and genomic microarrays.

6.7.1 OLIGONUCLEOTIDE ARRAYS

This involves the use of oligonucleotides, usually between 18 and 70 nucleotides long. They are often used for pathogen detection as they may be made up of multiple variations in the sequence of a target gene, thereby ensuring the sensitive detection of polymorphisms and differentiation among many strains. Oligonucleotide microarrays are also useful for detecting single-point mutations in genes, providing information for epidemiological and phylogenetic studies.

6.7.2 GENOMIC MICROARRAYS

This consists of a whole genome, fragments of a genome from a cDNA library, or the open reading frames/coding sequence from a strain of microorganism. They may be adapted to form a mixed genomic array, containing genomic DNA many related strains of a microorganism (Kostrzynska and Bachand, 2006; Foley and Grant, 2007).

Several applications of microarrays have been developed for studying and characterization of foodborne pathogens. An example is reported by Call et al. (2011) on enterohemorrhagic *E. coli* (EHEC) detection in chicken rinsate. The method was noted for its sensitivity and specificity in detecting and genotyping *E. coli* O157: H7. The PCR was used to amplify target DNA, and the products were hybridized to the glass microarray, which was composed of oligonucleotide probes. The results showed that the microarray was 32-fold more sensitive than agarose gel electrophoresis.

Similarly, *gyrB* gene is useful for the identification of closely related bacteria such as *E. coli*, *Salmonella*, and *Shigella* (Kakinuma et al., 2003). Thus, DNA microarray is an excellent tool for foodborne pathogens characterization and having practicality for both epidemiologic and phylogenetic studies.

6.8 RESTRICTION FRAGMENT LENGTH POLYMORPHISM (RFLP)

RFLP involves chromosomal DNA digestion with a restriction endonuclease and DNA fragments are separated by gel electrophoresis. It characterizes organisms by using the variations in homologous DNA sequences. Pathogen strain identification using RFLP fingerprinting technique is regarded as the most sensitive approach. DNA fragments in large numbers are generated, making it difficult to distinguish band patterns with those in standard gel electrophoresis. It has been eased with the application of a rare cutting restriction enzyme followed by pulsed-field gel electrophoresis (PFGE) method for DNA fragments separation. Another method involves the transfer of the large DNA fragments to membranes and subsequent hybridization to a labeled probe (Foley et al., 2009).

RFLP can be modified by the addition of PCR to the procedure. In this procedure, often abbreviated PCR-RFLP, PCR amplification is done, then restriction enzyme digestion and product differentiation by PFGE. Banding pattern helps in the differentiation of the pathogens. *Escherichia coli* and *Listeria* spp. could be characterized using PCR-RFLP (Fields et al., 1997; Paillard et al., 2003). In one report, the authors carried out the molecular characterization of *Listeria* up to the species level using PCR-RFLP, 23S rRNA gene is the amplified fragment. PCR-RFLP also revealed mixed cultures of *Listeria monocytogenes* and *Listeria innocua* in 22 isolates. The study showed the inexpensive, reliable, rapid, and easy-to-use nature of PCR-RFLP for the identification of *Listeria* species, and its adaptability to use in food microbiology (Paillard et al., 2003). For *E. coli* characterization, Fields et al. (1997) reported the characterization of *Escherichia coli* O157: H7 a notorious as pathogen behind several foodborne disease outbreaks. The pathogen H7 flagellar antigen is critical for diagnosis confirmation, but clinical isolates are frequently non-motile (NM) and do not produce detectable H antigen. As a result of this, the authors carried out a study to further characterize NM isolates and to characterize the H antigen in *E. coli* using PCR-RFLP. Amplification of the entire *fliC* coding sequence was carried out via PCR, restricted with *RsaI*, and fragment pattern was observed after gel electrophoresis. The experiment showed a single restriction pattern (pattern A) for O157: H7 isolates, O157: NM isolates that produced Shiga toxin and 16 of 18 O55: H7 isolates. The flagellar antigen group H7 isolates of non-O157 serotypes showed three patterns, one of which was unique from pattern A. Several patterns were found among isolates of the other 52 flagellar antigen groups tested. Around 13 out of 15 NM strains that did not produce the A pattern had patterns that matched those of other known H groups. Thus, the PCR-RFLP alongside O serogroup determination has potentials in identifying *E. coli* O157: H7 and related strains that do not express immunoreactive H antigen.

6.9 NUCLEIC ACID SEQUENCE-BASED AMPLIFICATION (NASBA)

This technology was developed in 1991 by J. Compton and described isothermal conditions, which are maintained during nucleic acids amplification. Three steps are generally involved in the complete amplification of RNA. First, a reverse transcriptase enzyme converts the target RNA into a cDNA, followed by reverse transcription. Secondly, RNase H, another enzyme, digests the target RNA sequence, the cDNA is bound with a second primer, and then reverse transcriptase catalyzes the conversion of the cDNA to its double-stranded form. Thirdly, the amplification process is used to produce RNA transcripts via the T7 RNA polymerase enzyme. The reaction involves two primers and three enzymes and takes place at a temperature of about 41°C, unlike the PCR, which requires repeated heating and cooling cycles during the reaction. Product recovery/detection for NASBA is usually labor-intensive, therefore, the coming of a novel real-time NASBA which uses fluorescently labeled probes to detect the single-stranded RNA (ssRNA) amplicons, thus, producing a homogenous NASBA assay. NASBA has been used for the detection of many foodborne pathogens including *Salmonella enterica*, *Vibrio cholerae*, *Staphylococcus aureus*, *Campylobacter jejuni* and *Campylobacter coli* (Shi et al., 2010; Law et al., 2015).

6.10 AMPLIFIED FRAGMENT LENGTH POLYMORPHISM (AFLP)

This technique was developed in the Netherlands in the 1990s by Zabeau and his colleagues. AFLP combines the principles of restriction enzyme digestion and PCR for the genotyping of microbes. It is a variation of random amplified polymorphic DNA (RAPD), able to detect restriction site polymorphisms using PCR amplification. The process can distinguish isolates based on the differences in restriction enzyme recognition sites in the genome. Bacterial DNA is digested with one or more restriction enzymes, for example, *EcoRI* or *MseI*. Afterward, short polynucleotide linker sequences are ligated to the free DNA ends, which serve as primer recognition targets for the initiation of PCR for the restriction fragments. After PCR amplification, the fragments are separated by electrophoresis, the amplification profiles are analyzed and are then compared to other isolates to distinguish among strains. Fluorescence AFLP (fAFLP) is a variant of AFLP, done using fluorescent PCR primers (Foley and Grant, 2007; Foley et al., 2009; Babalola, 2010). The technique is well suited for detecting differences among closely related strains. *Listeria monocytogenes* (Camargo et al., 2016; Lüth et al., 2018), *Bacillus cereus* (Ripabelli et al., 2000), and *Arcobacter* spp. (Oliveira et al., 2018) are some of the microorganisms that have been characterized using AFLP.

Moreover, Oliveira et al. (2018) reported the characterization of *Arcobacter* spp. in chicken. *Arcobacter butzleri*, *Arcobacter cryaerophilus*, and *Arcobacter skirrowii* are Gram-negative pathogenic microorganisms that cause watery diarrhea and septicemia in humans, thus having implications for food safety standards. The authors obtained the chicken meat from butcher shops in Sao Paulo and verified their virulence genes and genotypic profile using PCR and AFLP. Of the total 300 chicken cuts analyzed, 18.3% showed the presence of *Arcobacter* spp. identified as *A. butzleri* (63.6%) and *A. cryaerophilus* (36.3%). The virulence genes *ciaB* and *mviN* were present in all the strains. Other genes had a relative composition among the strains. The virulence genes were *cj1349* (98%), *pldA* (94.4%), *cadF* (72.7%), *tlyA* (92.7%), *hecA* (49%), *irgA* (47.2%), and *hecB* (34.5%). Further screening with AFLP showed 19 genotypic profiles for *A. butzleri*, and 17 for *A. cryaerophilus*. The study showed the effectiveness of PCR and AFLP for the characterization of microbes, the possible implications of *Arcobacter* contamination of meat products, and the need for awareness in the processing and packaging of those products.

Another study by Ripabelli et al. (2000) reported the characterization of *Bacillus cereus* from food as well as non-food sources using AFLP. Typical symptoms manifested in the food poisoning incidents included nausea, vomiting, and diarrhea. The study showed 16 different profiles for the *B. cereus* pathogens investigated and the distinguishing of three subtypes within serotype H1, the serotype responsible for the majority of the food poisoning associated with vomiting in England and Wales.

6.11 LOOP-MEDIATED ISOTHERMAL AMPLIFICATION (LAMP)

LAMP, first described in 2000 by Notomi and colleagues, utilized two primer sets and a DNA polymerase. The procedure generated a huge amount of amplified product within an hour which is usually 10^3-fold or higher as compared to simple PCR (Law et al., 2015). The procedure employs *Bst* polymerase (*Bacillus stearothermophilus* DNA polymerase) and is initiated by a forward inner primer with sequences of the sense strand of the target DNA which is used to hybridize a portion in the target and subsequently initiating complementary strand synthesis. Afterward, the outer primer hybridizes to another portion further along the initial position of hybridization in the target sequence resulting in the displacement of the just synthesized strand, which is released as a single-stranded DNA with a loop-like structure at one end. The inner primer-linked complementary strand acts as the template for a new DNA synthesis. The new synthesis is primed by a backward inner and outer primers that hybridize to the other end of the target, each primer attaching to a particular portion following the same steps as the forward primers. The synthesis also results in the formation of a loop-like structure at the end of the DNA. The overall structure of the produced double-stranded DNA looks like a dumbbell, having two loop-like structures at each end. After this, cycles comprising elongation and recycling steps lead to the final products constituted by a mixture of stem-loop DNAs having various stem lengths and cauliflower-like structures. Only the inner primers are used for strand displacement DNA synthesis during cycling reactions while all other four primers are used during the initial steps of the LAMP reaction. LAMP method unlike the PCR does not require thermocycling as the amplification is performed at a constant temperature between 60°C and 65°C. The detection of LAMP products can be done by electrophoresis, turbidimetry, or by using a SYBR Green stain (Jadhav et al., 2012; Zanoli and Spoto, 2013). Ueda and Kuwabara (2009) also used loop-mediated isothermal amplification (LAMP) method to detect *Salmonella* in human blood, feces, and food. The assay revealed the same specificity but a higher sensitivity of 100-fold when compared to PCR assay carried out. LAMP product obtained a detection limit of 10^2 CFU/mL with appropriate pre-enrichment culture.

6.12 MULTIPLEX LOOP-MEDIATED ISOTHERMAL AMPLIFICATION (mLAMP)

Xu et al. (2019) recently developed a multiplex LAMP assay for *Salmonella*, *Shigella*, and *Staphylococcus aureus* detection in fruit juice based on three primer sets and three varying restriction enzyme cleavage sites in both forward inner primer. Under the isothermal reaction condition of 64°C for 50 min, mLAMP assay was carried out after which digestion by separate restriction endonucleases *xhoI*, *kpnI*, and *BamHI* led the separation of the LAMP products followed by electrophoresis. Results of the mLAMP assay revealed high specificity of the primers and restriction digestion for the three pathogens with a detection limit of 100 fg DNA/25 μL.

6.13 RIBOTYPING

Kenneth and Stark (2006) investigated the potential of automated ribotyping approach for the identification of species of Clostridia in food using *EcoRI* digests in DuPont Qualicon RiboPrinter to ribotype 49 isolates of seven different *Clostridium* species and *EcoRV* digests to ribotype 17 *Clostridium botulinum* isolates to compare restriction enzyme efficacy. The result generated 26 ribogroups that were ribotype with *EcoRI* digests with 75.9% of *C. botulinum*. Automated ribotyping was confirmed to be promising for the detection of *C. botulinum* as well as other *Clostridium* species but can be greatly improved with high sensitivity when other restriction enzymes besides *EcoRI* are used.

Wiedmann et al. (2000) utilized automated ribotyping for molecular characterization of species of *Pseudomonas* isolated from milk using the restriction enzyme *EcoRI* and automated Qualicon RiboPrinter. Alkaline phosphatase-labeled antibodies were employed in detecting hybridization of the blotted nucleic acids with the DNA probe. The ribotype patterns yielded 38 different ribogroups out of 70 isolates of *Pseudomonas* sp. Evaluation of the discriminatory ability indicated automated ribotyping with *EcoRI* as a good characterization tool for strains of dairy pseudomonads.

Combina et al. (2005) described the involvement of a wide range of microbes in fermentation processes in food such as non-*Saccharomyces* yeasts: *P. membrefaciens*, *Pichia kluyveri*, *Metschnikowia pulcherrima*, *R. mucilaginosa*, *Hanseniaspora uvarum*, *Kloeckera apiculata*, *Rhodotorula*, *Zygosaccharomyces florentinus*, *Saccharomyces cerevisiae*, *Candida*, *Pichia*, *Hansenula*, and *Kluyveromyces*. Fusco and Quero (2012) revealed that one of the devastating public health issues is food and waterborne related diseases which have continued to rise yearly globally due to trade, globalization, economic development plus international travel. The authors pointed out that the risk is more in developing countries due to low levels of hygiene, sanitation, and socioeconomic conditions. Studies have shown that five major pathogens contaminate food and water: bacteria, viruses, fungi, helminths, and protozoa. Risk management and quality control strategies must be adopted to minimize the risk in the food production chain. Throughout the world, routine analysis is still very poorly done using conventional methods due to low reliability and inaccurate microbial numbers. Hence, considerable research must be done to explore the latest modern technology in molecular biology to enhance detection, control, and monitoring of foodborne pathogens.

D'Orazio (2011) showed that the goal of most food industries is to produce safe and quality food products for the consumers, hence the establishment of risk and quality control unit that will take care of various pathogenic microbes in the food production chain. Many have adopted various techniques in the detection and monitoring of foodborne pathogen such as ELISA bioassay and culture techniques. These techniques generally take a longer time to be processed, hence the need to adopt and fast and robust techniques that will generate both qualitative and quantitative results. The authors suggested that a biosensor is a novel approach that offers real-time rapid multiple analyzes for the detection of pathogens in food. Poultry industries have been highlighted to an area with an important application of biosensors such as enzyme-based biosensors, and immunosensors.

6.14 CONCLUSION

This chapter established that foodborne pathogens are currently the major causes of global public health concerns. The early detection of these microbes can reduce the risks associated with foodborne pathogens and sustain both human and environmental health. It has also been stated that foodborne pathogens are usually identified using various traditional culture-based approaches, but it was discovered that improvements in the techniques during the detection and diagnosis of pathogens in food have undergone remarkable developments. This chapter also validates the fact that the application of precise molecular diagnostic-based tests and real-time surveillance technologies are promising strategies for early detection of both known and unknown pathogens associated with human food. Therefore, there is a need for further developments of these molecular techniques for easy identification of several foodborne pathogens.

KEYWORDS

- **detection and diagnosis**
- **food chain**
- **foodborne pathogens**
- **molecular advances**
- **quarantine screening**

REFERENCES

Adzitey, F., Huda, N., & Ali, G. R. R., (2013). Molecular techniques for detecting and typing of bacteria, advantages and application to foodborne pathogens isolated from ducks. *Biotech, 3*, 97–107. doi: 10.1007/s13205-012-0074-4.

Arora, K., Chand, S., & Malhotra, B. D., (2006). Recent development in bio-molecular electronics techniques for food pathogens. *Analytical Chimica Acta, 568*, 259–274.

Babalola, O. O., (2003). Molecular techniques: An overview of methods for the detection of bacteria. *African Journal of Biotechnology, 2*, 710–713.

Bezirtzoglou, E., Maipa, V., Voidarou, C., Tsiotsias, A., & Papapetropoulou, M., (2000). Foodborne intestinal bacterial pathogens. *Microbial Ecology in Health and Disease, 12*(2), 96–104. doi: 10.1080/089106000750060350.

Boughattas, S., & Salehi, R., (2014). Molecular approaches for detection and identification of foodborne pathogens. *Journal of Food Quality and Hazards Control, 1*, 1–6.

Call, D. R., Brockman, F. J., & Chandler, D. P., (2001). Detecting and genotyping *Escherichia coli* O157:H7 using multiplexed PCR and nucleic acid microarrays. *International Journal of Food Microbiology, 67*, 71–80.

Camargo, A. C., Woodward, J. J., & Nero, L. A., (2016). The continuous challenge of characterizing the foodborne pathogen *Listeria monocytogenes. Foodborne Pathogens and Disease, 13*(8), 405–416. doi: 10.1089/fpd.2015.2115.

Cancino-Padilla, N., Fellenberg, M. A., Franco, W., Ibañez, R. A., & Vargas-BelloPérez, E., (2017). Foodborne bacteria in dairy products: Detection by molecular techniques. *Cien. Inv. Agr., 44*(3), 215–229.

Carr, J., Williams, D. G., & Hayden, R. T., (2010). Molecular detection of multiple respiratory viruses. In: Grody, W. W., Nakamura, R. M., Strom, C. M., & Kiechle, F. L., (eds.), *Molecular Diagnostics: Techniques and Applications for the Clinical Laboratory* (pp. 289–300). USA: Academic Press Inc.

Ceuppens, S., Li, D., Uyttendaele, M., et al., (2014). Molecular methods in food safety microbiology: Interpretation and implications of nucleic acid detection. *Comprehensive Reviews in Food Science and Food Safety, 13*, 551–577. doi: 10.1111/1541-4337.12072.

Chena, J. Q., Healeya, S., Regana, P., Laksanalamaib, P., & Hua, Z., (2017). PCR-based methodologies for detection and characterization of *Listeria monocytogenes* and *Listeria ivanovii* in foods and environmental sources. *Food Science and Human Wellness, 6*, 39–59.

Chun, S. R., Czajka, W. J., Lakamoto, M., & Benno, Y., (2001). Characterization of the *Lactobacillus casei* group and the *Lactobacillus acidophilus* group by automated ribotyping. *Microbiology and Immunology, 45*, 271–275.

Combina, M., Elia, A., Mercado, L., Catania, C., Ganga, A., & Martinez, C., (2005). Dynamics of indigenous yeast populations during spontaneous fermentation of wines from Mendoza, Argentina. *International Journal of Food Microbiology, 99*(3), 237–243.

Crerar, S. K., Dalton, C. B., Longbottom, H. M., et al., (1996). Foodborne disease: Current trends and future surveillance needs in Australia. *Medical Journal of Australia, 165*, 672–675.

D'Orazio, P., (2011). Biosensors in clinical chemistry – 2011 update. *Clinica Chimica Acta, 412*, 1749–1761.

Decallone, J., Delmee, M., Wauthoz, P., El Lioul, M., & Lambert, R., (1991). A rapid procedure for the identification of lactic acid bacteria based on the gas chromatographic analysis of cellular fatty acids. *Journal of Food Protection, 54*, 217–224.

Emerenini, E. C., Afolabi, O. R., Okolie, P. I., & Akintokun, A. K., (2013). Isolation and molecular characterization of lactic acid bacteria isolated from fresh fruits and vegetables using nested PCR analysis. *British Microbiology Research Journal, 3*(4), 368–377.

Fields, P. I., Blom, K., Hughes, H. J., Helsel, L. O., Feng, P., & Swaminathan, B., (1997). Molecular characterization of the gene encoding H antigen in *Escherichia coli* and development of a PCR-restriction fragment length polymorphism test for identification of *E. coli* O157:H7 and O157:NM. *Journal of Clinical Microbiology, 35*, 1066–1070.

Foley, S. L., & Grant, K., (2007). Molecular techniques of detection and discrimination of foodborne pathogens and their toxins. In: Shabbir, S., (ed.), *Foodborne Diseases*. Totowa, New Jersey: Humana Press Inc.

Foley, S. L., Lynne, A. M., & Nayak, R., (2009). Molecular typing methodologies for microbial source tracking and epidemiological investigations of Gram-negative bacterial foodborne pathogens. *Genetics and Evolution, 9*, 430–440. doi: 10.1016/j.meegid.2009.03.004.

Franco-Duarte, R., Cernákovấ, L., Kadam, S., et al., (2019). Advances in chemical and biological methods to identify microorganisms – from past to present. *Microorganisms, 7*, 130. doi: 10.3390/microorganisms7050130.

Fusco, V., & Quero, G. M., (2012). Nucleic acid-based methods to identify, detect and type pathogenic bacteria occurring in milk and dairy products. In: Eissa, A. A., (ed.), *Structure and Function of Food Engineering*. Rijeka: InTech.

Gatti, M., Fornasari, E., & Neviani, E. (1997). Cell-wall protein profiles of dairy thermophilic *Lactobacilli*. *Lett. Appl. Microbiol., 25*, 345–348.

Ghaffar, N. A., Ibrahim, A., Yossef, M., & Omar, M. N. A., (2014). Rapid detection of pathogenic bacteria in vegetables and fruits in Egyptian farms. *Journal of American Science, 10*(9), 242–252.

Giraffa, G., & Neviani, E., (2000). Molecular identification and characterization of food associated Lactobacilli. *Italian Journal of Food Science, 4*, 403–423.

Grimont, F., & Grimont, P. D. A., (1986). Ribosomal ribonucleic acid gene restriction as potential taxonomic tools. *Annales de l'Institut Pasteur Microbiologie, 137*(1), 165–175.

Holzapfel, W. H., Haberer, P., Geisen, R., Bjorkroth, J., & Schillinger, U., (2001). Taxonomy and important features of probiotic microorganisms in food and nutrition. *American Journal of Clinical Microbiology, 73*, 365S–73S.

Ivey, M. L., & Phister, T. G., (2011). Detection and identification of microorganisms in wine: A review of molecular techniques. *J. Ind. Microbiol. Biotechnol., 38*, 1619–1634. doi: 10.1007/s10295-011-1020-x.

Jadhav, S., Bhave, M., & Palombo, E. A., (2012). Methods used for the detection and subtyping of *Listeria monocytogenes*. *Journal of Microbiological Methods, 88*, 327–341. doi: 10.1016/j.mimet.2012.01.002.

Jamali, H., Paydar, M., Ismail, S., et al., (2015). Prevalence, antimicrobial susceptibility and virulotyping of *Listeria* species and *Listeria monocytogenes* isolated from open-air fish markets. *BMC Microbiology, 15*, 144. doi: 10.1186/s12866-–015-0476-7.

Kakinuma, K., Fukushima, M., & Kawaguchi, R., (2003). Detection and identification of *Escherichia coli*, *Shigella*, and *Salmonella* by microarrays using the *gyrB* gene. *Biotechnology and Bioengineering, 83*, 721–728.

Katzav, M., Isohanni, P., Lund, M., Hakkinen, M., & Lyhs, U., (2008). PCR assay for the detection of *Campylobacter* in marinated and non-marinated poultry products. *Food Microbiology, 25*, 908–914. doi: 10.1016/j.fm.2008.05.010.

Kenneth, C. A., & Stark, B., (2006). Automated ribotyping for the identification and characterization of foodborne clostridia. *Journal of Food Protection, 69*(12), 2970–2975.

Kimura, K., McCartney, A. L., McConnell, M. A., & Tannock, G. W., (1997). Analysis of fecal populations of *Bifidobacteria* and *Lactobacilli*, and investigation of the immunological responses of their human hosts to the predominant strains. *Appl. Environ. Microbiol., 63*, 3394–3398.

Klein, G., Pack, A., Bonaparte, C., & Reuter, G., (1998). Taxonomy and physiology of probiotic lactic acid bacteria. *International Journal of Food Microbiology, 41*, 103–125.

Kostrzynska, M., & Bachand, A., (2006). Application of DNA microarray technology for detection, identification, and characterization of foodborne pathogens. *Can. J. Microbiol., 52*, 1–8. doi: 10.1139/W05-105.

Law, J. W. F., Ab Mutalib, N. S., Chan, K. G., & Lee, L. H., (2015). Rapid methods for the detection of foodborne bacterial pathogens: Principles, applications, advantages and limitations. *Front. Microbiol., 5*, 770. doi: 10.3389/fmicb.2014.00770.

Li, B., Liu, H., & Wang, W., (2017). Multiplex real-time PCR assay for detection of *Escherichia coli* O157:H7 and screening for non-O157 Shiga toxin-producing *E. coli. BMC Microbiology, 17*, 215. doi: 10.1186/s12866-017-1123-2.

Liu, Y., Cao, Y., Wang, T., Dong, Q., Li, J., & Niu, C., (2019). Detection of 12 common foodborne bacterial pathogens by Taqman real-time PCR using a single set of reaction conditions. *Front. Microbiol., 10*, 222. doi: 10.3389/fmicb.2019.00222.

Lüth, S., Kleta, S., & Al Dahouk, S., (2018). Whole-genome sequencing as a typing tool for foodborne pathogens like *Listeria monocytogenes* – The way towards global harmonization and data exchange. *Trends in Food Science and Technology, 73*, 67–75.

Mafu, A. A., Pitre, M., & Sirois, S., (2009). Real-time PCR as a tool for detection of pathogenic bacteria on contaminated food contact surfaces by using a single enrichment medium. *Journal of Food Protection, 72*(6), 1310–1314.

Mazza, R., Piras, F., Ladu, D., Putzolu, M., Consolati, G. S., & Mazzette, R., (2015). Identification of *Listeria* spp. strains isolated from meat products and meat production plants by multiplex polymerase chain reaction. *Italian Journal of Food Safety, 4*, 5498.

McAuley, C. M., McMillan, K., Moore, S. C., Fegan, N., & Fox, E. M., (2014). Prevalence and characterization of foodborne pathogens from Australian dairy farm environments. *J. Dairy Sci., 97*, 7402–7412. http://dx.doi.org/ 10.3168/jds.2014-8735.

Mohania, D., Nagpal, R., Kumar, M., et al., (2008). Molecular approaches for identification and characterization of lactic acid bacteria. *Journal of Digestive Diseases, 9*(4), 190–198.

Morelli, L., (2001). *Taxonomy and Physiology of Lactic Acid Bacteria: Effects and Function on Nutrition.* Report of a Joint FAO/WHO expert consultation on evaluation on health and nutritional properties of probiotics in food including powder milk with lactic acid bacteria. Food and Agricultural Organization of the United Nations, New York.

Naravaneni, R., & Jamil, K., (2014). Rapid detection of foodborne pathogens by using molecular techniques. *Journal of Medical Microbiology, 54*, 51–54.

Nguyen, T. T., Giau, V. V., & Vo, T. K., (2016). Multiplex PCR for simultaneous identification of *E. coli* O157:H7, *Salmonella* spp., and *L. monocytogenes* in food. *3Biotech, 6*, 205. doi: 10.1007/s13205-016-0523-6.

Oliveira, M. G. X., Gomes, V. T. M., Cunha, M. P. V., Moreno, L. Z., Moreno, A. M., & Knöbl, T., (2018). Genotypic characterization of *Arcobacter* spp. isolated from chicken meat in Brazil. *Foodborne Pathogens and Disease, 15*, 293–299. doi: 10.1089/fpd.2017.2368.

Paillard, D., Dubois, V., Duran, R., et al., (2003). Rapid identification of *Listeria* species by using restriction fragment length polymorphism of PCR-amplified 23S rRNA gene fragments. *Applied and Environmental Microbiology, 69*, 6386–6392.

Razei, A., Sorouri, R., Mousavi, S. L., Nazarian, S. H., Amani, J., & Aghamollaei, H. (2017). Presenting a rapid method for detection of *Bacillus cereus*, *Listeria monocytogenes* and *Campylobacter jejuni* in food samples. *Iran J. Basic Med. Sci., 20*, 1050–1055. doi: 10.22038/IJBMS.2017.9275.

Ripabelli, G., McLauchlin, J., Mithani, V., & Threlfall, E. J., (2000). Epidemiological typing of *Bacillus cereus* by amplified fragment length polymorphism. *Letters in Applied Microbiology, 30*, 358–363.

Rodríguez-Lázaro, D., Oniciuc, E. A., García, P. G., et al., (2017). Detection and characterization of *Staphylococcus aureus* and methicillin-resistant *S. aureus* in foods confiscated in EU borders. *Front. Microbiol., 8*, 1344. doi: 10.3389/fmicb.2017.01344.

Roy, D., & Sirois, S., (2001). Molecular differentiation of *Bifidobacterium* species with amplified ribosomal DNA restriction analysis and alignment of short regions of the *idh* gene. *FEMS Microbiology Letters, 191*, 17–24.

Rudi, K., Naterstad, K., Drømtorp, S. M., & Holo, H., (2005). Detection of viable and dead *Listeria monocytogenes* on gouda-like cheeses by real-time PCR. *Letters in Applied Microbiology, 40*, 301–306.

Shin, H. H., Hwang, B. H., & Cha, H. J., (2016). Multiplex 16S rRNA-derived geno-biochip for detection of 16 bacterial pathogens from contaminated foods. *Biotechnol. J., 11*, 1405–1414. doi: 10.1002/biot.201600043.

Suzzi, G., & Corsetti, A., (2020). Food microbiology: The past and the new challenges for the next 10 years. *Front. Microbiol., 11*, 237. doi: 10.3389/fmicb.2020.00237.

Tsakalidou, E., Manolopoulou, E., Kabaraki, E., et al., (1994). The combined use of whole-cell protein extracts for the identification (SDS-PAGE) and enzyme activity screening of lactic acid bacteria isolated from traditional Greek dairy products. *Syst. Appl. Microbiol., 17*, 444–458.

Ueda, S., & Kuwabara, Y., (2009). The rapid detection of *Salmonella* from food samples by loop-mediated isothermal amplification (LAMP). *Biocontrol Science, 14*(2), 73–76.

Vo-Dinh, T., & Cullum, B., (2000). Biosensors and biochips: Advances in biological and medical diagnostics. *Fresenius' Journal of Analytical Chemistry, 366*, 540–551.

Walter, J., Hertel, C., Tannock, G. W., Lis, C. M., Munro, K., & Hammes, W. P., (2001). Detection of *Lactobacillus*, *Pediococcus*, *Leuconostoc*, and *Weissella* species in human feces by using group-specific PCR primers and denaturing gradient gel electrophoresis. *Appl. Environ. Microbiol., 67*, 2578–2585.

Wang, J., Li, Y., Chena, J., et al., (2018). Rapid detection of foodborne *Salmonella* contamination using IMBs-qPCR method based on *pagC* gene. *Brazilian Journal of Microbiology, 49*, 320–328.

William, R. A. D., & Sandler, S. A., (1971). Electrophoresis of glucose-6-phosphate dehydrogenase, cell wall composition and the taxonomy of heterofermentative lactobacilli. *J. Gen. Microbiol., 65*, 351–358.

Xu, C., Luo, H., & Zhang, Y., (2020). Development of multiplex loop-mediated isothermal amplification for three foodborne pathogens. *Food Sci. Technol. Campinas, 40*, 205–210. doi: https://doi.org/10.1590/fst.07319.

Zanoli, L. M., & Spoto, G., (2013). Isothermal amplification methods for the detection of nucleic acids in microfluidic devices. *Biosensors, 3*, 18–43. doi: 10.3390/bios3010018.

CHAPTER 7

Tracing Foodborne Pathogens Using Molecular-Based Approaches

CHANDRASHEKAR SRINIVASA, CHANDAN SHIVAMALLU, KOLLUR SHIVA PRASAD, SANTOSH KUMAR SHIMOGA RAJANNA, P. ASHWINI, and MURALI MAHADEVAMURTHY

ABSTRACT

Food safety has been considered an essential part of human life in the 21st century, which could be the reason for severe loss of life in case of inappropriate management. However, the challenges arise chiefly as a part of resistance offered by pathogenic bacteria, fungi, and viruses that exist with sources of food, i.e., plants and animals. As a part of their life cycle, these microorganisms release specific metabolites and toxins that are responsible for the contamination of the food. Tracing these prime components in identifying foodborne pathogens could be easily done with the help of molecular techniques like LAMP, NASBA, PCR, and DNA microarray. There have been several molecular tools that have been developed contemporarily with the advancement of technology. This chapter discusses the molecular techniques that are currently used in this regard and also discusses the development progress with respect to the materials, methods, and procedures used in these molecular techniques. The chapter also emphasizes on the intrinsic factors and impact of food processing on the stability of nucleic acids, variation between culture and molecular methods and counter-strategies to improve nucleic acid-based detection.

Global Food Safety: Microbial Interventions and Molecular Advancements. Saher Islam, PhD, Devarajan Thangadurai, PhD, Jeyabalan Sangeetha, PhD, & Natália Cruz-Martins, PhD (Editors)

7.1 INTRODUCTION

A wide array of diseases is associated with food and therefore food-borne pathogens pose greater challenges to the public health (Oliver et al., 2005). Several regions all over the world have seldom reported the incidence of foodborne diseases and therefore elevate the difficulty in quantifying the impact of these diseases on the global economy (Van de Venter, 2000). In addition, these diseases cost more than a billion dollars for hospitalization in the United States (Scallan et al., 2011; Scharff, 2012; Batz et al., 2014). Therefore, food safety has become the prime component for food processing industries. Consumption of contaminated food or water delivers pathogens within the body causing such diseases. The availability of ready-to-eat (RTE) foods as well as attractive foods sold by street vendors is the primary source of foodborne pathogens. Apart from these, even the fresh fruits, vegetables, seafood and meat are also at times prone to contamination (Wingstrand et al., 2006). The most found foodborne pathogens include *Listeria monocytogenes*, *Staphylococcus aureus*, *Campylobacter jejuni*, *Clostridium perfringens*, *Salmonella enterica*, Shiga toxin-producing *Escherichia coli* (STEC), *Escherichia coli* O157:H7, *Bacillus cereus* and *Vibrio* spp. (Zhao et al., 2014). Each microbe has a varying ability to cause diseases ranging from mild to life threatening. To ensure the safety of food, it is highly essential to assess these food sources for the presence of any contaminants and treat them.

Conventional methods of detecting these pathogens include culturing microorganisms on the agar plate and standard assessment using biochemical methods. They are simple and inexpensive, yet their use is often limited because of the tedious and time-consuming procedures (Mandal et al., 2011). Not only that, some of the microorganisms from the same species exhibit great variations between strains in turn augment to the difficulties in assessing them. For instance, there are more than 2,500 different types of *Salmonella*, which exhibit morphological similarities and cause difficulty in accurately identifying it (Foley et al., 2013; Li et al., 2013; Ricke, 2014). The challenges are further enhanced because of the physiological changes induced to the microorganisms by the environmental stress caused by various intrinsic and extrinsic factors such as disinfectants, preservatives, temperature variations, moisture content, etc. The exposure to these agents can render the pathogens into a dormant state, which can further relapse as a stronger and resistant species over the course of time. Additionally, false-negative results were also obtained because of viable but non-cultivable pathogens (VBNC). Viability of the pathogen cannot be differentiated in most of the conventional methods. To counter this issue, some methods of detection are using staining with acridine orange and BacLight fluorescence microscopy. Although these methods are relatively reliable, chances of identification of specific pathogens were often misleading. In addition, conventional methods require over a week for accurately confirming the pathogen and have reduced sensitivity (Lee et al., 2014; Zhao et al., 2014).

7.2 CHALLENGES OF THE FOOD PRODUCTION INDUSTRY

Growing population has drastically elevated the demand for sustainable foods. Technological advances have led to the growth of food production industry as a global market and therefore, to ensure the sustainability of food during its transport and export advanced technologies are required (Windhorst, 2009). Right from the raw materials to the end products, a constant threat of microbial contamination exists, which can potentially render these food sources undesirable. The central authority, US Food and Drug Administration (USFDA) has set certain standards for the manufacturing and packaging of processed as well as non-processed foods. Entry of pathogens can occur via unhygienic practices among the workers of the processing industry, use of contaminated water, or through the contaminated equipments such as screens, fermenters, etc. (Ashbolt, 2004; Green and Selman, 2005; Alum et al., 2016).

Bacteria, fungi, and viruses cause varying types and intensity of infections, thus making the food sources harmful for consumption. Contamination of meat by bacteria belonging to the genera such as *Bacillus*, *Leuconostoc*, *Acinetobacter*, *Streptococcus*, and some species of *Lactobacillus* are known to grow on the meat surface in the form of slime and thus rendering it unfit for human consumption (Borch et al., 1996). This can change the color of the meat in accordance with the pigment produced by the specific type of bacteria contaminating it. Similarly, some of the aerobic bacteria cause an undesirable stench and flavor to the meat because of the process known as souring caused by *Clostridium* and *Streptococcus* species (Ingram, 1952). Furthermore, certain type of bacteria form biofilms, a term used to refer to the attachment of active bacterial cells and extracellular complexes to the solid matrix. This kind of contamination can occur if the equipments are not washed and sterilized thoroughly before every process (Van Houdt and Michiels, 2010). In a study carried out by Joseph et al. (2001), the biofilms from *Salmonella* spp. were found to be resistant even against the use of strong sanitizers such as chlorine and iodine, suggesting the enhanced resistance exerted by the pathogen when in the form of biofilms. A fungus is also one of the major contaminants of the food industry. In a study by Njobeh et al. (2009), among the 95 food samples tested in Cameroon, all of them were contaminated with either *Aspergillus* or *Penicillium* spp. Fungi can produce certain mycotoxins that have shown harmful effects upon consumption of such contaminated foods. These mycotoxins are basically secondary metabolites produced by these fungi to acclimatize to the external environment. Among them, aflatoxins produced by *Aspergillus*, ochratoxins produced by *Aspergillus* and *Penicillium* spp., fumonisins produced by *Fusarium oxysporum* are commonly known. These toxins are commonly found in beans and nuts (Kumi et al., 2014) and grains and cereals (Lawley, 2013a, b). To ensure the safety of food, prior detection of the presence of such contaminants is essential to isolate such products thereby reducing its entry into the market.

In the agricultural sector, owing to the limited availability of land, livestock grazing, crop damage due to environmental adversities and pollution, food safety has now become a global issue (Dunkley and Dunkley, 2013; Hoekstra and Wiedmann, 2014). Therefore, lifecycle assessment (LCA) is among the most popular to carry out a thorough qualitative assessment of all the input and output materials in the agricultural systems (Stoessel et al., 2012; Hellweg and Milài Canals, 2014; Pelletier et al., 2014). One measure to ensure safety is by growing food using organic farming with less harmful chemicals. The complicated mosaic of food industry focuses on where and how the product is obtained and how it is marketed. Therefore, advocating organic farming ensures a greater degree of safety to the food products.

Poultry is yet another field that has received less attention in terms of safety measures. The government policies on animal welfare have led to the cage-free growth of poultry often resulting in infections arising from the water and feeds leading to implications with regard to its health and in turn the health of the consumers (De Reu et al., 2008; Windhorst, 2008; Mench et al., 2011). Some of the social and cultural specifications create demand of one type of meat over the other, resulting in a deficit state for that meat. This in turn leads to excessive cost for the authentic type as well as poor or mediocre quality meat which are non-authentic types (O'Bryan et al., 2008). All these have resulted in a greater degree of fraud in terms of authentic brands and disease outbreaks like the one that has very recently occurred in Eastern China (Wuhan) and led to the outbreak of a deadly disease caused by a new severe acute respiratory syndrome Coronavirus-2 (SARS-CoV-2) (Li et al., 2020). Therefore, stringent policies need to be in place for the type of meat allowed for consumption, certification, accurately labeling the product and other traceability. Traceability is one such component that requires sophisticated tools for data mining, processing, and analyzing.

7.3 TRACEABILITY AS A PRIME COMPONENT IN IDENTIFYING FOODBORNE PATHOGENS

The enormous variety of foods available in the market has indeed enhanced the complexities for food safety. In terms of foodborne pathogens, a lot of study has been carried out to accurately identify and isolate the contaminated foods, which has in turn paved the way for tracking of such organisms to its point of origin and its epidemiological risks pertaining to these organisms focusing on each case separately. Although enormous information is available but still not many studies have been carried out regarding the authenticity of the food material itself. Therefore, focusing on developing biological methods for the generation of a large data repository containing all the data available will assist better detection of pathogens specific to certain type of food systems (Beatson and Walker, 2014). Knowledge on the microbial genetics, which has recently been booming has opened newer avenues in terms of enforcing food safety measures and the detection of pathogens present in food. Molecular detection tools using the DNA have enlarged the understanding of not just the pathogen but also the presence of pathogen in the food system. Therefore, rather than mere understanding of the overall stress response exerted by the pathogen, all the virulence-related gene expressions are also well studied as the factors affecting the foods (Lungu et al., 2009; Soni et al., 2011; Spector and Kenyon, 2012; Callaway et al., 2013; Ferreira et al., 2014). These components of the pathogen and its interaction with its external environment have led to obtain a plethora of data which can be used for various studies. With this background, the following sections discusses on the use of molecular methods for the detection of pathogens.

7.4 ADVANTAGES OF MOLECULAR METHODS USED IN FOOD SAFETY

Microbiological examinations traditionally have relied upon the use of culture-based methods for the detection of pathogens present in any given sample, including food sources. The fact that these methods are time-consuming and expensive has led to the quest for alternative methods for detection. Among the various methods, molecular approaches have been most promising for the detection of pathogens. Studies have shown that for the detection of non-cultivable pathogens (such as the viral pathogens), molecular-based approach is the only choice. Foodborne pathogens, such as human Norovirus, were detected only after the advent of molecular methods (Payne et al., 2012). Additionally, as mentioned in the previous sections, the VBNC cells enter this state due to the environmental variations and are often impossible to detect using conventional techniques. Such pathogens, *Campylobacter* spp. from the food and stool samples, for instance, can only be detected using DNA-based methods. Most infections often cause disease conditions that vary in intensity from mild to severe. In most cases, detection of the type of pathogen involved in the disease is of utmost importance to design optimal treatment methods. This can only be achieved using molecular methods, which detect the pathogen within a span of a day or two, while the conventional methods take close to 15 days to accurately identify the pathogen. For instance, *Mycobacterium avium* subsp. *paratuberculosis* (MAP) culture conditions have not been optimized and therefore take over 10 weeks for mere detection of the pathogen. In this course, the pathogen would successfully establish itself in the host and cause severe infection and thus, speedy detection protocol (by molecular methods) is essential (Slana et al., 2008; Douarre et al., 2010). Also, most enrichment media and selective media are specific for a certain class of microbes but not specific to a single organism. This leads to false-negative results, which can only be prevented using molecular methods. Therefore, species-specific detection of the pathogen and the differentiation of different species from a genus can all be achieved using molecular methods. These methods are highly specific to a particular genotype, pathotype as well as serotype. For example, classical O- and H-serotyping of *E. coli* can also be performed by PCR (Wang et al., 2003; DebRoy et al., 2011). The various molecular methods that are available for the detection of foodborne pathogens are detailed in subsections.

7.4.1 POLYMERASE CHAIN REACTION (PCR)

PCR has now become one of the most common molecular techniques for the detection of foodborne pathogens. Although its invention dates to over 30 years, it still is recognized as a tool for every laboratory for quick detection and diagnostic assays. A short sequence of DNA gets amplified with the help of primers and polymerase enzyme. This system operates in three steps, namely, denaturation of the double stranded DNA to single strands, annealing of the primer to its complementary sequence and the last step of primer extension to synthesize complementary sequence in multiple copies. This is one of the most effective tools for the detection of even a small amount of the pathogen in the food sources (Mandal et al., 2011). Foodborne pathogens such as *E. coli* O157:H7, *Listeria monocytogenes*, *Campylobacter jejuni*, *Salmonella* spp. *Staphylococcus aureus* and *Shigella* have been detected by PCR (Cheah et al., 2008; Alves et al., 2012; Chiang et al., 2012; Zhou et al., 2013).

7.4.2 MULTIPLEX PCR

This is yet another routinely used method that is rapid and gives more precision in terms of detection. This method employs the use of different sets of primers targeting different regions of the genomic DNA. In this method, proper care should be taken while designing primers since all the primers should have similar annealing temperature. Any variations in the annealing temperatures between primers will lead to missing product (Zhao et al., 2014). Additionally, the concentration of the primers also plays a key role in the efficiency of the product formed because it can form primer-dimer between the primers itself and not anneal to the genomic DNA. This also can lead to a failed reaction and missing product. Because of its high efficiency and reliability, it is now used to detect many pathogens in the food system. In a study carried out by Chen et al. (2012), it was possible to detect the presence of *Salmonella enteritidis*,

Escherichia coli O157:H7, *Shigella flexneri*, *Staphylococcus aureus* and *Listeria monocytogenes* simultaneously using five pairs of primers namely invasion protein (invA), intimin (eaeA) gene, invasion plasmid antigen H (ipaH), 16S rDNA, and listeriolysin O (hlyA), respectively. In another study, Ryu et al. (2013) detected the presence of six *Listeria* spp. using multiplex PCR method.

As an improvement to the multiplex PCR technique, Gene Expression Profiler (GeXP)-PCR technology was developed by Zhou et al. (2013) for the simultaneous detection of six foodborne pathogens namely *Listeria monocytogenes*, *Salmonella enterica*, *Shigella* spp., *Staphylococcus aureus*, *Escherichia coli* O157:H7, and *Campylobacter jejuni*. This method involves PCR amplification followed by capillary electrophoretic separation of the amplified product rather than agarose gel electrophoresis that is normally performed. The primers used are chimeric primers which contain universal tags at the 5′ end, which are also rendered to the amplified product. Further, the universal primers are used that are complementary to the universal tags. This will facilitate the amplification of the desired sequence. The forward sequence of these universal primers is attached to a fluorescent label that facilitates the detection of the amplified product during capillary electrophoretic separation. This was a highly sensitive method for high-throughput analysis with a detection limit of 420, 310, 270, 93, 85, and 66 colony-forming unit/mL (CFU/mL) for *Salmonella*, *Listeria*, *Staphylococcus*, *Escherichia*, and *Shigella* spp., respectively.

7.4.3 REAL-TIME OR QUANTITATIVE PCR (qPCR)

Real-time PCR or qPCR, as it is popularly known, offers a more advanced technology in terms of its detection system, which offers a continuous assessment of the amplification process. In this method, the dideoxy-nucleotide triphosphates (ddNTPs) are labeled with fluorescent probes that are detected continuously by the monitor and the concentration of the product formed is directly proportional to the dye detected (Omiccioli et al., 2009). The fluorescent systems available for the detection in qPCR include TaqMan probes, SYBR green and molecular beacons. TaqMan probes, the double-dye probes, are a short sequence of nucleotides containing fluorophore (reporter dye) at the 5′ end and a quenching dye at the 3′ end (Levin, 2005). Both the dyes are present near each other preventing the emission of the fluorescence of the reporter. The 5′ to 3′ exonuclease activity of the polymerase separates both the dyes during the polymerization process and this in turn is detected as signals during the chain extension step in the PCR reaction (Patel et al., 2006). Conversely, SYBR green is a non-specific intercalating agent that as such emits mild fluorescence, which is in turn enhanced by its binding to the minor groove of the double-helical DNA structure (Law et al., 2015). Contrary to these two dyes, molecular beacons are much different from the above two. It contains oligonucleotide probes possessing hairpin loop and stem structures with complementary regions to the target sequence. Herein, the loop contains the complementary sequence whereas the stem is formed by the annealing of two complementary sequences. Like TaqMan probes, in the case of molecular beacons also there are two dyes namely reporter and quencher dyes which are present near each other and therefore does not fluoresce. When it hybridizes with the complementary sequence in the amplicons, a spontaneous conformational change is induced to the probe that separates the two dyes and thus facilitates fluorescence (Law et al., 2015).

Virtually all the recent studies on pathogenic contamination of food are now carried out using real-time PCR because it gives real-time information of the amplification process. The presence of *Salmonella* in freshly cut fruits and vegetables were detected first using molecular beacons as the qPCR probes in a study carried out by Liming and Bhagwat (2004). Furthermore, using SYBR green probes the tropical shellfish were tested positive for the presence of thermostable direct hemolysin (TDH) gene (tdh-positive) *Vibrio parahaemolyticus* with a detection limit of 10^2 CFU/ml for spiked culture with shrimp homogenate (Tyagi et al., 2009). Similarly, *Staphylococcus aureus* (enterotoxin gene cluster harboring strain) detection was possible in raw milk using SYBR green (10^3 CFU/ml) and TaqMan probes (10^4 CFU/ml).

qPCR assay also has a multiplex option for the detection of the presence of multiple pathogens in a particular food source. Presence of *Shiga* toxin produced by *E. coli* serogroups O26, O45, O103, O111, O121, and O145 were detected by multiplex qPCR in the ground beef samples (Fratamico et al., 2011). Kim et al. (2012) developed multiplex qPCR primers for the detection of *Vibrio parahaemolyticus*, *Vibrio cholerae*, and *Vibrio vulnificus* simultaneously. In a single multiplex qPCR assay, *Salmonella enterica* subsp. *enterica*, *Listeria monocytogenes*, *Escherichia coli* O157,

Vibrio parahaemolyticus, *Vibrio vulnificus*, *Campylobacter jejuni*, *Enterobacter sakazakii* and *Shigella* spp. were detected using specific target primers using molecular beacon probes with a detection limit ranging from 1.3×10^3–1.6×10^4 CFU/g stool (Hu et al., 2014).

The greatest advantage of using qPCR method is the reduced time required for the diagnostic procedure. The laborious and time-consuming steps of assessing the PCR products on the gel are eliminated in the real-time PCR and therefore fasten the process. This also provides the benefit of reduced risks of cross-contamination and automation (Fricker et al., 2007). This has allowed the evolution of various detection kits for the specific detection of pathogens. *Salmonella* BAX™ PCR (DuPont Qualicon), AnDiaTec® *Salmonella* real time PCR Kit (AnDia Tec), Probelia™ *Salmonella* spp. (Sanofi Diagnostics Pasteur), and TaqMan™ *Salmonella* detection kit (PE Applied Biosystems) (Park et al., 2014) are a few of the popular detection kits available in the market. The commercial kits available for the detection of foodborne pathogens are summarized in Table 7.1.

7.4.4 NUCLEIC ACID-BASED AMPLIFICATION (NASBA)

In the year 1991, Compton developed the NASBA based on the amplification of nucleotide sequence under isothermal conditions, rather than the thermocycling for PCR-based amplification. In this procedure, a single-stranded RNA (ssRNA) is first converted into a complementary DNA (cDNA) by the action of reverse transcriptase enzyme at a temperature of around 41°C. The primary enzymes involved in this reaction are avian myeloblastosis virus (AMV) reverse transcriptase, T7 RNA polymerase and RNase H. The amplified product obtained through this reaction can be observed using agarose gel electrophoresis (Zhao et al., 2014). This method is time consuming, labor intensive and not economical owing to the high cost of RNA isolation and cDNA preparation. This has now led to the design of real-time detection systems using fluorescent labeling molecular beacons for the detection of ssRNA itself (Leone et al., 1998). In practice, real-time NASBA has been used for the detection of all those pathogens that are also effectively detected using real-time PCR. Yet, this method has more advantages, it can detect only living cells because it detects the RNA rather than DNA as in the case of PCR-based methods.

TABLE 7.1 List of Commercially Available Kits Used in the Detection of Foodborne Pathogens

Commercial Kit	Assay Format	Target Pathogens	Pre-Enrichment (h)	Total Assay Time	LOD	Company
Assurance®	Solid-phase enzyme immunoassay	*E. coli*, *Listeria*	18–48	2 h	–	Biocontrol
DuPont ™ BAX® System	Real-time PCR methodology	*E. coli*, *Salmonella*, *Vibrio*, *Campylobacter*	24	~1 h	1 CFU per sample, from 25 g up to 375 g	DuPont
GeneQuence	DNA hybridization assay	Simultaneous detection of *Salmonella*, *Listeria* spp., *L. monocytogenes*	24	2 h	1 CFU in 25 g	Neogen
Mericon DNA bacteria kit and *Mericon* bacteria plus kit	Real-time PCR methodology	*S. aureus*, *Campylobacter* spp., *Cronobacter* spp., *Shigella* spp.	20	73 min	10 copies per reaction	Quiagen
Assurance GDS® MPX	IMS+multiplex PCR methodology	*E. coli* O157:H7, O26, O45, O103, O111, O121, O145	10–18	~1 h	–	Biocontrol
DuPont™ Lateral Flow System	Sandwich immunoassay	*Salmonella*, *Listeria*, *E. coli*	20–40	10 min	1 CFU in 25 g	DuPont
VIP® Gold	Sandwich immunoassay	EHEC, *Listeria*, *Salmonella*	8–18	10 min	–	Biocontrol
Atlas® System	Transcription-mediated application	*Listeria*, *Salmonella*, *E. coli*	24	2 h	–	Roka Bioscience
Pathatrix®	IMS+PCR methodology	*E. coli*, *Salmonella*, *Listeria*	7–10	15 min	–	Life Technologies

It should be noted that DNA can be detected in both living and dead cells, whereas RNA is detected in the living cells only and therefore provide the advantage of better detection (Dwivedi and Laykus, 2011). Commercial kits have been made available for NASBA method from manufacturers such as KIT Biomedical Research, Life Sciences, bioMerieux NS GenProbe (Gracias and McKillip, 2007). One of the most popularly available kits for the detection of *Listeria monocytogenes* and *Vibrio cholera* is NucliSENS EasyQ® Basic Kit (Nadal et al., 2007).

7.4.5 LOOP-MEDIATED ISOTHERMAL AMPLIFICATION (LAMP)

LAMP is a novel method based on the auto-cycling strand displacement DNA synthesis using Bst DNA polymerase under a controlled temperature (isothermal) ranging between 59–65°C for a period of 60 min. In this technique, four primers are designed targeting two inner and two outer regions of the genome amplifying six specific regions. The resulting structure appears like a cauliflower having multiple loops and stems varying in size. The advantage of this method is the production of a large amplicons within a span of 60 min. This size is close to 10^3 folds greater than the conventional PCR amplification. The product can be detected using the fluorescent probes such as SYBR green dye (Zhao et al., 2014). The first detected pathogen using this technique is the $stxA_2$ gene of the foodborne pathogen *E. coli* O157:H7 by Maruyama et al. (2003). Ever since, this has gained popularity in the field of food safety and diagnostics of foodborne pathogen. Studies have proved that LAMP results are highly sensitive and rapid therefore well accepted worldwide (Wang et al., 2008). Mori and Motomi (2009) reported the availability of commercial kits for the detection of well-known pathogens such as *Salmonella*, *Campylobacter*, *Listeria*, verotoxin producing *E. coli* and *Legionella* spp. Loopamp is the detection kit available for *Shigella* spp. (Song et al., 2005), *Salmonella enterica* (Ohtsuka et al., 2005), *Campylobacter* (Yamazaki et al., 2009) and enteroinvasive *E. coli* (EIEC) (Song et al., 2005). Further advances in the LAMP assay have led to the design of reverse-transcription, real-time, multiplex, and in situ variants (Law et al., 2015).

7.4.6 OLIGONUCLEOTIDE DNA MICROARRAY

Microarray technology is a recent development in the field of diagnostics and has gained a widespread acceptance for its greater advantage over all other methods. This employs a multi-gene detection system made-up of glass slides that are coated with specific, chemically synthesized oligonucleotide probes of 25–80 bp length (Severgini et al., 2011). Each microarray contains up to hundreds of oligonucleotides that are specific for a segment of the target DNA. The sample nucleic acids (DNA, RNA, or cDNA) are previously labeled using fluorescent dyes and then hybridized with the oligonucleotides coated on the microarray chips. The nucleic acid samples hybridize to the specific and complementary oligonucleotides, sequence of which is already known. Fluorescent signals are detected using specific detectors post hybridizing like the real-time PCR detection system (Lauri and Mariani, 2009).

Using this technology, the first detection of the foodborne pathogen *Shigella* and *E. coli* serotypes were reported by Li et al. (2006). The varying levels of pathogenicity in *E. coli*, harmless strain *Escherichia coli* K-12 to deadly strain *Escherichia coli* O157:H7 (Lauri and Mariani, 2009), suggests the need for the specific identification of the serotype that can further provide information on the type of treatments required. In a study carried out by Wang et al. (2007), about 22 different foodborne pathogens were screened at a single run using this technology. Those pathogens are *Clostridium perfringens*, *Staphylococcus aureus*, *Shigella* spp., *Listeria monocytogenes*, *Vibrio cholerae*, *Campylobacter jejuni*, *Salmonella* spp., *Vibrio parahaemolyticus* and *Bacillus cereus*. Many microarray chips are commercially available but mostly used for gene expression studies. These commercial kits have improved sensitivity and specificity by increasing the number of probes per target sequence, and therefore a single target can bind to different probes (Rasooly and Herold, 2008; Severgnini et al., 2011). Most commercially available microarrays do not satisfy all the requirements of diagnostics because they possess low to medium density probes producing reliable information. In case of specific requirements, customization is offered by many organizations such as the Department of Bioresources at Seibersdorf that design microarray systems according to our needs. StaphyChips® developed by Affymetrix and in collaboration with Advanced Array Technology (ATT, Eppendorf Array Technologies) is one such microarray that can detect close to 15 *Staphylococcus* species including the methicillin-resistant strain (Mothershed and Whitney, 2006). The applications of nucleic acid-based methods are collectively summarized in Table 7.2 (Law et al., 2015).

TABLE 7.2 The Applications of Nucleic Acid-based Methods

Detection Method	Foodborne Pathogens	Detection Limit	Food Matrix	Assay Time
Multiplex PCR	*Salmonella* spp., *Salmonella enteritidis*	103 CFU/mL	Artificially and naturally contaminated chicken carcasses, minas cheese and fresh pork sausages	24 h
	STEC O26, O103, O111, O145, sorbitol fermenting O157 non-sorbitol fermenting O157	5×10^4 CFU/mL in minced beef and sprouted seeds. 5×10^3 CFU/mL in raw-milk cheese.	Artificially contaminated minced beef, sprouted seed (soy, alfalfa, and leek) and raw-milk cheese	24 h
	E. coli O157:H7, *Listeria monocytogenes*, *S. aureus*, *Yersinia enterocolitica*, *Salmonella*	10^3 CFU/mL	Artificially contaminated pork	–
Real-time PCR	*S. enterica*	41.2 fg/PCR for *S. typhimurium* genomic DNA. 18.6 fg/PCR for *S. enteritidis* genomic DNA.	Artificially contaminated chicken, liquid egg, and peanut butter	10 h
	L. monocytogenes, *E. coli* O157, *Salmonella* spp.,	<18 CFU/10 g	Artificially contaminate ground beef; naturally contaminated beef, pork, turkey, and chicken	24 h
	L. monocytogenes, *E. coli* O157:H7, *Salmonella* spp.	2×102 CFU/mL	Artificially contaminated ground pork	24 h
	Salmonella spp., *L. monocytogenes*	5 CFU/25 g	Artificially and naturally contaminated meat, fish, fruits, vegetables, dairy products, eggs, chocolate bar, omelet, lasagna, and various cooked dishes	<30 h

TABLE 7.2 *(Continued)*

Detection Method	Foodborne Pathogens	Detection Limit	Food Matrix	Assay Time
	S. aureus, *Salmonella*, *Shigella*	9.6 CFU/g for *S. aureus*, 2.0 CFU/g for *Salmonella* and 6.8 CFU/g for *Shigella*	Fresh pork	<8 h
NASBA	*E. coli*	40 cells/mL	Drinking water	4 h
	S. enteritidis	101 CFU/ reaction	Artificially contaminated fresh meats, poultry, fish, ready-to-eat salads and bakery products	26 h
	L. monocytogenes	400 CFU/mL	Artificially contaminated cooked ham and smoked salmon slices	72 h
	Bacillus amyloliquefaciens, *B. cereus* and *B. circulens*	–	Artificially contaminated milk	–
	S. enteritidis and *S. typhimurium*	<10 CFU/mL	–	<90 min
LAMP	*Vibrio vulnificus*	5.4 CFU/ reaction for avirulent *V. vulnificus* strain in pure culture. 2.5×10^3 CFU/g for a virulent *V. vulnificus* strain in spiked raw oyster, no enrichment. 1 CFU/g for a virulent *V. vulnificus* strain in spiked raw oyster, after 6 h enrichment.	Artificially contaminated raw oysters	8 h

TABLE 7.2 *(Continued)*

Detection Method	Foodborne Pathogens	Detection Limit	Food Matrix	Assay Time
	Salmonella spp. and *Shigella* spp.	5 CFU/10 mL	Artificially contaminated milk	<20 h
	V. parahaemolyticus	10 CFU/ reaction	Naturally contaminated seafood samples: fish, shrimp, and mussel	16 h
	STECO26, O45, O103, O111, O121, O145 and O157	1–20 cells/ reaction in pure culture and 10^5–10^6 CFU/25 g in produce	Artificially contaminated lettuce, spinach, and sprouts	-
Oligonucleotide DNA microarray	*E. coli* O157:H7, *S. enterica*, *L. monocytogenes* and *C. jejuni*	1×10^{-4} ng for each genomic DNA	Naturally contaminated fresh meat samples: chicken, beef, pork, and turkey	-
	L. monocytogenes	8 log CFU/mL	Artificially contaminated milk	-
	E. coli, *Shigella* spp., *Salmonella* spp., *Proteus* spp., *C. jejuni*, *L. monocytogenes*, *E. faecalis*, *Y. enterocolitica*, *V. parahaemolyticus*, *V. fluvialis*, β-hemolytic *Streptococcus*, *S. aureus*	10 CFU/mL of pure culture	Artificially and naturally contaminated pork, chicken, fish and milk	-

7.5 INTRINSIC FACTORS AFFECTING THE STABILITY OF NUCLEIC ACIDS

Nucleic acids are most vulnerable to spontaneous degradation by slight change in the pH, temperature or even enzymatically. Especially RNA is easily degraded by enzymatic hydrolysis much faster than DNA primarily due to its phosphodiester bonds being less stable (Lindahl, 1993). Yet, DNA undergoes depurination owing to the less stable N-glycosyl bonds and to an extent depyrimidination. Hydrolytic deamination, oxidation, and alkylation are also prospective DNA damages causing aberrant bases leading to mutation of bases. Microbial DNases and RNases secreted by the living or dead cells also cause significant damage to the nucleic acids and thereby hamper detection (Palchevskiy and Finkel, 2009). In environmental samples, the microbial metabolism contributes to a great extent for the degradation of the nucleic acids and therefore is a major factor deciding its stability during detection procedures (Tsai et al., 1995; Deere et al., 1996; Wetz et al., 2004). DNA stability increases under dry conditions, low temperatures and in the absence of microbial activity and oxygen (Lorenz and Wackernagel, 1994). Likewise, detection of foodborne pathogens (inactivated during food processing) using molecular methods depends on the initial concentration of DNA, degree of protection exerted by the dead microbial cells, temperature, food matrix and the interaction between the various microorganisms (both pathogenic and non-pathogenic) present in the food source (Herman, 1997). In the same way, RNA is also prospective of microbial degradation upon treatment under lethal conditions such as high temperature. Yet, the type of RNA determines its susceptibility to degradation. For instance, ribosomal RNAs (rRNAs) are more stable, but the degradation of messenger RNA (mRNA) is determined by the type of gene that it is transcribed from (Cenciarini et al., 2008).

7.6 IMPACT OF FOOD PROCESSING ON THE STABILITY OF NUCLEIC ACIDS

Most foods undergo a few steps of processing for improving its shelf-life. Most of the food processing steps affect the stability of nucleic acids and thus affect detection procedures. These processing methods directly affect the membrane potential of the pathogens by eliminating the structural integrity of the cells thus ensuring the elimination of pathogenic contaminants (Dock and Floros, 2000). Such treatments often lead to significant damage to the macromolecules of the microbial cells including its DNA and RNA composition. In processed foods, a substantial level of nucleic acids is released into the food medium due to the damage to the microbial cells in case of contamination. However, some amount of nucleic acids is degraded by the activity of enzymes, and therefore the quantification needs to be monitored with care to ensure optimal results (Goncalves et al., 2016).

The main intention of food processing is to eliminate the microbial contamination in food sources. To ensure the efficiency of processing, it is validated by culturing the food source on enriched media followed by molecular detection of the pathogen, if any. This culturing using enrichment media is essential to eliminate the false-positive results that can otherwise be encountered owing to the presence of dead cells. Such inactivated cells may possess nucleic acids that can give misleading results and therefore should be eliminated. Therefore, studies carried out using molecular detection in combination with viability assessment are the most optimized methods for detection of foodborne pathogens. However, this can be true for pathogens such as fungal and bacterial contaminants. In the case of viral pathogens, it is often impossible to culture those using conventional methods and its inactivation is also very difficult. In such conditions, the only method of choice is the qPCR for the viral RNA molecules that is sure to provide information on the live cells (Postollec et al., 2011; Knight et al., 2013). Table 7.3 (Ceuppens et al., 2014) provide details of the various food processing methods and their effect on the nucleic acid stability.

TABLE 7.3 Various Food Processing Methods and Their Effect on the Nucleic Acid Stability

Treatment Type	Treatment Conditions	Storage Conditions	Nucleic Acid Detection
DNA	Autoclaving of soybean (15 μg/mL)	–	Three-fold decrease in the abundance of 80 bp target fragments after 20 min of autoclaving, 30-fold decrease after 40 min, 400-fold decrease after 60 min and 80,000-fold decrease after 80 min
	Heating of soybean DNA (15 μg/mL) to 99°C for 1–7 h	–	Severe DNA degradation, but no decrease in the abundance and thus detection of 80 bp target fragments (the mean size of the DNA segments was about 400 bp)

TABLE 7.3 *(Continued)*

Treatment Type	Treatment Conditions	Storage Conditions	Nucleic Acid Detection
	Heating of *E. coli* plasmid DNA pSG100 (480 µg/mL) at 37 and 65°C for 90 min in 10 mM Tris-HCl with pH 4.0 and in tomato serum with pH 4.3	–	Degradation to fragments of <1,754 bp by all treatments
	Heating of *E. coli* plasmid DNA pSG100 (480 µg/mL) at 85°C in 10 mM Tris-HCl with pH 8.4 for 90 min	–	DNA fragments of 1,754 bp persist
	Heating of genomic DNA of transgenic maize Bt176 (48 µg/mL) at 85°C – pH 8.4	–	Degradation to fragments of <1,416 bp after 30 min and 60 min and to fragments of <1,255 bp after 90 min
Micro-waves	Exposure of soybean DNA (15 ng/µL) to microwaves of 800 W for 0–15 min	–	Severe DNA degradation, but no decrease in the abundance of 80 bp target fragments
Sonication	Exposure of soybean DNA (15 ng/µL) to sonication at 170 W for 5 min – 8 h	–	Three-fold decrease in the abundance of 80 bp target fragments and 20-to 30-fold decrease of 170 bp target fragments
Storage	–	Incubation in chicken rinse at 4 and 20°C	Degradation of 90% of the 300 bp fragments after 8 h at 20°C and after 9.5 h at 4°C and degradation of 90% of the 600 bp fragments after 0.5 h at 20°C and after 1.5 h at 4°C
	Genomic and plasmid DNA of *Y. enterocolitica* (between 1 mg/mL and 1 µg/mL)	Incubation in chicken homogenate at 4 and 20°C	Degradation of 90% of the 300 bp fragments after 1 h at 20 and 4°C and degradation of 90% of the 600 bp fragments after 0.5 h at 20°C and after 1.5 h at 4°C

TABLE 7.3 *(Continued)*

Treatment Type	Treatment Conditions	Storage Conditions	Nucleic Acid Detection
	–	Incubation in pork rinse at 4 and 20°C	Degradation of 90% of the 300 bp fragments after 120.5 h at 20°C and degradation of 90% of the 600 bp fragments after 26.5 h at 20°C and after 35 h at 4°C
	–	Incubation in pork homogenate at 4 and 20°C	Degradation of 90% of the 300 bp fragments after 38.5 h at 20°C and degradation of 90% of the 600 bp fragments after 74.5 h at 20°C and after 93.5 h at 4°C
	Degradation of *E. coli* pUC18 plasmid DNA (250 μg/mL) and sugar beet chromosomal DNA (250 μg/mL)	Incubation in raw beet juice at 4, 37, and 70°C	DNA was completely degraded at 70°C within 10 min, while intact DNA persisted for 10 min at 37°C and for ≥20 min at 4°C
	Chromosomal DNA of maize (20 μg/mL)	Incubation in maize silage effluent at ambient temperature	DNA fragments of 1,914 bp persist for 5 min and 684 bp fragments for 60 min
	–	Incubation in ovine saliva at 39°C	Fragments of 850 bp and 1,914 bp persist for 1 hand fragments of 350 bp and 684 bp persist for 24 h
	–	Incubation in ovine rumen fluid at 39°C	Fragments of 1,914, 850, and 350 bp persist for ≤1 min
	Legionella pneumophila DNA (180 ng/mL)	Incubation in sewage effluent at 16°C	DNA fragments of 186 and 108 bp persist for ≥4 d
	Salmonella typhimurium DNA (14–22 ng/mL, corresponding to 5–6 log CFU/mL)	Incubation in seawater at 10 and 20°C	DNA fragments of 284 bp persist for 3 to 8 d at 10°C and for 2 to 4 d at 20°C
	Legionella pneumophila DNA (180 ng/mL)	Incubation in ocean water at 16°C	Degradation of DNA to fragments <186 bp after 2 d and <108 bp after 3 d

TABLE 7.3 (Continued)

Treatment Type	Treatment Conditions	Storage Conditions	Nucleic Acid Detection
	Aeromonas salmonicida DNA (10 ng/mL)	Incubation in freshwater with sandy and loamy sediment at 13°C	Degradation of DNA to fragments of <423 bp after 10 d in water, but DNA fragments of 423 bp persist for 13 weeks in the sandy and loamy sediment
	–	Incubation in loamy sand soil at 23°C	Fragments of 1,029 bp persist for 60 d in loamy sand soil at 3.3% of the initial concentration
	E. coli plasmid pUC8-ISP DNA (0.2 µg/g)	Incubation in clay soil at 23°C	Fragments of 1,029 bp still present after 60 d in clay soil at 11.2% of the initial concentration
	–	Incubation in silty clay soil at 23°C	Degradation of DNA to fragments of <1,029 bp in the silty clay soil after 10 d
	–	Incubation in sterilized (ethylene oxide-treated) soils at 23°C	Fragments of 1,029 bp still present after 60 d in sterilized soil sat 100% of the initial concentration
RNA storage	Free poliovirus RNA (30 ng/mL)	Incubation in seawater at 4 and 23°C	Degradation of RNA to fragments of <394 bp after 3 d at 23°C and after 14 d at 4°C
	–	Incubation in filter-sterilized seawater at 4 and 23°C	Degradation of RNA to fragments of <394 bp after 35 d at 4 and 23°C

7.7 DEALING WITH THE VARIATION BETWEEN CULTURE AND MOLECULAR METHODS

7.7.1 HURDLES IN THE WAY OF ACCURATE CONTAMINATION

In an ideal scenario, if all the false positive and false negative results are eliminated then results from the culture-based method will correspond with that of the molecular methods. However, such an ideal condition is highly unlikely because most definitely under physiological conditions, some amount of negative or failed results can be expected. Generally, unequivocal positive results are obtained when infectious viruses and live bacteria are cultured and assessed using molecular tools. But this does not happen when the desired microorganism is present within a microenvironment, thus challenging the nucleic acid extraction protocol. For instance, when a type of bacteria forms biofilms, internalized as spores or nodule formation within the plant system will interfere with the nucleic acid extraction and thus result in false negatives (DiCaprio et al., 2012). Certain microorganisms also pose difficulty in extracting the nucleic acids because of its rigid cell wall structure like the cell wall of MAP (Donaghy et al., 2008). The best method to counter these setbacks is by optimizing the sample preparation procedure, especially for the Gram negative or positive bacteria present in suspension or in food sources. Therefore, most of the molecular detection methods employ pre-enrichment step to ensure the pathogen concentration at detectable levels. Extremely low concentration of pathogen load can often go undetected and therefore this enrichment step will facilitate in enhancing the detectability of the foodborne pathogen (Jasson et al., 2009).

In addition to the intrinsic difficulties exerted by the pathogen itself, the food processing methods also add to the issues causing injury to the cells, if not its death. Such an injury will lead to elicitation of stress responses and often exhibit a prolonged lag phase on its way towards recovery. In its course of recovery, it might require more stringent conditions such as nutrients and therefore, even the enrichment step before performing the molecular detection may fail to quantify the presence of the pathogen accurately (Wang and Duncan, 2017). For instance, although qPCR quantifies the RNA samples of viruses, it does not specifically describe the infectious, defective, or free RNA samples leading to inconsistencies in the quantification assays. Some other pathogens are tough to be cultured irrespective of their physiological conditions, and therefore the molecular data does not match with the culture results (Joris et al., 2011).

Yet another issue is that the bacterial or viral DNA remains within the system until complete degradation, which may take up to several days. Any assay carried out during this period will end up showing results even for this nucleic acid, which although is ineffective in causing infection but is quantified. For instance, in southern California urban rivers, the water samples were tested for the presence of the pathogenic viruses. About 16% of the samples with non-infectious adenoviruses were reported (Choi and Jiang, 2005). Further, several food processing steps employ bacteriophages to aid in processing to reduce surface contamination and these introduce a large amount of nucleic acids and interfere in accurate quantification (Anon, 2009).

7.7.2 COUNTER STRATEGIES TO IMPROVE NUCLEIC ACID-BASED DETECTION

As stated earlier, all the microorganisms cannot be cultured and hence culturing cannot provide conclusive evidence during detection. This is further exemplified by the presence of human Norovirus which show up differently in different samples such as 3.9% to 59.1% in shellfish (Lowther et al., 2010; Woods and Burkhardt, 2010), 23.3% in cherry tomatoes (Stals et al., 2011), 6.6% to 40.0% in soft red fruits (Baert et al., 2011; Stals et al., 2011), 28.2% to 50.0% in leafy greens (Baert et al., 2011) and 4.2% to 45% in water (Lodder et al., 2010; Borchardt et al., 2012; Allmann et al., 2013). These disparities are fortunately resolved owing to the adaptations available for real-time PCR. One such modification is the amplification of the whole genome and compared with the transfection assays (Baert et al., 2008). In addition, stains that can distinguish between the viable and dead cells such as ethidium monoazide (EMA) or propidium monoazide (PMA) can be tested before the PCR procedure to eliminate the damaged cells or viral capsids from being detected. Spores of yeast and other fungi and several bacterial cells have been accurately quantified using this method (Vesper et al., 2008; Rawsthorne et al., 2009; Taskin et al., 2011; Elizaquivel et al., 2012; Radulovic et al., 2012; Dinu and Bach, 2013; Gensberger et al., 2013). Because these dyes require double-stranded DNA/RNA or single strands with extensive secondary structures, EMA/PMA pre-treatment has been applied with variable success to detect infectious virus particles (Fittipaldi et al., 2010; Parshionikar et al., 2010). Specific receptors that bind the viruses can also be used before the RT-PCR step to eliminate freely present nucleic acids from the detection system. Li et al. (2011) conducted a study using Caco-2 cells, bovine serum albumin and pig gastric mucin before the PCR step, which reduced the human Noroviruses signal up to 1,000 folds. Epidemiological data was generated using these assays in order to set the standard for the allowed concentration of the viral genome that can be present in order to certify its safety (Stals et al., 2013). In yet another study, immuno-capture of the specific viral or bacterial receptors using antibody was also tested. This was indeed useful to eliminate nucleic acids from the detection assays, but this method was not successful in distinguishing infectious and non-infectious types. Stratmann et al. (2006) established peptide-mediated magnetic separation (PMS) prior to PCR detection. Detection of 2.7 log

CFU/mL *Mycobacterium avium* subsp. *paratuberculosis* (MAP) in bulk milk samples by PMS-PCR was successful using a conserved surface protein followed by PCR detection with MAP specific primers. Although all these methods described in this section have proven to be effective, the efficiency of the method is entirely subjective and can be optimal for a particular type of pathogen but fail to successfully detect some others. Therefore, the samples should be individually optimized to effectively quantify the foodborne pathogens using molecular methods.

7.8 CONCLUSION

To ensure stringent compliance and regulatory guidelines, the food and water quality need to be maintained. For this, two major strategies are being employed: improvement of previous existing methods as per the needs for a particular pathogen and development of newer technologies in the field of diagnostics. Food recalls owing to contaminated samples has led to greater economic loss and its associated damage to the reputation of the brand. The various steps at which contamination can likely occur are farming, production, processing, packaging, distribution, and consumption. The disease outbreaks that have caused hospitalization and mortality of affected patients has raised great concerns regarding the suitability of these food sources for consumption. Therefore, methods that can detect the presence of foodborne pathogen are extremely important for assessment of the food quality. Although culture-based methods have proven a gold standard for the detection of specific pathogens, they have not been widely used for various drawbacks. Availability of nucleic acid-based detection methods for the assessment of foodborne pathogens without the need to culture the microorganisms has paved newer avenues in food microbiology for a faster and more economical diagnostics. Conventional PCR, multiplex PCR, qPCR is available for the detection of the pathogen but required high skills for the handling and execution of the assays. NASBA and LAMP are relatively easier to be executed and helpful in the detection of the pathogens as well as their toxins and therefore provide a deeper knowledge.

KEYWORDS

- **detection and diagnosis**
- **DNA microarray**
- **food safety**
- **foodborne pathogens**
- **loop-mediated isothermal amplification**
- **molecular techniques**
- **nucleic acid-based amplification**
- **nucleic acids**
- **polymerase chain reaction**

REFERENCES

Allmann, E., Pan, L., Li, L., Li, D., Wang, S., & Lu, Y., (2013). Presence of enteroviruses in recreational water in Wuhan, China. *Journal of Virological Methods, 193*(2), 327–331.

Alum, E. A., Urom, S. M. O. C., & Ben, C. M. A., (2016). Microbiological contamination of food: The mechanisms, impacts and prevention. *International Journal of Science and Technology and Research, 5*(3), 65–78.

Alves, J., Marques, V. V., Pereira, L. F. P., Hirooka, E. Y., & De Oliveira, T. C. R. M., (2012). Multiplex PCR for the detection of *Campylobacter* spp. and *Salmonella* spp. in chicken meat. *Journal of Food Safety, 32*(3), 345–350.

Anon, A., (2009). The use and mode of action of bacteriophages in food production; scientific opinion of the panel on biological hazards. The *EFSA Journal*, 1076, 26.

Ashbolt, N. J., (2004). Microbial contamination of drinking water and disease outcomes in developing regions. *Toxicology, 198*(1–3), 229–238.

Baert, L., Mattison, K., Loisy-Hamon, F., et al., (2011). Norovirus prevalence in Belgian, Canadian and French fresh produce: A threat to human health? *International Journal of Food Microbiology, 151*(3), 261–269.

Baert, L., Wobus, C. E., Van, C. E., Thackray, L. B., Debevere, J., & Uyttendaele, M., (2008). Detection of murine norovirus 1 by using plaque assay, transfection assay, and real-time reverse transcription-PCR before and after heat exposure. *Applied and Environmental Microbiology, 74*(2), 543–546.

Baraketi, A., Salmieri, S., & Lacroix, M., (2018). Foodborne pathogens detection: Persevering worldwide challenge. In: Rinken, T., & Kivirand, K., (eds.), *Biosensing Technologies for the Detection of Pathogens: A Prospective Way for Rapid Analysis*, IntechOpen, London.

Batz, M., Hoffmann, S., & Morris, Jr. J. G., (2014). Disease-outcome trees, EQ-5D scores, and estimated annual losses of quality-adjusted life years (QALYs) for 14 foodborne pathogens in the United States. *Foodborne Pathogens and Disease, 11*(5), 395–402.
Beatson, S. A., & Walker, M. J., (2014). Tracking antibiotic resistance. *Science, 345*(6203), 1454–1455.
Borch, E., Kant-Muermans, M. L., & Blixt, Y., (1996). Bacterial spoilage of meat and cured meat products. *International Journal of Food Microbiology, 33*(1), 103–120.
Borchardt, M. A., Spencer, S. K., Kieke, Jr. B. A., Lambertini, E., & Loge, F. J., (2012). Viruses in non-disinfected drinking water from municipal wells and community incidence of acute gastrointestinal illness. *Environmental Health Perspectives, 120*(9), 1272–1279.
Brandao, D., Liebana, S., & Pividori, M. I., (2015). Multiplexed detection of foodborne pathogens based on magnetic particles. *New Biotechnology, 32*(5), 511–520.
Callaway, T. R., Edrington, T. S., Loneragan, G. H., Carr, M. A., & Nisbet, D. J., (2013). Shiga toxin-producing *Escherichia coli* (STEC) ecology in cattle and management based options for reducing fecal shedding. *Agriculture, Food and Analytical Bacteriology, 3*(2), 39.
Cenciarini-Borde, C., Courtois, S., & La Scola, B., (2009). Nucleic acids as viability markers for bacteria detection using molecular tools. *Future Microbiology, 4*(1), 45–64.
Ceuppens, S., Li, D., Uyttendaele, M., et al., (2014). Molecular methods in food safety microbiology: interpretation and implications of nucleic acid detection. *Comprehensive Reviews in Food Science and Food Safety, 13*(4), 551–577.
Cheah, Y. K., Salleh, N. A., Lee, L. H., Radu, S., Sukardi, S., & Sim, J. H., (2008). Comparison of PCR fingerprinting techniques for the discrimination of *Salmonella enterica* subsp. *enterica* serovar Weltevreden isolated from indigenous vegetables in Malaysia. *World Journal of Microbiology and Biotechnology, 24*(3), 327.
Chiang, Y. C., Tsen, H. Y., Chen, H. Y., et al., (2012). Multiplex PCR and a chromogenic DNA macroarray for the detection of *Listeria monocytogenes*, *Staphylococcus aureus*, *Streptococcus agalactiae*, *Enterobacter sakazakii*, *Escherichia coli* O157: H7, *Vibrio parahaemolyticus*, *Salmonella* spp. and *Pseudomonas fluorescens* in milk and meat samples. *Journal of Microbiological Methods, 88*(1), 110–116.
Choi, S., & Jiang, S. C., (2005). Real-time PCR quantification of human adenoviruses in urban rivers indicates genome prevalence but low infectivity. *Applied and Environmental Microbiology, 71*(11), 7426–7433.
De Reu, K., Messens, W., Heyndrickx, M., Rodenburg, T. B., Uyttendaele, M., & Herman, L., (2008). Bacterial contamination of table eggs and the influence of housing systems. *World's Poultry Science Journal, 64*(1), 5–19.
DebRoy, C., Roberts, E., Valadez, A. M., Dudley, E. G., & Cutter, C. N., (2011). Detection of Shiga toxin-producing *Escherichia coli* O26, O45, O103, O111, O113, O121, O145, and O157 serogroups by multiplex polymerase chain reaction of the wzx gene of the O-antigen gene cluster. *Foodborne Pathogens and Disease, 8*(5), 651–652.
Deere, D., Porter, J., Pickup, R. W., & Edwards, C., (1996). Survival of cells and DNA of *Aeromonas salmonicida* released into aquatic microcosms. *Journal of Applied Bacteriology, 81*(3), 309–318.
DiCaprio, E., Ma, Y., Purgianto, A., Hughes, J., & Li, J., (2012). Internalization and dissemination of human norovirus and animal caliciviruses in hydroponically grown romaine lettuce. *Applied and Environmental Microbiology, 78*(17), 6143–6152.

Dinu, L. D., & Bach, S., (2013). Detection of viable but non-culturable *Escherichia coli* O157: H7 from vegetable samples using quantitative PCR with propidium monoazide and immunological assays. *Food Control, 31*(2), 268–273.

Dock, L. L., & Floros, J. D., (2000). Thermal and non-thermal preservation of foods. In *Essentials of Functional Foods*. 49–89. Aspen Publishers, Inc., Gaithersburg, MD.

Donaghy, J. A., Rowe, M. T., Rademaker, J. L. W., et al., (2008). An inter-laboratory ring trial for the detection and isolation of *Mycobacterium avium* subsp. *paratuberculosis* from raw milk artificially contaminated with naturally infected faeces. *Food Microbiology, 25*(1), 128–135.

Douarre, P. E., Cashman, W., Buckley, J., Coffey, A., & O'Mahony, J. M., (2010). Isolation and detection of *Mycobacterium avium* subsp. *paratuberculosis* (MAP) from cattle in Ireland using both traditional culture and molecular based methods. *Gut. Pathogens, 2*(1), 11.

Dunkley, C. S., & Dunkley, K. D., (2013). Greenhouse gas emissions from livestock and poultry. *Agriculture, Food and Analytical Bacteriology, 3*(1), 17–29.

Dwivedi, H. P., & Jaykus, L. A., (2011). Detection of pathogens in foods: The current state-of-the-art and future directions. *Critical Reviews in Microbiology, 37*(1), 40–63.

Elizaquível, P., Sánchez, G., & Aznar, R., (2012). Quantitative detection of viable foodborne *E. coli* O157: H7, *Listeria monocytogenes* and *Salmonella* in fresh-cut vegetables combining propidium monoazide and real-time PCR. *Food Control, 25*(2), 704–708.

Ferreira, V., Wiedmann, M., Teixeira, P., & Stasiewicz, M. J., (2014). *Listeria monocytogenes* persistence in food-associated environments: Epidemiology, strain characteristics, and implications for public health. *Journal of Food Protection, 77*(1), 150–170.

Fittipaldi, M., Rodriguez, N. J. P., Codony, F., Adrados, B., Peñuela, G. A., & Morató, J., (2010). Discrimination of infectious bacteriophage T4 virus by propidium monoazide real-time PCR. *Journal of Virological Methods, 168*(1, 2), 228–232.

Foley, S. L., Johnson, T. J., Ricke, S. C., Nayak, R., & Danzeisen, J., (2013). *Salmonella* pathogenicity and host adaptation in chicken-associated serovars. *Microbiology and Molecular Biology Reviews, 77*(4), 582–607.

Fratamico, P. M., Bagi, L. K., Cray, Jr. W. C., et al., (2011). Detection by multiplex real-time polymerase chain reaction assays and isolation of Shiga toxin-producing *Escherichia coli* serogroups O26, O45, O103, O111, O121, and O145 in ground beef. *Foodborne Pathogens and Disease, 8*(5), 601–607.

Fricker, M., Messelhäußer, U., Busch, U., Scherer, S., & Ehling-Schulz, M., (2007). Diagnostic real-time PCR assays for the detection of emetic *Bacillus cereus* strains in foods and recent foodborne outbreaks. *Applied and Environmental Microbiology, 73*(6), 1892–1898.

Gensberger, E. T., Sessitsch, A., & Kostić, T., (2013). Propidium monoazide–quantitative polymerase chain reaction for viable *Escherichia coli* and *Pseudomonas aeruginosa* detection from abundant background microflora. *Analytical Biochemistry, 441*(1), 69–72.

Gonçalves, H. M., Moreira, L., Pereira, L., et al., (2016). Biosensor for label-free DNA quantification based on functionalized LPGs. *Biosensors and Bioelectronics, 84*, 30–36.

Gracias, K. S., & McKillip, J. L., (2007). Nucleic acid sequence-based amplification (NASBA) in molecular bacteriology: A procedural guide. *Journal of Rapid Methods and Automation in Microbiology, 15*(3), 295–309.

Green, L. R., & Selman, C., (2005). Factors impacting food workers' and managers' safe food preparation practices: A qualitative study. *Food Protection Trends, 25*(12), 981–990.

Hellweg, S., & Canals, L. M., (2014). Emerging approaches, challenges and opportunities in life cycle assessment. *Science, 344*(6188), 1109–1113.

Herman, L., (1997). Detection of viable and dead *Listeria monocytogenes* by PCR. *Food Microbiology, 14*(2), 103–110.

Hoekstra, A. Y., & Wiedmann, T. O., (2014). Humanity's unsustainable environmental footprint. *Science, 344*(6188), 1114–1117.

Hu, Q., Lyu, D., Shi, X., et al., (2014). A modified molecular beacons–based multiplex real-time PCR assay for simultaneous detection of eight foodborne pathogens in a single reaction and its application. *Foodborne Pathogens and Disease, 11*(3), 207–214.

Ingram, M., (1952). Internal bacterial taints ('bone taint' or 'souring') of cured pork legs. *Epidemiology and Infection, 50*(2), 165–181.

ISO, (2012). Microbiology of food and animal feed. Real-time polymerase chain reaction (PCR)-based method for the detection of foodborne pathogens. Horizontal method for the detection of Shiga toxin-producing *Escherichia coli* (STEC) and the determination of O157, O111, O26, O103 and O145 serogroups. *ISO Norm. ISO/TS 13136*. https://www.iso.org/standard/53328.html (accessed on 20 December 2021).

Jasson, V., Rajkovic, A., Debevere, J., & Uyttendaele, M., (2009). Kinetics of resuscitation and growth of *L. monocytogenes* as a tool to select appropriate enrichment conditions as a prior step to rapid detection methods. *Food Microbiology, 26*(1), 88–93.

Joris, M. A., Verstraete, K., De Reu, K., & De Zutter, L., (2011). Loss of vtx genes after the first subcultivation step of verocytotoxigenic *Escherichia coli* O157 and non-O157 during isolation from naturally contaminated fecal samples. *Toxins, 3*(6), 672–677.

Joseph, B., Otta, S. K., Karunasagar, I., & Karunasagar, I., (2001). Biofilm formation by *Salmonella* spp. on food contact surfaces and their sensitivity to sanitizers. *International Journal of Food Microbiology, 64*(3), 367–372.

Knight, A., Li, D., Uyttendaele, M., & Jaykus, L. A., (2013). A critical review of methods for detecting human noroviruses and predicting their infectivity. *Critical Reviews in Microbiology, 39*(3), 295–309.

Kumi, J., Mitchell, N. J., Asare, G. A., et al., (2014). Aflatoxins and fumonisins contamination of home-made food (weanimix) from cereal-legume blends for children. *Ghana Medical Journal, 48*(3), 121–126.

Lauri, A., & Mariani, P. O., (2009). Potentials and limitations of molecular diagnostic methods in food safety. *Genes and Nutrition, 4*(1), 1–12.

Law, J. W. F., Ab Mutalib, N. S., Chan, K. G., & Lee, L. H., (2015). Rapid methods for the detection of foodborne bacterial pathogens: Principles, applications, advantages and limitations. *Frontiers in Microbiology, 5*, 770.

Lawley, R., (2013a). Aflatoxins. *Food Safety Watch*. Scientific Research.

Lawley, R., (2013b). Ochratoxins. *Food Safety Watch*. Scientific Research.

Learn-Han, L., Yoke-Kqueen, C., Salleh, N. A., et al., (2008). Analysis of *Salmonella agona* and *Salmonella* Weltevreden in Malaysia by PCR fingerprinting and antibiotic resistance profiling. *Antonie Van Leeuwenhoek, 94*(3), 377.

Lee, N., Kwon, K. Y., Oh, S. K., Chang, H. J., Chun, H. S., & Choi, S. W., (2014). A multiplex PCR assay for simultaneous detection of *Escherichia coli* O157: H7, *Bacillus cereus*, *Vibrio parahaemolyticus*, *Salmonella* spp., *Listeria monocytogenes*, and

Staphylococcus aureus in Korean ready-to-eat food. *Foodborne Pathogens and Disease, 11*(7), 574–580.

Leone, G., Van, G. B., Schoen, C. D., Van, S. H., & Kramer, F. R., (1998). Molecular beacon probes combined with amplification by NASBA enable homogeneous, real-time detection of RNA. *Nucleic Acids Research, 26*(9), 2150–2155.

Levin, R. E., (2004). The application of real-time PCR to food and agricultural systems: A review. *Food Biotechnology, 18*(1), 97–133.

Li, D., Baert, L., Van, C. E., & Uyttendaele, M., (2011). Critical studies on binding-based RT-PCR detection of infectious noroviruses. *Journal of Virological Methods, 177*(2), 153–159.

Li, R., Lai, J., Wang, Y., et al., (2013). Prevalence and characterization of *Salmonella* species isolated from pigs, ducks and chickens in Sichuan Province, China. *International Journal of Food Microbiology, 163*(1), 14–18.

Li, X., Zai, J., Zhao, Q., et al., (2020). Evolutionary history, potential intermediate animal host, and cross-species analyses of SARS-CoV-2. *Journal of Medical Virology, 92*(6), 602–611.

Li, Y., Liu, D., Cao, B., et al., (2006). Development of a serotype-specific DNA microarray for identification of some *Shigella* and pathogenic *Escherichia coli* strains. *Journal of Clinical Microbiology, 44*(12), 4376–4383.

Lindahl, T., (1993). Instability and decay of the primary structure of DNA. *Nature, 362*(6422), 709–715.

Lodder, W. J., Van, D. B. H. H. J. L., Rutjes, S. A., & De Roda, H. A. M., (2010). Presence of enteric viruses in source waters for drinking water production in The Netherlands. *Applied and Environmental Microbiology, 76*(17), 5965–5971.

Lorenz, M. G., & Wackernagel, W., (1994). Bacterial gene transfer by natural genetic transformation in the environment. *Microbiology and Molecular Biology Reviews, 58*(3), 563–602.

Lowther, J. A., Avant, J. M., Gizynski, K., Rangdale, R. E., & Lees, D. N., (2010). Comparison between quantitative real-time reverse transcription PCR results for norovirus in oysters and self-reported gastroenteric illness in restaurant customers. *Journal of Food Protection, 73*(2), 305–311.

Lungu, B., Ricke, S. C., & Johnson, M. G., (2009). Growth, survival, proliferation and pathogenesis of *Listeria monocytogenes* under low oxygen or anaerobic conditions: A review. *Anaerobe, 15*(1, 2), 7–17.

Mandal, P. K., Biswas, A. K., Choi, K., & Pal, U. K., (2011). Methods for rapid detection of foodborne pathogens: An overview. *American Journal of Food Technology, 6*(2), 87–102.

Maruyama, F., Kenzaka, T., Yamaguchi, N., Tani, K., & Nasu, M., (2003). Detection of bacteria carrying the stx2 gene by in situ loop-mediated isothermal amplification. *Applied and Environmental Microbiology, 69*(8), 5023–5028.

Mench, J. A., Sumner, D. A., & Rosen-Molina, J. T., (2011). Sustainability of egg production in the United States - The policy and market context. *Poultry Science, 90*(1), 229–240.

Mori, Y., & Notomi, T., (2009). Loop-mediated isothermal amplification (LAMP): A rapid, accurate, and cost-effective diagnostic method for infectious diseases. *Journal of Infection and Chemotherapy, 15*(2), 62–69.

Mothershed, E. A., & Whitney, A. M., (2006). Nucleic acid-based methods for the detection of bacterial pathogens: Present and future considerations for the clinical laboratory. *Clinica Chimica Acta, 363*(1, 2), 206–220.

Nadal, A., Coll, A., Cook, N., & Pla, M., (2007). A molecular beacon-based real time NASBA assay for detection of *Listeria monocytogenes* in food products: Role of target mRNA secondary structure on NASBA design. *Journal of Microbiological Methods, 68*(3), 623–632.

Njobeh, P. B., Dutton, M. F., Koch, S. H., Chuturgoon, A., Stoev, S., & Seifert, K., (2009). Contamination with storage fungi of human food from Cameroon. *International Journal of Food Microbiology, 135*(3), 193–198.

O'Bryan, C. A., Crandall, P. G., & Ricke, S. C., (2008). Organic poultry pathogen control from farm to fork. *Foodborne Pathogens and Disease, 5*(6), 709–720.

Oliver, S. P., Jayarao, B. M., & Almeida, R. A., (2005). Foodborne pathogens in milk and the dairy farm environment: food safety and public health implications. *Foodborne Pathogens and Disease, 2*(2), 115–129.

Omiccioli, E., Amagliani, G., Brandi, G., & Magnani, M., (2009). A new platform for real-time PCR detection of *Salmonella* spp., *Listeria monocytogenes* and *Escherichia coli* O157 in milk. *Food Microbiology, 26*(6), 615–622.

Palchevskiy, V., & Finkel, S. E., (2009). A role for single-stranded exonucleases in the use of DNA as a nutrient. *Journal of Bacteriology, 191*(11), 3712–3716.

Park, S. H., Aydin, M., Khatiwara, A., et al., (2014). Current and emerging technologies for rapid detection and characterization of *Salmonella* in poultry and poultry products. *Food Microbiology, 38*, 250–262.

Parshionikar, S., Laseke, I., & Fout, G. S., (2010). Use of propidium monoazide in reverse transcriptase PCR to distinguish between infectious and non-infectious enteric viruses in water samples. *Applied and Environmental Microbiology, 76*(13), 4318–4326.

Patel, J. R., Bhagwat, A. A., Sanglay, G. C., & Solomon, M. B., (2006). Rapid detection of *Salmonella* from hydrodynamic pressure-treated poultry using molecular beacon real-time PCR. *Food Microbiology, 23*(1), 39–46.

Payne, D. C., Sharapov, U., Hall, A. J., & Hu, D. J., (2012). Foodborne viruses. In: Oyarzabal, O. A., & Backert, S., (eds.), *Microbial Food Safety: An Introduction* (pp. 73–92). New York: Springer Science+Business Media.

Pelletier, N., Ibarburu, M., & Xin, H., (2014). Comparison of the environmental footprint of the egg industry in the United States in 1960 and 2010. *Poultry Science, 93*(2), 241–255.

Postollec, F., Falentin, H., Pavan, S., Combrisson, J., & Sohier, D., (2011). Recent advances in quantitative PCR (qPCR) applications in food microbiology. *Food Microbiology, 28*(5), 848–861.

Priyanka, B., Patil, R. K., & Dwarakanath, S., (2016). A review on detection methods used for foodborne pathogens. *The Indian Journal of Medical Research, 144*(3), 327.

Radulović, Z., Mirković, N., Bogović-Matijašič, B., Petrušić, M., Petrović, T., Manojlović, V., & Nedović, V., (2012). Quantification of viable spray-dried potential probiotic lactobacilli using real-time PCR. *Archives of Biological Sciences, 64*(4), 1465–1472.

Rasooly, A., & Herold, K. E., (2008). Food microbial pathogen detection and analysis using DNA microarray technologies. *Foodborne Pathogens and Disease, 5*(4), 531–550.

Rawsthorne, H., Dock, C. N., & Jaykus, L. A., (2009). PCR-based method using propidium monoazide to distinguish viable from nonviable *Bacillus subtilis* spores. *Applied and Environmental Microbiology, 75*(9), 2936–2939.

Ricke, S. C., (2014). Application of molecular approaches for understanding foodborne *Salmonella* establishment in poultry production. *Advances in Biology, 2014*. https://doi.org/10.1155/2014/813275.

Ryu, J., Park, S. H., Yeom, Y. S., et al., (2013). Simultaneous detection of *Listeria* species isolated from meat processed foods using multiplex PCR. *Food Control, 32*(2), 659–664.

Scallan, E., Hoekstra, R. M., Angulo, F. J., et al., (2011). Foodborne illness acquired in the United States - major pathogens. *Emerging Infectious Diseases, 17*(1), 7–15.

Scharff, R. L., (2012). Economic burden from health losses due to foodborne illness in the United States. *Journal of Food Protection, 75*(1), 123–131.

Severgnini, M., Cremonesi, P., Consolandi, C., De Bellis, G., & Castiglioni, B., (2011). Advances in DNA microarray technology for the detection of foodborne pathogens. *Food and Bioprocess Technology, 4*(6), 936–953.

Slana, I., Paolicchi, F., Janstova, B., Navratilova, P., & Pavlik, I., (2008). Detection methods for *Mycobacterium avium* subsp. *paratuberculosis* in milk and milk products: A review. *Veterinární Medicína, 53*(6), 283.

Soni, K. A., Nannapaneni, R., & Tasara, T., (2011). The contribution of transcriptomic and proteomic analysis in elucidating stress adaptation responses of *Listeria monocytogenes*. *Foodborne Pathogens and Disease, 8*(8), 843–852.

Spector, M. P., & Kenyon, W. J., (2012). Resistance and survival strategies of *Salmonella enterica* to environmental stresses. *Food Research International, 45*(2), 455–481.

Stals, A., Baert, L., Jasson, V., Van, C. E., & Uyttendaele, M., (2011). Screening of fruit products for norovirus and the difficulty of interpreting positive PCR results. *Journal of Food Protection, 74*(3), 425–441.

Stals, A., Van Coillie, E., & Uyttendaele, M., (2013). Viral genes everywhere: Public health implications of PCR-based testing of foods. *Current Opinion in Virology, 3*(1), 69–73.

Stoessel, F., Juraske, R., Pfister, S., & Hellweg, S., (2012). Life cycle inventory and carbon and water foodprint of fruits and vegetables: Application to a Swiss retailer. *Environmental Science and Technology, 46*(6), 3253–3262.

Stratmann, J., Dohmann, K., Heinzmann, J., & Gerlach, G. F., (2006). Peptide aMptD-mediated capture PCR for detection of *Mycobacterium avium* subsp. *paratuberculosis* in bulk milk samples. *Applied and Environmental Microbiology, 72*(8), 5150–5158.

Taskin, B., Gozen, A. G., & Duran, M., (2011). Selective quantification of viable *Escherichia coli* bacteria in biosolids by quantitative PCR with propidium monoazide modification. *Applied and Environmental Microbiology, 77*(13), 4329–4335.

Tsai, Y. L., Tran, B., & Palmer, C. J., (1995). Analysis of viral RNA persistence in seawater by reverse transcriptase-PCR. *Applied and Environmental Microbiology, 61*(1), 363–366.

Tyagi, A., Saravanan, V., Karunasagar, I., & Karunasagar, I., (2009). Detection of *Vibrio parahaemolyticus* in tropical shellfish by SYBR green real-time PCR and evaluation of three enrichment media. *International Journal of Food Microbiology, 129*(2), 124–130.

Van De, V. T., (2000). Emerging foodborne diseases: A global responsibility. *Food, Nutrition and Agriculture, 26*, 4–13.

Van, H. R., & Michiels, C. W., (2010). Biofilm formation and the food industry, a focus on the bacterial outer surface. *Journal of Applied Microbiology, 109*(4), 1117–1131.

Vesper, S., McKinstry, C., Hartmann, C., Neace, M., Yoder, S., & Vesper, A., (2008). Quantifying fungal viability in air and water samples using quantitative PCR after treatment with propidium monoazide (PMA). *Journal of Microbiological Methods, 72*(2), 180–184.

Vidic, J., Vizzini, P., Manzano, M., et al., (2019). Point-of-need DNA testing for detection of foodborne pathogenic bacteria. *Sensors, 19*(5), 1100.

Wang, X. W., Zhang, L., Jin, L. Q., et al., (2007). Development and application of an oligonucleotide microarray for the detection of foodborne bacterial pathogens. *Applied Microbiology and Biotechnology, 76*(1), 225–233.

Wang, Y., & Duncan, T. V., (2017). Nanoscale sensors for assuring the safety of food products. *Current Opinion in Biotechnology, 44*, 74–86.

Wetz, J. J., Lipp, E. K., Griffin, D. W., et al., (2004). Presence, infectivity, and stability of enteric viruses in seawater: Relationship to marine water quality in the Florida Keys. *Marine Pollution Bulletin, 48*(7, 8), 698–704.

Windhorst, H. W., (2008). A projection of the regional development of egg production until (2015). *World's Poultry Science Journal, 64*(3), 356–376.

Wingstrand, A., Neimann, J., Engberg, J., et al., (2006). Fresh chicken as main risk factor for campylobacteriosis, Denmark. *Emerging Infectious Diseases, 12*(2), 280.

Woods, J. W., & Burkhardt, W., (2010). Occurrence of norovirus and hepatitis A virus in US oysters. *Food and Environmental Virology, 2*(3), 176–182.

Yamazaki, W., Taguchi, M., Kawai, T., et al., (2009). Comparison of loop-mediated isothermal amplification assay and conventional culture methods for detection of *Campylobacter jejuni* and *Campylobacter coli* in naturally contaminated chicken meat samples. *Applied and Environmental Microbiology, 75*(6), 1597–1603.

Zhao, X., Lin, C. W., Wang, J., & Oh, D. H., (2014). Advances in rapid detection methods for foodborne pathogens. *Journal of Microbiology and Biotechnology, 24*(3), 297–312.

Zhou, B., Xiao, J., Liu, S., et al., (2013). Simultaneous detection of six foodborne pathogens by multiplex PCR with a GeXP analyzer. *Food Control, 32*(1), 198–204.

Part III

Detecting and Tracing Foodborne Pathogens Using NGS and WGS

CHAPTER 8

Foodborne Pathogen Detection Using Next-Generation Sequencing

CHARLES OLUWASEUN ADETUNJI, OLUGBEMI TOPE OLANIYAN, DEVARAJAN THANGADURAI, OLUGBENGA SAMUEL MICHAEL, BENJAMIN EWA UBI, EUGENE AYODELE AYENI, MUHAMMAD AKRAM, JULIANA BUNMI ADETUNJI, UMME LAILA, AJAYI KOLAWOLE TEMIDAYO, WADZANI DAUDA PALNAM, SAHER ISLAM, KAYODE ADEBISI AROWORA, HINA ANWAR, and DANIEL INGO HEFFT

ABSTRACT

Next-generation sequencing (NGS) has been identified as a powerful molecular approach in revolutionizing food microbiology, most especially for effective identification of foodborne pathogens. This technology is very effective to reduce the burden of pathogenic infections from the health of humans. NGS shows highly throughputs which synthesize millions of sequences in a single time or same time. The role of these sequences is very effective for the detection of taxa of microbes or those organisms which are un-cultivable or exist in less amount. The use of NGS in several fields is vast including identification of various foodborne pathogenic microorganisms. Therefore, this chapter provides detailed information on the application of NGS as a powerful bioinformatic technique for detection and maintenance of food safety and assurance of high level of food quality that could be used in feeding the ever-increasing population globally.

Global Food Safety: Microbial Interventions and Molecular Advancements. Saher Islam, PhD, Devarajan Thangadurai, PhD, Jeyabalan Sangeetha, PhD, & Natália Cruz-Martins, PhD (Editors)

8.1 INTRODUCTION

First-generation sequencing included a number of approaches that were used to obtain sequences of DNA with a low throughput. A part of this was Sanger sequencing, developed by Frederick Sanger whereby the exact nucleotide sequence of a desired DNA sample could be deciphered. The method was first described in 1977 and consisted of a chain termination strategy using dideoxynucleotide triphosphates (ddNTPs) (Sanger et al., 1977; Heather and Chain, 2016). The basis of the procedure was employed for DNA sequencing in the Human Genome Project, which was obviously time consuming and had a poor quality of beginning sequences. Due to these setbacks, there was a need for more economically friendly and rapid approaches in clinical laboratories and in research settings, which has led to the development of next generation sequencing (NGS). NGS is an approach that allows the sequencing of up to millions of DNA fragments via a high-throughput approach (Lee et al., 2019).

One of the important and effective methods which are used for the analysis, identification, and detection of pathogenic microorganisms is high throughput sequencing or NGS technology. The beneficial effects compared to conventional methods are the production of sequences which are not only used for the precise identification and characterization of pathogens. Moreover, it may also be applied for the screening of mutation resistant genes and plays a role in pathogenicity and virulence factors (Wensing et al., 2005; Lemey et al., 2014; Salje et al., 2017; Schmidt et al., 2017).

Generally, in all over the world the application of bioinformatics and NGS methods have become valuable and effective in research most especially those related to public health due to their lower rates of error. Therefore, in order to achieve good, effective, and accurate information, there is a need to acquire a lot of expertise because NGS technique is one of the complex technologies (Gargis et al., 2016). Moreover, the use of NGS has played a greater role towards a successful reduction in the level of foodborne illness (FBI) that might affect human health (Motro and Moran-Gilad, 2017). Therefore, this chapter intends to provide detailed information on the application of NGS as a powerful molecular approach in revolutionizing food microbiology.

8.2 GENERAL OVERVIEW ON NEXT GENERATION SEQUENCING (NGS)

NGS has been applied for several routine applications such as diagnostics, investigation of disease outbreaks, forensics, antimicrobial resistance (AMR), and food safety (Jagadeesan et al., 2019). It has been discovered that the application of NGS coupled with single nucleotide polymorphisms (SNPs), could be applied for effective detection of virulence genes, resistance to antibiotics, or other important features can be elucidated. Therefore, NGS is rapidly replacing other molecular techniques utilized for the characterization of foodborne outbreaks as well as for surveillance purposes (Ronholm, 2018). Whole-genome sequencing (WGS) approaches via NGS techniques could enable researchers to characterize a microbe for epidemiological investigations with a remarkable level of speed and accuracy.

For food applications, NGS is the first method that enables the complete detection of the microbiota of a food specimen with millions of reads of heterogeneous DNA fragments, unlike the PCR that detects a few species among the entire microbiota. The reads are basically composed of 16S rRNA gene fragments of bacteria that are afterwards taxonomically assigned and analyzed using bioinformatics (Ceuppens, 2014). Several studies have employed this approach for the better understanding of food microbial ecology, improving fermentation activities, and screening for potential pathogens to ensure food safety (Quigley et al., 2012; Loman et al., 2013; Miralles et al., 2019; Tatsika et al., 2019).

NGS technologies allow sequencing of isolates via high-throughput sequencing technology either by the WGS of a particular microbe strain/ isolate or the gene-specific identification of the several microbes present in a specimen for investigation. Several commercially available technologies that are used include Illumina, Life Technologies, which may be represented as Thermo Fisher or Ion Torrent, Oxford Nanopore Technologies, and Pacific Biosciences. At present, Illumina produces a collection of sequencers (MiSeq, NextSeq 500 and HiSeq series) optimized for varying levels of throughputs and turnaround times. MiSeq is a fast, targeted sequence and small genomes sequencing technology with run times are as low as 4 hours. On the other hand, HiSeq 2500, a newer model of the HiSeq, is adapted for applications that are capable of yielding up to 1 Tb in 6 days. The sequencing and bioinformatics analysis of a study using HiSeq 2500 involving Shiga toxin-producing *E. coli* O157:H7 was done by sequencing whole genomes of single isolates and subsequently mapping to a reference genome and analysis for SNP. NextSeq 500 is a fast benchtop sequencer for individual labs that is capable of producing 120 Gb, or a single 30× genome, in less than 30 hours. It employs a novel sequencing strategy of two channels which reduces data processing times and increases throughput.

In the Ion Torrent system, clonal amplification of adapter-ligated DNA fragments is carried out by emulsion PCR on beads. Afterwards, a sequencing-by-synthesis reaction is carried out in microwells with the beads. The sequencing approach measures pH changes brought about by the release of hydrogen ions during DNA extension and detection by a sensor positioned at the bottom of the microwell. In a study using this approach for characterizing multiple Shiga toxin producing *Escherichia coli* (STEC) serotypes, the genome of single isolates was sequenced, followed by multi-locus sequence typing (MLST), *k*-mer, and phylogenetic analysis against 5,029 bacterial genomes. Pacific Biosciences (PacBio) employs the use of real-time sequencing of a single molecule, where a capped DNA template can be sequenced multiple times, using a strand-displacing polymerase as is used in loop-mediated isothermal amplification (LAMP). The polymerase is immobilized at the bottom of a chamber called zero-mode waveguides (ZMW), where DNA synthesis occurs and can be monitored in real-time as fluorescent signals are produced and recorded in a video. The last, nanopore-based sequencing employs a single molecule strategy which has been commercialized by Oxford Nanopore Technologies. The system depends on the transition of DNA through a small channel, as sequencing is accomplished by measuring distinct changes in current that are induced as the bases threaded through biological nanopores by a molecular motor protein (Reuter et al., 2015; Sekse et al., 2017).

NGS approaches are useful for the detection of viruses, bacteria, and fungal communities in food. Attention should be paid to differentiate viable, active, and inactive cells as dead cells may contribute to the DNA pool during sequencing assays. Although further studies are required along this line, RNA is believed to be an indicator of viability since it is thought that RNA degrades more quickly than DNA and therefore would be found only in metabolically active cells. Due to this reason RNA is usually used for the detection of viable cells (Adzitey et al., 2013). RNA can also be used for viral detection, for example, the Norovirus (Ronholm,

2018). For the analysis of NGS data, abundant reads of high quality are important to achieve accurate classification for taxonomy and diversity assessment. In targeted metagenomics, sequence reads are usually classified into operational taxonomic units (OTUs). This classification is dependent on how similar the reads are or by taxonomic assignment based on how similar the sequences are with those present in a database. The several databases employed include the ribosomal database project (RDP), SILVA, and Greengenes projects, which provide information for small and large subunits of rDNA sequences. The UNITE database focuses mainly on internal transcribed spacer (ITS) sequences from eukaryotes (Mayo et al., 2014).

8.3 CLINICAL APPLICATIONS OF NEXT GENERATION SEQUENCING (NGS)

NGS shows various applications related to clinical environment and they are applied at different level in clinical microbiology which includes the tracking outbreaks around the country, effective in the surveillance of hospital infection, plays role in the discovery of pathogen, detection of mutation in foodborne isolates and multiple pathogens detection in a single sample. The role of NGS is very effective, most especially in the diagnosis of infectious diseases in a clinical microbiology. Many efforts are done by the medical community to establish the NGS at various levels. Therefore, NGS is one of the best technologies which could give lots of information regarding to the health status of mankind. This includes information related to characteristics of microbe's response of host for disease. For example, some of the microorganism which are potentially pathogenic present in cerebrospinal fluid (CSF), synovial fluids, mouth, gut, and skin. One of the positive points of NGS is that it directly detects the microorganism from the sample such as parasites, fungi, yeast, and bacteria (Goldberg et al., 2015). Next-generation sequencing has been observed and proposed to change the phase of clinical diagnosis and microbiology research due to its more efficient and faster way of samples analysis. Studies have shown that the traditional methods are not reliable and consume a lot of time in identifying mycobacteria and other pathogenic microbes. Neuroleptospiros is a condition described by Wilson et al. (2014) revealed the various investigations which were carried on diverse samples such as CSF, urine, brain, stool, blood, sputum, serum, oropharyngeal/nasopharyngeal swab plus plasma utilizing NGS. The data generated were utilized for treatment or diagnostic protocol for patients care and improved outcomes. It was also stated that the knowledge of NGS techniques is to widen our understanding of the role of microorganisms in relation to the ecology, health, and disease state. Also, studies have revealed that NGS has been extensively utilized in food safety and risk management to investigate various foodborne pathogens in the food production chain in order to prevent outbreaks. Worthey et al. (2011) utilized whole-exome sequencing as a diagnostic tool to identify immunological defect in a patient, which further justified the utilization of this technique in precision medicine. Studies have demonstrated that the genomic revolution with the discovery of NGS has revitalized microbial physiology and genomic science as it could be referred to as the state-of-the-art genomic technique. It is believed that clinical microbiology will allow improved detection of potential pathogen and lineage of bacteria utilizing NGS. Hendriksen et al. (2019) revealed the role of NGS techniques in microbial surveillance and diagnosis. Ansorge (2016) revealed that the advancement in next generation systems and its applications for DNA sequencing have greatly reshaped the pathogenic detection in complex samples.

8.4 TYPES OF "OMICS" TERMINOLOGIES UTILIZED IN FOOD MICROBIOLOGY

Several "omics" terminologies have been employed in the application of NGS most especially in food microbiology. Among these are metagenomics, which involves the analysis of the diverse microbial communities present in a food sample. This involves whole DNA sequencing or gene-specific sequencing and metatranscriptomics. This is usually achieved using cDNA fragments and gives technical reliability, sensitivity, and clear maps of the transcribed regions of the genome. In metatranscriptomics analysis, the non-ribosomal fraction of RNA is given special attention, which includes messenger RNA (mRNA), micro-RNA (miRNA), transfer RNA (tRNA), small nuclear RNA (snRNA), small interfering RNA (siRNA) and other non-coding RNAs (Kulski, 2016; Seske et al., 2017).

8.5 DETECTION OF FOODBORNE PATHOGEN USING NEXT GENERATION SEQUENCING (NGS)

Levy et al. (2016) revealed that next-generation sequencing is a powerful approach that could be applied in investigating genome in a large data scale which has been growing exponentially since its launch many years ago. The authors highlighted the relevance of NGS technique most especially in genomics research. Also, new modern, cheap, and efficient sequencers continue to play a role in research through the high resolution of complex genome is being unraveled. There is also a robust growth among commercial companies and research communities for the advancement in genomic research. Food microbial science studies the role of microorganisms in food production line for beneficial as well as harmful effects on human health and the environment. Hence, many fast, efficient, and low-cost NGS methods have reorganized microbial taxonomy and analysis of the entire landscape of genome sequencing, particularly in relations with food microbial species.

Baltasar et al. (2014) also highlighted the recent developments in the application of NGS techniques as an efficient tool for detection of microbial taxa, microbial operons, microbial ecology plus food ecosystem and uncultured pathogens. Bonomo et al. (2019) highlighted the role and application of next generation systems in the understanding of microbial identification and taxonomy in food microbiota. This technology has been deployed for increasing surveillance for food safety and monitoring especially food chain processes. Many of these applications are very useful in applied microbiology for conservation, food production and safety as well as in probiotic physiology. The authors suggested that through the knowledge of bioinformatics and sequencing, a lot of new discoveries have been made in food microbial ecosystem, thereby enhancing the holistic screening of many food matrices for pathogenic microbes. DNA barcode fragments as global markers aligning with bacteria phyla in food with the understanding of categorizing all microbial spectrum using high resolution throughput sequencer have equally been unraveled. Many strains of food microbial species and their molecular mechanisms have been uncovered using high throughput genomic screening. It should be noted that NGS technology can be utilized in two broad ways: targeted sequencing and shotgun sequencing of microbial DNA or gene structure offering a brief biodiversity plus phylogeny of diverse components of microbial populations and phylobiome of food ecological constituents especially in food microbiology.

Shad et al. (2012) revealed that the field of biological science has witnessed tremendous advancement due to the increased utilization of next-generation sequencing technologies in environmental DNA along with DNA barcodes for ecological research. Solieri et al. (2012) revealed that researchers have deeply uncovered the mechanisms involved in NGS of biological samples. The introduction of this technology to foodborne pathogens for microbial genomics sequencing resulted in unraveling the links associated with growth, survival, and pathogenesis involved in the molecular basis of food microbiome such as yeasts and lactic acid bacteria (LAB) found in fermented food, hence capacity to correlate phenotype with genotype.

Ruppé et al. (2017) revealed NGS as an improved sequencing technology that has transformed the research landscape in microbial genomics, particular those existing in the gut microbiota and environment. This technology is adapted to study or identify resistance genes in AMR dynamics, risk management, surveillance in the environment and food microbiota. Balamurugan et al. (2019) discovered that bioinformatics in combination with NGS possess a great capacity to change food microbiology and to uncover the mechanism of possible pathophysiology in microbial strains. The use of SNP analysis and genomic multi-locus sequence typing are powerful approaches for studying phylogenetic clustering. Metagenomics science and DNA barcoding of food samples provide information about biochemical and physiological activity of microbial ecosystem.

Urooj et al. (2020) revealed that gut physiology has witnessed a substantial research progress in the study of microbiota due to the revolution in sequencing technology. Most of the research is now focusing on the relationship between gut microbiota ecology and health progress through the adoption of metagenomics, NGS and WGS analysis for the development of possible probiotic strains in food technology to improve yield and quality. Several microflora isolates are subjected to molecular techniques and computational physiology to identify unique strains of probiotic bacteria in order to promote good health through diet. Bello Ortí et al. (2015) revealed that the field of life science has witnessed change in large scale transcriptomics and genomic sequencing due to advancement from first to third-generation sequencing in data production. This has enabled broader understanding of physiology and structural concept in conservation and biodiversity of microbial food science.

Durand et al. (2013) revealed that many pathogenic organisms cause serious havoc to agricultural and economic sectors due to increased contamination of food products witnessed recently. The authors suggested that the development of modern biotechnology tools will help in monitoring and controlling many of these microbial pathogens particularly mycotoxin-producing species. DNA based biomarkers using NGS are capable of identifying real time microbial species in food together with increasing the knowledge of gene assembly and modulatory pathways linked with the control of mycotoxins gene expression.

Tillmar et al. (2013) revealed that many foods analytic methods have been demonstrated over the last decades in an attempt by food industries to produce quality food products to the expectation of the consumers. Fingerprinting technologies such as foodomics and molecular biology approaches are novel technologies gaining tremendous attention among food producers to analyze food diversity and pathogenic food detection. The authors suggested that with the increased availability of next-generation sequencing will change the food industry by offering an excellent opportunity to uncover food microbial flora.

Pightling et al. (2015) developed a core-genome sequence typer (CGST) for *in silico* molecular characterization of *Listeria monocytogenes* employing next-generation sequence data which achieved greater discriminatory power than other molecular typing methods such as PFGE, ribotyping, and multilocus sequence typing (MLST). The *L. monocytogenes* core-genome sequence typing (LmCGST) method utilizes the high-confidence core (HCC) genomes profiles for high resolution identification and development of evolutionary relevant nomenclature. Analysis of 29 high quality chromosome sequences of *L. monocytogenes* were performed resulting in the identification of 4766 and 2114 ORFs (open reading frames) of the pan-genome and core-genome of *L. monocytogenes* respectively. This information was used to calculate and identify 1013 ORFs that make up the HCC genome and HCC profiles were further employed to develop a bioinformatics pipeline with accurate ortholog identification for core-genome sequence typing of *L. monocytogenes* showing evolutionary relationships with easy interpretation and without possible errors of missing SNPs and ORFs (Pightling et al., 2015).

In the article of Lewis et al. (2020), the ability of NGS approaches to characterize and discriminate *Salmonella* sp. and Bacteriophage MS2 contamination in fresh iceberg lettuce was assessed. The detection limit of different NGS tools applied for the analysis of *Salmonella* and phage MS2 were compared. Non-pathogenic phage MS2 was used as a surrogate for Norovirus and propagated in *E. coli* since the phage showed the structural characteristics of the same kind. Both DNA and RNA were extracted concurrently using the following NGS methods: ScriptSeq RNA-seq, NEBNext ultra II RNA-seq and 16S amplicon sequencing. Using the bioinformatics protocols, reads were mapped to the *Salmonella* and MS2 complete genomes and alignments were analyzed. Results of comparison of sensitivity based on detection limits revealed Script Seq RNA-seq as the most sensitive with microbial loads of 10^4 CFU/reaction and 10^5 CFU/reaction for *Salmonella* sp. and MS2, respectively. It was however observed that the assessed NGS technologies with the bioinformatics pipeline do not match the conventional molecular and culture-based techniques with regards to sensitivity for use as a screening tool for pathogens.

Kim et al. (2018) employed metagenomics for the identification and discrimination of the microbial communities in Chinese cabbage to gain proper insight into bacterial pathogens present on fresh vegetables whose consumption rate was preferred by the people in this region. Illumina Miseq technology was used to perform 16S rRNA gene amplicon sequencing targeting the V5 to V6 region for 54 extracted DNA samples of the Chinese cabbage. Using bioinformatics pipeline, merging of raw paired-end reads, barcode trimming, chimera removal, operational taxonomic unit (OTU) picking, diversity analysis, visualization, and phylogenetic classification were performed. Results of the bacterial community analysis revealed a dominance of Proteobacteria and Bacteroidetes, at Phylum levels. The common genera identified, when mapped at 97% similarity cut-off, were *Chryseobacterium*, *Sphingomonas*, and *Pseudomonas*. It was then inferred that the metagenomic approach proves to be of high specificity in the detection of foodborne bacterial pathogens on fresh vegetables without ruling out the need for more epidemiological data to realize source attribution.

8.6 CONCLUSIONS

This chapter has provided detailed information on the relevance of NGS technology in the detection of foodborne pathogen. This chapter also supports that NGS technology is a rapidly developing technique that could be applied to establish the food quality and reduction in sequencing error rates. NGS technology has been documented to play an effective role in different fields, but their main role in clinical settings could not be over-emphasized because it is also used for the identification of foodborne pathogens to prevent illness. Moreover, detailed specific case studies have applied the NGS approach as an effective tool in the detection of foodborne pathogens.

KEYWORDS

- **bioinformatics**
- **detection**
- **food pathogens**
- **next-generation sequencing**
- **operational taxonomic unit**

REFERENCES

Adzitey, F., Huda, N., & Ali, G. R. R., (2013). Molecular techniques for detecting and typing of bacteria, advantages and application to foodborne pathogens isolated from ducks. *Biotech, 3*, 97–107. doi: 10.1007/s13205-012-0074-4.

Ansorge, W. J., (2016). Next-generation DNA sequencing (II): Techniques, applications. *Next Gener. Seq. Appl., S1*, 005. doi: 10.4172/2469-9853.S1-005.

Bankevich, A., Nurk, S., Antipov, D., et al., (2012). SPAdes: A new genome assembly algorithm and its applications to single-cell sequencing. *J Comput Biol., 19*, 455–477. https://doi.org/10.1089/cmb.2012.0021.

Bello-Ortí, B., Howell, K. J., Tucker, A. W., Maskell, D. J., & Aragon, V., (2015). Metatranscriptomics reveals metabolic adaptation and induction of virulence factors by *Haemophilus parasuis* during lung infection. *Veterinary Research, 46*(1), 102. doi: 10.1186/s13567-015-0225-9.

Bonomo, M. G., Calabrone, L., & Salzano, G., (2019). Next-generation sequencing in food microbiota: Biotechnological and food safety benefits. *Microbiology, 15*(8), 713–715.

Ceuppens, S., Li, D., Uyttendaele, M., Renault, P., Ross, P., et al., (2014). Molecular methods in food safety microbiology: Interpretation and implications of nucleic acid detection. *Comprehensive Reviews in Food Science and Food Safety, 13*, 551–577. doi: 10.1111/1541–4337.12072.

Dover, N., Barash, J. R., Hill, K. K., Davenport, K. W., Teshima, H., Xie, G., et al., (2013). *Clostridium botulinum* strain Af84 contains three neurotoxin gene clusters: Bont/A2, bont/F4 and bont/F5. *PLoS One, 8*, e61205. https://doi.org/10.1371/journal.pone.0061205.

Durand, N., El Sheikha, A. F., Suarez-Quiros, M. L., et al., (2013). Application of PCR-DGGE to the study of dynamics and biodiversity of yeasts and potentially OTA producing fungi during coffee processing. *Food Control, 34*(2), 466–471.

Gargis, A. S., Kalman, L., & Lubin, I. M., (2016). Assuring the quality of next-generation sequencing in clinical microbiology and public health laboratories. *J Clin Microbiol., 54*(12), 2857–2865.

Gill, D. M., (1982). Bacterial toxins: A table of lethal amounts. *Microbiol Rev., 46*, 86–94.

Goldberg, B., Sichtig, H., Geyer, C., Ledeboer, N., & Weinstock, G. M., (2015). Making the leap from research laboratory to clinic: Challenges and opportunities for next-generation sequencing in infectious disease diagnostics. *mBio, 6*(6), e01888-15. doi: 10.1128/mBio.01888-15.

Heather, J. M., & Chain, B., (2016). The sequence of sequencers: The history of sequencing DNA. *Genomics, 107*, 1–8. doi: 10.1016/j.ygeno.2015.11.003.

Hendriksen, R. S., Bortolaia, V., Tate, H., Tyson, G. H., Aarestrup, F. M., & McDermott, P. F., (2019). Using genomics to track global antimicrobial resistance. *Front Public Health, 7*, 242. doi: 10.3389/fpubh.2019.00242.

Jagadeesan, B., Gerner-Smidt, P., Allard, M. W., et al., (2019). The use of next-generation sequencing for improving food safety: Translation into practice. *Food Microbiology, 79*, 96–115.

Jagadeesan, B., Gerner-Smidt, P., Allard, M. W., Leuillet, S., Winkler, A., Xiao, Y., et al., (2019). The use of next-generation sequencing for improving food safety: Translation into practice. *Food Microbiology, 79*, 96–115.

Kim, D., Hong, S., Kim, Y. T., Ryu, S., Kim, H. B., & Lee, J. H., (2018). Metagenomic approach to identifying foodborne pathogens on Chinese cabbage. *J. Microbiol. Biotechnol., 28*(2), 227–235. https://doi.org/10.4014/jmb.1710.10021.

Kulski, J. K., (2016). Next-generation sequencing - an overview of the history, tools, and "omic" applications. In: *Next Generation Sequencing - Advances, Applications and Challenges*. http://dx.doi.org/10.5772/61964.

Lee, S. I., Kim, S. A., Park, S. H., & Ricke, S. C., (2019). Molecular and new-generation techniques for rapid detection of foodborne pathogens and characterization of microbial communities in poultry meat. In: Venkitanarayanan, K., Thakur, S., & Ricke, S. C., (eds.), *Food Safety in Poultry Meat Production*. Springer Nature, Switzerland. https://doi.org/10.1007/978-3-030-05011-5_11.

Lemey, P., Rambaut, A., Bedford, T., et al., (2014). Unifying viral genetics and human transportation data to predict the global transmission dynamics of human influenza H3N2. *PLoS Pathog., 10*(2), e1003932.

Levy, S. E., & Myers, R. M., (2016). Advancements in next-generation sequencing. *Annu Rev Genomics Hum Genet., 17*, 95–115.

Lewis, E., Hudson, J. A., Cook, N., Barnes, J. D., & Haynes, E., (2020). Next-generation sequencing as a screening tool for foodborne pathogens in fresh produce. *Journal of Microbiological Methods, 171*, 105840.

Loman, N. J., Constantinidou, C., Christner, M., et al., (2013). A culture-independent sequence-based metagenomics approach to the investigation of an outbreak of Shiga-toxigenic *Escherichia coli* O104:H4. *JAMA, 309*, 1502–1510.

Mayo, B., Rachid, C. T., Alegría, A., Leite, A. M., Peixoto, R. S., & Delgado, S., (2014). Impact of next generation sequencing techniques in food microbiology. *Current Genomics, 15*(4), 293–309.

Miralles, M. M., Maestre-Carballa, L., Lluesma-Gomez, M., & Martinez-Garcia, M., (2019). High-throughput 16S rRNA sequencing to assess potentially active bacteria and foodborne pathogens: A case example in ready-to-eat food. *Foods, 8*, 480. doi: 10.3390/foods8100480.

Motro, Y., & Moran-Gilad, J., (2017). Next-generation sequencing applications in clinical bacteriology. *Biomol. Detect. Quantif., 14*, 1–6. doi: 10.1016/j.bdq.2017.10.002.

Pightling, A. W., Petronella, N., & Pagotto, F., (2015). The *Listeria monocytogenes* core-genome sequence typer (LmCGST): A bioinformatic pipeline for molecular characterization with next-generation sequence data. *BMC Microbiology, 15*, 224–234. doi: 10.1186/s12866-015-0526-1.

Quigley, L., O'Sullivan, O., Beresford, T. P., Ross, R. P., Fitzgerald, G. F., & Cotter, P. D., (2012). High-throughput sequencing for detection of subpopulations of bacteria not previously associated with Artisanal Cheeses. *Applied and Environmental Microbiology, 78*, 5717–5723.

Reuter, J. A., Spacek, D., & Snyder, M. P., (2015). High-throughput sequencing technologies. *Mol Cell, 58*, 586–597. doi: 10.1016/j.molcel.2015.05.004.

Ronholm, J., (2018). Editorial: Game Changer – next-generation sequencing and its impact on food microbiology. *Front. Microbiol., 9*, 363. doi: 10.3389/fmicb.2018.00363.

Ruppé, E., Greub, G., & Schrenzel, J., (2017). Messages from the first international conference on clinical metagenomics (ICCMg). *Microbes Infect., 19*(4, 5), 223–228.

Salje, H., Lessler, J., Maljkovic, B. I., et al., (2017). Dengue diversity across spatial and temporal scales: Local structure and the effect of host population size. *Science, 355*(6331), 1302–1306.

Sanger, F., Nicklen, S., & Coulson, A. R., (1977). DNA sequencing with chain-terminating inhibitors. *Proc. Nati. Acad. Sci. USA, 74*, 5463–5467.

Schmidt, K., Mwaigwisya, S., Crossman, L. C., et al., (2017). Identification of bacterial pathogens and antimicrobial resistance directly from clinical urines by nanopore-based metagenomic sequencing. *J. Antimicrob. Chemother., 72*(1), 104–114.

Sekse, C., Holst-Jensen, A., Dobrindt, U., Johannessen, G. S., Li, W., Spilsberg, B., & Shi, J., (2017). High throughput sequencing for detection of foodborne pathogens. *Front. Microbiol., 8*, 2029. doi: 10.3389/fmicb.2017.02029.

Shokralla, S., Spall, J. L., Gibson, J. F., & Hajibabaei, M., (2012). Next-generation sequencing technologies for environmental DNA research. *Mol. Ecol., 21*(8), 1794–1805. doi: 10.1111/j.1365-294X.2012.05538.x.

Solieri, L., Dakal, T. C., & Giudici, P., (2013). Next-generation sequencing and its potential impact on food microbial genomics. *Ann. Microbiol., 63*, 21–37. https://doi.org/10.1007/s13213-012-0478-8.

Tatsika, S., Karamanoli, K., Karayanni, H., & Genitsaris, S., (2019). Metagenomic characterization of bacterial communities on ready-to-eat vegetables and effects of household washing on their diversity and composition. *Pathogens, 8*, 37. doi: 10.3390/pathogens8010037.

Tillmar, A. O., Dell'Amico, B., Welander, J., & Holmlund, G., (2013). A universal method for species identification of mammals utilizing next generation sequencing for the analysis of DNA mixtures. *PLoS One, 8*(12), e83761. doi: 10.1371/journal.pone.0083761.
Urooj, B., Sohail, M., Khan, A. N., et al., (2020). Role of next generation sequencing (NGS) in searching for promising lactic acid strains used as animal probiotics. *Acta Scientific Nutritional Health, 4*(4), 179–183.
Wensing, A. M., Van De, V. D. A., Angarano, G., et al., (2005). Prevalence of drug-resistant HIV-1 variants in untreated individuals in Europe: Implications for clinical management. *J. Infect. Dis., 192*(6), 958–966.
Wentz, T. G., Hu, L., Hammack, T. S., Brown, E. W., Sharma, S. K., & Allard, M. W., (2019). Next-generation sequencing for the detection of foodborne microbial pathogens. In: Singh, S., & Kuhnm, J., (eds.), *Defense Against Biological Attacks*. Springer, Cham. https://doi.org/10.1007/978-3-030-03071-1_14.

CHAPTER 9

Next-Generation Sequencing for Strain Tracking and Community Profiling of Food Microbes

GOLLAPALLI PAVAN, MANJUNATH CHAVADI,
MANJUNATHA HANUMANTHAPPA, and MANJULA RAMU

ABSTRACT

The availability of genomic information in recent years has been explored rapidly due to the impact of next-generation sequencing (NGS) techniques, leading to an increased number of sequenced genomes from a broad spectrum of organism. At a time, thousands or millions of sequences can be produced by using high-throughput technology like NGS. Additionally, the limitations implied by the short-read sequencing platforms, where the sequenced genomes lack these techniques, have overcome entire genomic content. This sequencing technology allows identifying microbial taxa from unculturable or even from organisms present in small numbers more accurately. Further, NGS technology also identifies all microbial genes and operons or differential expressed genes at various conditions. The introduction of NGS strategies has revolutionized microbial ecology, which allowed us to investigate several food ecosystems. This chapter discusses standard NGS techniques and platforms focusing on the study of food microbiota. Furthermore, several *in silico* approaches implemented to optimize the genome assemblies that facilitate on finishing process

Global Food Safety: Microbial Interventions and Molecular Advancements. Saher Islam, PhD, Devarajan Thangadurai, PhD, Jeyabalan Sangeetha, PhD, & Natália Cruz-Martins, PhD (Editors)

of genome sequencing and the contribution of NGS to a systems-level understanding of food microorganisms were discussed.

9.1 INTRODUCTION

Bioinformatics has evolved as an interdisciplinary field in the modern scientific era applying various methodologies from computer science, applied mathematics, and statistics to study biological phenomena. The application of bioinformatics is ubiquitous in the contemporary research field. Food science is also a well-developed field that involves food, causes of deterioration, food processing principles, and food quality improvement. The effect of microorganisms on food can be predicted by applying bioinformatics techniques, and genomics, proteomics, and systems biology approaches. These computational techniques will assist in identifying food production requirements, food processing, quality and nutritional value improvement, and others. Several methods have emerged for developing and producing high-yielding crops with good quality and resistance to diseases (pathogenic microbes).

A wide variety of microbial effects on the food can be predicted and assessed using bioinformatics applications (Garrigues et al., 2013). The current food science research has been moved to using 'Omics' methods from the earlier classical methodologies in physical, biological, and chemical makeup and processing of the food (Cellini et al., 2004). The accelerated progress of high-throughput sequencing, i.e., next-generation sequencing (NGS) technologies, has dramatically impacted microbial research, where the whole genome of an organism is processed in a single step. These advances have replaced the traditional molecular methods for detecting the microorganisms and their genome draft typing.

The developments in computational algorithms and software provide a wide variety of appreciable beneficiaries in food science, viz. taste and flavor improvement. The molecular modeling and simulation methods can model the structure of taste receptors and develop more intense testing compounds as food additives (Talevi et al., 2012). Few examples of such taste receptors have been listed in Table 9.1. It is found that lactic acid is essential for flavor formation in dairy production. This flavor profile of various dairy products is due to multiple compounds like lactic acid bacteria (LAB) and their intermolecular interactions. These interactions can be identified and understood through bioinformatics applications to further improve flavor (Liu et al., 2008).

TABLE 9.1 Type of Taste and Their Corresponding Taste Receptors

Type of Taste	Taste Receptor
Sour taste	Ion channel identified to degeneration
Bitter	~50 G-protein coupled receptor
Umami	Glutamate receptor
Sweet	G-protein coupled receptor (Tas1r3)
Salt	The epithelial ion channel (ENaC)

Similarly, food quality and safety also have a significant dependency on bioinformatics. As mentioned earlier, the NGS techniques are being used to find the whole genome sequence of the foodborne pathogens (Brul et al., 2006). These approaches will also help identify the source of food-borne illness, developing molecular markers to identify the occurrence of food spoilage and prediction of stress resistance.

This chapter details different sequencing techniques and bioinformatics approaches used in food microbiology, concentrating on how these techniques help improve the quality of fermented food products by microbial production. We also discussed genome sequence data retrieval, predicting the genomic functional elements, genome-scale metabolic modeling and its application in complex fermentation, identifying the food properties like flavor and texture, food safety, and risk factor assessment. The different application areas and approaches used in bioinformatics are outlined in a schematic representation of Figure 9.1.

9.2 SEQUENCING TECHNOLOGY

Genome technologies are most widely used in the food processing industries in setting food processing parameters. Using an empirical combination of appropriate preservation hurdles has been replaced by the use of knowledge-based approaches. The predictive modeling of spoilage and pathogenic microorganisms' behavior upon and after food processing is one of the prime areas where genomics plays a dominant role in food microbiology (Brul et al., 2006). A variety of next-generation sequencing tools exist which have replaced the traditional Sanger sequencing method.

The high-throughput nucleotide sequencing mainly determines targeted sequencing, which generates relatively shorter read lengths and single-molecule sequencing, producing relatively more extensive read lengths. A large amount of raw sequence data is generated, requiring further processing (Mathew et al., 2015). Several new high-throughput technologies like genomics, metagenomics, transcriptomics, and metatranscriptomics have emerged from the onset of NGS techniques (Table 9.2).

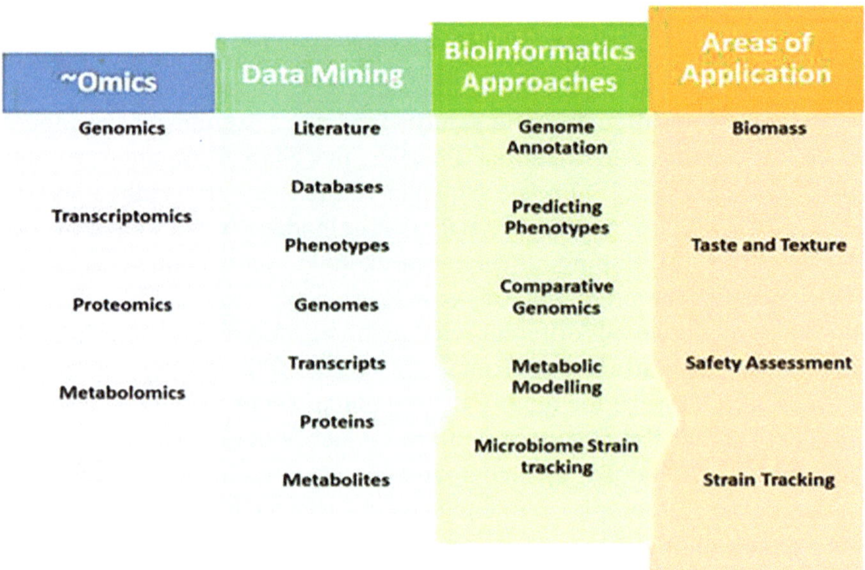

FIGURE 9.1 Schematic representation of different sequencing, bioinformatics applications, and areas of applications in food microbiology.

Two different substantial ways are implemented in this technology: shotgun sequencing (total microbial nucleic acid sequencing) and targeted sequencing (gene-specific). Shotgun sequencing considers polygenetic composition within a community to provide a number and potential function of genes (Solieri et al., 2013). But in targeted sequencing, highly conserved DNA segments or cDNAs are first amplified using group-specific or universal primers by PCR. The microbiology of an array of foods and fermented foods has been studied using both of the above two techniques (Liu et al., 2011; Ercolini, 2013).

TABLE 9.2 Comparison of First, Second, and Third-Generation Sequencing Platforms based on Features, Chemistry, and Performance

Platform	Chemistry	PCR Amplification	Disadvantages	Applications	References
Sanger sequencing	Asynchronous with base-specific terminator	Standard PCR	PCR biases; low degree of parallelism; high cost of sequencing	Gene/genome sequencing	Sanger et al. (1977); Margulies et al. (2005)
Roche 454	Sequencing-by synthesis (pyrosequencing)	EmPCR	PCR biases; asynchronous synthesis; homopolymer run; base insertion and deletion errors; emPCR is cumbersome and technically challenging	De novo genome sequencing, RNA-seq, resequencing/targeted re-sequencing	Metzker (2010)
Illumina	Polymerase-based sequencing-by-synthesis	Bridge amplification	PCR biases; low multiplexing capability of samples	De novo genome sequencing, RNA-seq, resequencing/ targeted re-sequencing, metagenomics, ChIP	Metzker (2010)
SOLiD	Ligation-based sequencing	EmPCR	EmPCR is cumbersome and technically challenging PCR biases; long run time	Transcript counting, mutation detection, ChIP, RNA-seq, etc.	Pandey et al. (2008); Metzker (2010)
HeliScope	Polymerase (asynchronous extension)	SM; no PCR	Asynchronous synthesis; homopolymer run; high instrument cost; short read lengths; high error rates compared with other reversible terminator chemistries	Resequencing, transcript counting, ChIP, RNA-seq	Metzker (2010)

TABLE 9.2 *(Continued)*

Platform	Chemistry	PCR Amplification	Disadvantages	Applications	References
Polonator	Synchronous controlled synthesis	Em PCR	Low read length; emPCR is cumbersome and technically challenging	Bacterial genome, resequencing, SNPs, and structural variants detection	Metzker (2010)
PacBio	Phospho-linked fluorescent nucleotides	SMRT	High instrument cost; low number of sequences read per run; highest error rates compared with other NGS chemistries	De novo genome sequencing, RNA-seq, resequencing/targeted re-sequencing, metagenomics, SNPs, and structural variants detection	Travers et al. (2010)
CMOS non-optical sequencing	Template-directed DNA polymerase synthesis	–	–	De novo genome sequencing	Rothberg et al. (2011)

Large genome-scale datasets are generated using different technologies on NGS platforms, which differ from each other based on their engineering, sequencing chemistry, result output (number and length of sequence reads), accuracy, and cost (Glenn, 2011). The most commonly used platforms for NGS (second-generation techniques) include 454 (Roche), Illumina (Illumina), SOLiD, and Ion Torrent (Life Technologies), and PacBio (Pacific Biosciences) systems. Liu et al. (2012) and Quail et al. (2012) have pertinently compared the pros and cons of these above techniques.

Further, the currently under development third-generation techniques include DNA nano ball sequencing, helioscope single-molecule sequencing, nanopore DNA sequencing, tunneling current DNA sequencing, sequencing with mass spectrometry, and microscopy-based techniques (Schadt et al., 2010).

9.2.1 SEQUENCING DATA (FASTQ)

The most common format to store both the biological sequence and its corresponding quality score is the FASTQ format. For brevity, a single ASCII character is used to encode both the sequence letters and quality score. The Wellcome Trust Sanger Institute has developed this FASTQ format to assort sequences with FASTA format along with quality data. Later the high-throughput sequencing devices like Illumina Genome Analyzer made this format as de facto standard for storing the output of sequencing (Cock et al., 2009). The FASTQ file is comprised of four lines per sequence, as shown below:

```
@SEQ_ID
GATTTGGGGTTCAAAGCAGTATCGATCATAAAGGAACTCAC-
GATTT
+
!”*((((***+))%%%++)(%%%%).1***-+*”))**55CCF>>>>>>CCC
CCCC65
```

where;

- **Line 1:** Includes ‘@’ character at the beginning, followed by a sequence identifier. Like a FASTA title line, here also the sequence description is optional.

- **Line 2:** Contains the raw sequence data with only letters for nucleotide bases.
- **Line 3:** Includes '+' character at the beginning. They are further followed by the same sequence identifier and any description, which is also optional.
- **Line 4:** The value for the sequence quality and contain the same number of symbols corresponding to the letters present.

The quality runs from 0×21 (lowest quality; '!' in ASCII) to 0×7e (highest quality; '~' in ASCII) is represented in a byte. As illustrated below, the quality is defined in the left-to-right increasing order of quality value of characters.

```
!"#$%&'()*+,-./0123456789:;<=>?@ABCDEFGHIJKLMNOPQRSTUVWXYZ[\]^_`abcdefghijklmnopqrstuvwxyz{|}~
```

This format of wrapping both the sequence and its quality strings is also allowed by original Sanger FASTQ files. The unfortunate choice of "@" and "+" as markers has made parsing complications and thus generally discouraged the use of Sanger FASTQ files. There are different FASTQ formats available, as listed below:

- Phred scores and the qual format
- NCBI sequence read archive
- Illumina 1.3+ FASTQ format
- Solexa FASTQ format
- Sanger FASTQ format
- ABi SOLiD color space FASTQ

9.2.2 SEQUENCE ASSEMBLY

Genome reconstruction using relatively short DNA fragments that can be sequenced, which is arduous primarily by genomic repeats, can be performed by a computational method called genomic assembly (Medvedev et al., 2007). The occurrence of two or more identical copies of DNA segments within the genomes will induce ambiguity in genome reconstruction. The information contained in these reads alone cannot resolve this ambiguity. Additionally, the presence of unusual base-pair compositions will also lead to difficulty in sequencing the genome. Thus, highly fragmented contigs comprising tens of thousands of contiguous genomic segments are observed in a typical genome assembly of eukaryotic genomes. Because of this reason, the techniques that can generate complementary information to that contained in the sequence reads were developed by genomic scientists. *Haemophilus influenzae* is the first living organism, where the sequence assembly was dependent upon paired-read data. This links relatively distant segments of the genome together and allows the assembled contigs to be ordered into a "scaffold" of the organism's main chromosome.

9.2.3 SCAFFOLDING

A portion of the reconstructed genome sequence from end-sequenced whole-genome shotgun clones is called a scaffold composed of contigs and gaps. A high confidence level of ordered bases in a contiguous length of genomic sequence is called a contig. The scaffolding process can be driven by using any type of information that hits the relative location of genomic segments along a chromosome. Most of the time in genome scaffolding, the chromosome structure is interrogated based on the genomic technologies specifically designed over indirect inference based on evolutionary arguments (Ghurye and Pop, 2019). Linkage information for scaffolding can be provided by different sequencing technologies, which can span anywhere from several hundred to tens of thousands of base pairs (Illumina, Pacific Biosciences, and Oxford Nanopore) to hundreds of thousands of base pairs (linked reads and optical maps) to millions of base pairs (Chicago and Hi-C). The best results are usually obtained from a mixture of technologies or from technologies that yield data spanning a broad range of distances (such as third-generation sequencing reads, optical maps, or Hi-C links) from a given wide range of the lengths of repeats in most organisms. The valuable information for repeats of a broad range of size is provided by sequencing reads that can be viewed as linking information spanning any distance within the read length. The genomic contigs can be organized into chromosome-wide scaffolds based on critical information types, as illustrated in Table 9.3 (Ghurye and Pop, 2019).

TABLE 9.3 Comparison of Different Sequencing and Mapping Technologies

Category	Description	Scaffolding Data	Separation on the Genome	Orientation	Ordering	Distance
Physical mapping	This technologies endeavors to access the location of specific loci along genomic chromosomes.	Restriction maps	10–100 Kb	Yes	No	Yes
		Optical maps	10–100 Kb	Yes	Yes	Yes
Subcloning	This method involves the generation of large fragments from the genome which are sequenced separately, but they retain a link between the sequencing reads generated from the same fragment.	10X Genomics	100 Kb	Yes	Yes	Yes
		Illumina TSLR (TruSeq Synthetic Long Read)	100 Kbp	Yes	Yes	Yes
Long-read data	Pacific Bioscience and Oxford Nanopore, like sequencing technologies generate long sequencing reads, that can be found as a special case of subcloning.	Pacific Bioscience	10–15 Kb	Yes	Yes	Yes
		Oxford Nanopore	15–20 Kb	Yes	Yes	Yes
Paired read	A technology that reveals about relative placement of pairs of reads in a genome-wide sequencing and be a common source of information for scaffolding.	Paired-end reads	100–500 bp	Yes	Yes	Yes
		Mate pairs	1,000–10,000 bp	Yes	Yes	Yes
Chromosome conformation	The three-dimensional structure of chromosomes inside a cell can be studied by generating paired-read data.	Hi-C	30–100 Mb	Yes	Yes	No
		Chicago	3–100 Mb	Yes	Yes	No
Synteny	The colocalization of genes or genomic loci along a chromosome is referred to as synteny.	Reference genome(s)	Up to genome size	Yes	Yes	Yes

9.2.4 GAP CLOSURE STRATEGIES

The genome sequences most often contain gaps after scaffolding, which needs to be processed by filling such gaps. Various strategies are for this purpose, of those generating new sequencing data is the common one that includes PacBio's long reads (Koren and Phillippy, 2015). Other methods involve predicting the most likely order and orientation of the contigs using bioinformatics tools like Projector 2 (van Hijum et al., 2005) or Mauve (Darling et al., 2004). These tools infer contig order by comparing them to one or more reference sequences. The complete genome sequence is obtained by sequencing across the gaps between contigs, while gap filling has occupied a significant portion of other genome sequencing projects' time and expense. This is primarily due to (1) use of the relatively high (~8.0X) coverage in the random phase of the project; (2) the use of a large insert (15–23 kb) libraries in addition to the small insert library; (3) long sequence read lengths; and (4) use of sequence from both ends of all clones.

Sequence gaps are those regions for which a template is available (i.e., the unsequenced region in the middle of a clone) will be closed by designing primers pointing outward from the ends of contigs (toward the center of the unread segment) and performing Dye-Terminator sequencing reactions on the appropriate template. Physical gaps are those for which no template is available (i.e., no clones identified which cross the gap) and result in contigs whose order concerning one another is unknown. There should be no physical gaps in the sequence, but only the sequencing gaps must persist if the libraries are truly random. However, if due to non-random selection, there are physical gaps, then they can be closed using the methods summarized below (Table 9.4).

TABLE 9.4 The Strategies of Gap Closure for Whole Genome Sequencing Projects

Contig Ordering	Description
Clone links	Forward and reverse ends of clones (2 kb and 18 kb libraries)
Split peptides	Protein matches at ends of potentially adjacent contigs
Southern blots	Oligonucleotide fingerprint comparisons
PCR	Combinatorial reactions with all contig end-oligonucleotides
Sequence Gaps (DNA Template Available)	
Dye terminator reactions	Primer walking to cover both strands of the gap

TABLE 9.4 *(Continued)*

Contig Ordering	Description
Physical Gaps (No DNA Template Available)	
PCR	Regular or long-range PCR products sequenced directly
Large insert libraries	Clones isolated from large insert libraries sequenced directly

9.3 GENOMICS AND BIOINFORMATICS APPLICATIONS IN FOOD MICROBIOLOGY

9.3.1 FUNCTIONAL GENOMICS

The most functional role of bioinformatics is in predicting the function of a gene from its sequence information. Due to the enormous increase in the sequence data in present days, it requires careful handling of such data in identifying the functional elements of those sequences. Various types of analysis can identify the function of a gene, such as prediction of laccases (Weirick et al., 2014); gene function prediction from the bacterial genome (Overbeek et al., 2014; Seemann, 2014); and propose the bacterial strain-specific properties through predicting gene functions on the pathway database (Yamada et al., 2011; Kanehisa et al., 2012) or even predicting functionalities of complex microbial communities (Aziz et al., 2008; Langilla et al., 2013). The genetic elements which potentially have a positive impact in some strains (like flavor enhancement) and the other hand, detrimental (like food spoilage) species, such as, *Brethanomyces bruxcellenis* can be identified by comparative genomic sequence analysis (Crauwels et al., 2014). PhenolLinck (Bayjanov et al., 2012) and DuctApe (Galardini et al., 2014) are the two most commonly used omics tools, which require genome sequence information. The reads obtained from single-cell sequencing are assembled by multiple displacement amplification technique (Lasten, 2009; Gurevich et al., 2013). Bioinformatics applications can help identify mobile elements such as transposons, plasmids, or phages since they show great importance in carrying functionality from one bacterial strain to another, for example, galactose utilizing operon transfer between *Lactococcus lactis* strains. The transposon insertion finder (Machielsen et al., 2011; Nakagome et al., 2014) is dependent on the approaches like next-generation sequencing and bioinformatics tools.

9.3.2 BIOMASS PRODUCTION

Genome-scale metabolic modeling is the technique that has been extensively used in recent years to improve food production. The metabolic potential of the food-related microbial strain can be enhanced using the whole-genome sequence in this approach (Marrleveld et al., 2013). The techniques allow completing the metabolic model's missing metabolic reactions, besides genome sequence quality, as a limiting factor (Benedict et al., 2014). An integrated silico approach, such as flux-balance analysis, analyzes the genome-scale metabolic data and algorithms to mimic organism growth under predetermined conditions of substrate availability in the medium. Based on the organism's requirements, the medium composition can be optimized using this growth simulation (Wegkamp et al., 2010).

The robustness of strain can also improve the overall yield of biomass after harvesting, significantly influencing the changing fermentation conditions under which starter culture is preferred. Additionally, for a given phenotype, the gene expression levels of a strain are also significant, along with the gene content. In *L. lactis*, the number of potential causative genes related to its survival was identified using the transcriptome-trait matching (TTM) approach. Indeed, it was observed that the survivability of *L. lactis* to heat, oxidative stress and spray drying was improved by preconditioning using GTM (gene trait matching) and TTM (DeSantis et al., 2014; Dijkstra et al., 2014).

Furthermore, information about low-cost substrate for fermentation, improvement in productivity of compounds (amino acids or succinic acids), flavor enhancement or texture activity of strain can be understood and analyzed better by using these metabolic models. These modeling techniques can also be implemented in complex fermentation processes, which provide information on species/strain interactions' insights.

9.3.3 FOOD FLAVOR AND TEXTURE IMPROVEMENT

The flavor profile of many food products is due to many different molecules and their interactions, but not due to any single compound. The fermentation process also influences the flavor and texture of the food product, and it is microorganism-specific which can be changed. For instance, when added to cheese fermentations, adjunct strains, exopolysaccharide-producing organisms to enhance the texture of yogurt (Robitaille et al., 2009; Yilmaz et al., 2015). Meanwhile, altering the conditions of fermentation or altering the wine starter culture will assist in altering the flavor profile of a wine. The flavor formation of *L. lactis* MG1363 was predicted by metabolic modeling and subsequently verified experimentally (Flahaut et al., 2013).

Similarly, *Lactobacillus delbrueckii* subsp. *bulgaricus* genome sequence (Hao et al., 2011) revealed milk fermentation and yogurt production. During fermentation, produced or converted metabolites are mainly responsible for the taste and texture of food products. The last sensory characteristic of the taste and texture can be directly predicted by metabolic patterns rather than associating with microorganisms' gene content. Hundreds of metabolites present in food samples can be simultaneously measured using metabolomic profiling techniques (Alonso et al., 2015).

9.3.4 RISK ASSESSMENT

The selective screening of microbial genome sequences for the genes' specific functionality is encouraged instead of whole bacterial genes function prediction. It also ensures a computationally efficient way to analyze probable health or safety risks of microbial strains present in the sample. Genomic comparison of the specific bacterium will aid in investigating antibiotic resistance and virulence of selected strains (Bennedsen et al., 2011).

9.3.5 STRAIN TRACKING IN THE MICROBIOME

A database with gene-specific relations and approach to realm insight on genes' specific functionality is essential to obtain valuable reports out of metagenomics experiments. The mixed-culture fermentation containing DNA can be profiled by shotgun metagenomics but find it difficult to deduce due to strain-level diversity (Erkus et al., 2013). But mRNA-derived sequences of complex fermentations can be profiled by meta-transcriptome approaches, where gene expression measurement allows to determine expression levels of genes in a mixed culture. It has been reported that the meta-genome and -transcriptome sequencing of bacterial communities is involved in cheese rind fermentation (Wolfe et al., 2014). In industrial fermentations, phage prediction of the bacteriophages has great importance because of their biodiversity maintenance anomaly (Rodriguez-Valera et al., 2009). The microbes in mixed-culture fermentation interact between them, which is time-consuming; hence, it can be predicted, which can help save cost and time. Further exploration of the microbial consortia's metabolic requirements during fermentation could lead to knowledge-based advancement in the strategies of fermentation stability.

9.3.6 GENOTYPE-PHENOTYPE ASSOCIATION MAPPING

The genetic basis of phenotypes that include plasmid-based genomic library analysis and gene-disruption libraries were identified based upon the genetic selection and Sanger sequencing-based screening methods. Besides, the whole-genome array, comparative genome hybridization, or single-nucleotide polymorphism (SNP) array is used to capture causal loci. Being labor-intensive and cost concern, the advances in NGS technologies have increased the feasibility of genotype-phenotype association mapping even for complex microbial phenotypes. The genotype-phenotype relationships can be elucidated in three-level by these NGS applications (Figure 9.2):

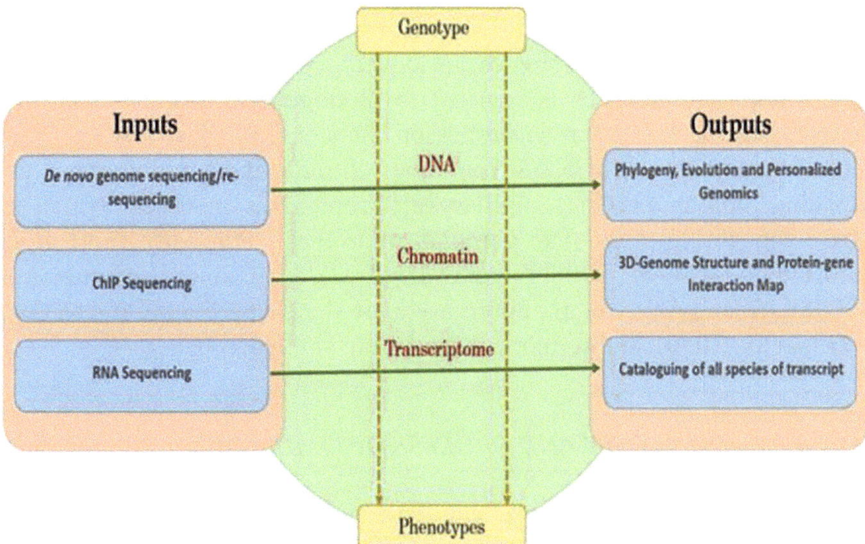

FIGURE 9.2 Three levels of NGS applications used in the genotype-phenotype mapping.

- Identification of individual genetic variation within a population for which a reference genome is available (re-sequencing). Sequence reads mapped to a reference genome were used to detect SNPs, small insertions, or deletions (indels), and large-scale structural variations, such as copy number variations (CNV), thereby improving our knowledge on evolutive mechanisms shaping genome and phenotype diversity within species.
- Adopting high-throughput studies by hybridization-based methods like ChIP-seq for genome-wide analysis of DNA/protein interactions, gene-protein interactions, and 3D-genome structures.
- The use of NGS for transcriptomics (RNA-Seq) facilitates gene identification, expression, and splicing analysis.

9.3.7 COMMUNITY PROFILING

Understanding the relationship between community composition, microbial functions, and their impact on the food's sensorial properties is a fundamental goal of community profiling. Evaluating each introduced or indigenous species in food processing is critical while exploring microbial diversity determines microbial role as food pathogens, spoilage of food, or a potential starter culture. To overcome traditional approaches, currently culture-free strategies are developed where the sequencing of PCR-amplified libraries is targeted to phylogenetic barcodes (like 16S rRNA genes). This approach relies on the steps, as illustrated in Figure 9.3. The application of NGS techniques eliminated all traditional *in vivo* cloning, colony picking, capillary electrophoresis methods, which are time-consuming steps. The complex microbial community from global soil (Leininger et al., 2006; Fierer et al., 2007), deep mines (Edwards et al., 2006), ocean (Angly et al., 2006; Sogin et al., 2006), human microbiome (Qin et al., 2010), and fermented food were studied.

9.3.8 FOOD SAFETY AND PROCESS OPTIMIZATION

The food industry's principal challenge is to produce safe foods with desired functionalities using minimal processing techniques. The source of foodborne illness is determined more innovatively by implementing the genome sequencing projects that are now focusing on foodborne pathogens (Brul et al., 2006). The implementation of NGS technologies in the whole-genome characterization of the food-related microbes is the first step towards food spoilage prevention. The raw material quality, bacterial combinations selection used as starter cultures, controlled fermentation conditions, shape population structures and dynamics are the parameters required for environmental and food processing. The rapid and cost-effective NGS approach like 16S tagged NGS involves properly monitoring these above fluctuations during the fermentation process.

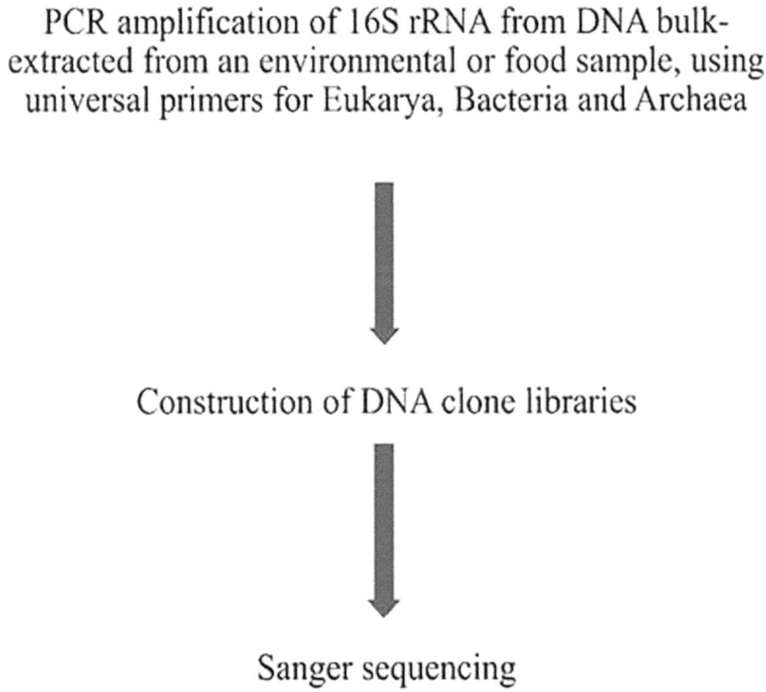

FIGURE 9.3 Sequencing of PCR-amplified libraries targeted to phylogenetic barcodes (like 16S rRNA genes) in community profiling.

A link between the microbial composition to environmental and process parameters was established by NGS-assisted studies. For instance, the pyrosequencing and mass spectroscopy (MS) or gas chromatography (GC) techniques are used to check the quality of the meat, which is dependent on a complex shift in microbiota and metabolites secreted under different conditions, respectively (Ercolini et al., 2013). The data generated by NGS and other "Omics" technologies like transcriptomics and metabolomics has facilitated constructing a system model at species and even at meta-species levels by integrating mathematical algorithms. This allows predicting the species composition in the food by considering the environmental and processing parameters. The occurrence of food spoilage, pathogenic bacteria and thermal preservation stress resistance can be identified using molecular markers.

9.3.9 DATA ANALYSIS

The key factors in achieving accurate taxonomy assignment and assessment of the diversity of microbial species are high-quality read length. In the targeted metagenomics, the sequence reads can be clustered into operational taxonomic units (OTUs) based on the similarity or compositional approaches (Sul et al., 2011). Thus, the relationship of the sequence reads with known microbial groups can be predicted by the taxonomic assignment, which further interprets ecological and/or functional features. A wide variety of user-friendly tools and web-based databases can be used with minimum bioinformatics experience. The vast majority of the microbial species from some ecosystems have never been characterized or taxonomically classified. The most commonly used tools and online databases used for data analysis are listed in Table 9.5.

TABLE 9.5 Some Tools and Databases That can be Used for Accurate Assignment of Taxonomy and Assessment of Diversity of Microbial Species

Databases/ Tools	Description	URL	References
The ribosome database Greengenes project	Updated databases for the small (16S and 18S) and large (23S and 28S) subunits of rDNA sequences	https://rdp.cme.msu.edu/ http://greengenes.lbl.gov https://www.arb-silva.de/	DeSantis et al. (2006); Cole et al. (2009); Quast et al. (2013)
UTITE database	Focuses mainly on ITS sequences from eukaryotes	https://unite.ut.ee/	Nilsson et al. (2019)

TABLE 9.5 *(Continued)*

Databases/ Tools	Description	URL	References
FunGene database	Phylogenetic markers (like rpoB, gyrG, recA, etc.), and functional genes (such as those coding for antibiotic resistance, biodegradation activities, etc.	http://fungene.cme.msu.edu	Fish et al. (2013)
RAST and MG-RAST	Provide databases designed to help in the interpretation of genomic and metagenomic data, respectively	http://www.kbase.us/applist/rast https://www.mg-rast.org/	Aziz et al. (2008)
IMG database	Based on a large number of genome and metagenomic datasets available	https://img.jgi.doe.gov/	Markowitz et al. (2013)
PICRUSt software	Infer the number of genes shared by different microbial taxa. This program provides information on the phenotypic relationships within a microbial community based solely on its phylogenetic composition	http://picrust.github.io/	Rachid et al. (2013)

9.4 CONCLUSION

The field of food microbiology is revolutionized by the advancement in genomics, especially the techniques like NGS. These high-through technologies will assist in various applications like microbial profiling, genetic data mining, genotype-phenotype linking, pathogen detection, determining the fate of starter cultures, food manufacturing, and ripening, food spoilage and shelf-life, strain tracking, and several others. The development of various database resources, algorithms, and experimental advancements to integrate the bioinformatics approaches has promoted exponential microbial sequence data. Substantial knowledge of both the microbes and the operation of an organism's matrix physical parameters is essential for these studies. Finally, combining these sequencing with other 'omics' techniques and systems biology approaches will help analyze microbial diversity and evolution, contributing to the overall quality of food products with their useful and/or adverse activities.

KEYWORDS

- **community profiling**
- **food safety**
- **next-generation sequencing**
- **scaffolding**
- **sequence assembly**
- **strain tracking**

REFERENCES

Abarenkov, K., Nilsson, R. H., Larsson, K. H., et al., (2010). The UNITE database for molecular identification of fungi – recent updates and future perspectives. *New Phytologist, 186*(2), 281–285.

Alonso, A., Marsal, S., & Julia, A., (2015). Analytical methods in untargeted metabolomics: State of the art in (2015). *Frontiers in Bioengineering and Biotechnology, 3*, 23.

Angly, F. E., Felts, B., Breitbart, M., et al., (2006). The marine viromes of four oceanic regions. *PLoS Biology, 4*(11), e368.

Aziz, R., Bartels, D., Best, A., et al., (2008). The RAST server: Rapid annotations using subsystems technology. *BMC Genomics, 9*, 75.

Bayjanov, J. R., Molemaar, D., Tzeneva, V., Siezen, R. J., & Van, H. S. A., (2012). PhenoLink - a web tool for linking phenotype to ~omics data for bacteria: Application of gene-trait matching for *Lactobacillus pantarum* strain. *BMC Genomics, 13*, 170.

Benedict, M. N., Mundy, M. B., Heny, C. S., Chia, N., & Price, N. D., (2014). Likelihood-based gene annotations for gap filling and quality assessment in genome-scale metabolic model. *PLoS Computational Biology, 10*, e1003882.

Bennedsen, M., Stuer-Lauridsen, B., Danielsen, M., & Johansen, E., (2011). Screening for antimicrobial resistance genes and virulence factors via genome sequencing. *Applied and Environmental Microbiology, 77*, 2785–2787.

Bokulich, N. A., & Mills, D. A., (2012). Next-generation approaches to the microbial ecology of food fermentations. *BMB Reports, 45*, 377–389.

Brul, S., Schuren, F., Montijn, R., Keijser, B. J., Van, D. S. H., & Oomes, S. J., (2006). The impact of functional genomics on microbiological food quality and safety. *International Journal of Food Microbiology, 112*(3), 195–199.

Cellini, F., Chesson, A., Colquhoun, I., et al., (2004). Unintended effects and their detection in genetically modified crops. *Food and Chemical Toxicology, 42*(7), 1089–1125.

Cock, P. J. A., Fields, C. J., Goto, N., Heuer, M. L., & Rice, P. M., (2009). The Sanger FASTQ file format for sequences with quality scores, and the Solexa/Illumina FASTQ variants. *Nucleic Acids Research, 38*(6), 1767–1771.

Cole, J. R., Wang, Q., Cardenas, E., et al., (2009). The ribosomal database project: Improved alignments and new tools for rRNA analysis. *Nucleic Acids Research, 37*, D141–145.

Crauwels, S., Zhu, B., Steensels, J., et al., (2014). Assessing genetic diversity among *Brethanomyces* yeasts by DNA fingerprinting and whole-genome sequencing. *Applied and Environmental Microbiology, 80*, 4398–4313.

Darling, A. C., Mau, B., Blattner, F. R., & Perna, N. T., (2004). Mauve: Multiple alignment of conserved genomic sequence with rearrangements. *Genome Research, 14*, 1394–1403.

DeSantis, T. Z., Hugenholtz, P., Larsen, N., et al., (2006). Greengenes, a chimera-checked 16S rRNA gene database and workbench compatible with ARB. *Applied and Environmental Microbiology, 72*, 5069–5072.

Dijkstra, A. R., Alkema, W., Starrenbury, M., Hugenholtz, J., Van, H. S. A., & Bron, P. A., (2014). Fermentation-induced variation in host and oxidative stress phenotypes of *Lactococcus lactis* MG1363 reveal transcriptome signatures for robustness. *Microbial Cell, 13*, 148.

Dijkstra, A. R., Setyawati, M. C., Bayjanov, J. R., et al., (2014). Diversity in robustness of *Lactococcus lactis* strains during heat stress, oxidative stress and spray drying stress. *Applied and Environmental Microbiology, 80*, 603–611.

Edwards, R. A., Rodriguez-Brito, B., Wegley, L., et al., (2006). Using pyrosequencing to shed light on deep mine microbial ecology. *BMC Genomics, 7*, 57.

Ercolini, D., (2013). High-throughput sequencing and metagenomics: Moving forward in the culture-independent analysis of food microbial ecology. *Applied and Environmental Microbiology, 79*, 3148–3155.

Erkus, O., De Jager, V. C., Spus, M., et al., (2013). Multifactorial diversity sustains microbial community stability. *The ISME Journal, 7*, 2126–2136.

Fierer, N., Breitbart, M., Nulton, J., et al., (2007). Metagenomic and small-subunit rRNA analyses reveal the genetic diversity of bacteria, archaea, fungi, and viruses in soil. *Applied and Environmental Microbiology, 73*, 7059–7066.

Fish, J. A., Chai, B., Wang, Q., et al., (2013). FunGene: The functional gene pipeline and repository. *Frontiers in Microbiology, 4*, 291.

Flahaut, N. A., Wiersma, A., Van De, B. B., et al., (2013). Genome-scale metabolic model for *Lactococcus lactis* MG1363 and its application to the analysis of flavor formation. *Applied Microbiology and Biotechnology, 97*, 8729–8739.

Galardini, M., Mengoni, A., Biondi, E. G., et al., (2014). DuctApe: A suite for the analysis and correlation of genomic and Omnilog™ phenotype microarray data. *Genomics, 103*, 1–10.

Garrigues, C., Johansen, E., & Critteuden, R., (2013). Pangenomics - an avenue to improved industrial starter cultures and probiotics. *Current Opinion in Biotechnology, 24*, 187–191.

Ghurye, J., & Pop, M., (2019). Modern technologies and algorithms for scaffolding assembled genomes. *PLoS Computational Biology, 15*(6), e1006994.
Glenn, T. C., (2011). Field guide to next generation DNA sequencers. *Molecular Ecology Resources, 11*, 759–769.
Gurevich, A., Saveliev, V., Vyahhi, N., & Tesler, G., (2013). QUAST: Quality assessment tool for genome assemblies. *Bioinformatics, 29*, 1072–1075.
Hao, P., Zheng, H., Yu, Y., et al., (2011). Complete sequencing and pangenomic analysis of *Lactobacillus delbrueckii* subsp. *bulgaricus* reveal its genetic basis for industrial yogurt production. *PLoS One, 6*, e15964.
Kanehisa, M., Goto, S., Sato, Y., Furumichi, M., & Tanabe, M., (2012). KEGG for integration and interpretation of large-scale molecular data set. *Nucleic Acids Research, 40*, D109–114.
Kececioglu, J. D., & Myers, E. W., (1995). Combinatorial algorithms for DNA sequence assembly. *Algorithmica, 13*, 7.
Koren, S., & Phillippy, A. M., (2015). One chromosome, one contig: Complete microbial genomes from long-read sequencing and assembly. *Current Opinion in Microbiology, 23*, 110–120.
Langilla, M. G., Zaneveld, J., Caporase, J. G., et al., (2013). Predictive functional profiling of microbial communities using 16S rRNA marker gene sequences. *Nature Biotechnology, 31*, 815–821.
Lasten, R. S., (2009). Genomic DNA amplification by the multiple displacement amplification (MDA) method. *Biochemical Society Transactions, 37*, 450–453.
Leininger, S., Urich, T., Schloter, M., et al., (2006). Archaea predominate among ammonia-oxidizing prokaryotes in soils. *Nature, 442*, 806–809.
Liu, G. E., (2011). Recent applications of DNA sequencing technologies in food, nutrition and agriculture. *Recent Patents on Food, Nutrition and Agriculture, 3*, 187–191.
Liu, L., Li, Y., Li, S., et al., (2012). Comparison of next-generation sequencing systems. *Journal of Biomedicine and Biotechnology, 2012*, 251364.
Liu, M., Nauta, A., Francke, C., & Siezen, R. J., (2008). Comparative genomics of enzymes in flavor-forming pathways from amino acids in lactic acid bacteria. *Applied and Environmental Microbiology, 74*(15), 4590–4600.
Machielsen, R., Siezen, R. J., Van, H. S. A., & Van, H. V. J. E., (2011). Molecular description and industrial potential of Tn6098 conjugative transfer conferring alpha-galacyoride metabolism in *Lactococcus lactis*. *Applied and Environmental Microbiology, 77*, 555–563.
Margulies, M., Egholm, M., Altman, W. E., et al., (2005). Genome sequencing in microfabricated high-density picolitre reactors. *Nature, 437*(7057), 376–380.
Markowitz, V. M., Chen, I. M., Palaniappan, K., et al., (2012). IMG: The integrated microbial genomes database and comparative analysis system. *Nucleic Acids Research, 40*, D115–122.
Marrleveld, T. R., Khandelwal, R. A., Olivier, B. G., Teusink, B., & Bruggeman, F. J., (2013). Basic concepts and principles of stoichiometric modelling of metabolic networks. *Biotechnology Journal, 8*, 997–1008.
Matthew, J. S., Den, B. H. C., & Martin, W., (2015). Genomics tools in microbial food safety. *Current Opinion in Food Science, 4*, 105–110.

Medvedev, P., Georgiou, K., Myers, G., & Brudno, M., (2007). Computability of models for sequence assembly. In: Giancarlo, R., &. Hannenhalli, S., (eds.), *Algorithms in Bioinformatics* (pp. 289–301). Berlin: Springer.
Metzker, M. L., (2010). Sequencing technologies - the next generation. *Nature Reviews Genetics, 11*(1), 31–46.
Meyer, F., Paarmann, D., D'Souza, M., et al., (2008). The metagenomics RAST server - a public resource for the automatic phylogenetic and functional analysis of metagenomes. *BMC Bioinformatics, 9*, 386.
Nakagome, M., Solovieva, E., Takahashi, A., Yasue, H., Hirochika, H., & Miyao, A., (2014). Transposon insertion finder (TIF): A novel program for detection of de novo transpositions of transposable elements. *BMC Bioinformatics, 15*, 71.
Nilsson, R. H., Larsson, K. H., Taylor, A. F. S., et al., (2019). The UNITE database for molecular identification of fungi: Handling dark taxa and parallel taxonomic classifications. *Nucleic Acids Research, 47*(D1), D259–264.
Overbeek, R., Olson, R., Pusch, G. D., et al., (2014). The SEED and the rapid annotation of microbial genomes using subsystems Technology (RAST). *Nucleic Acids Research, 42*, D206–214.
Pandey, V., Nutter, R. C., & Prediger, E., (2008). Applied biosystems SOLID system: Ligation-based sequencing. In: Janitz, M., (ed.), *Next-Generation Genome Sequencing: Towards Personalized Medicine* (pp. 431–444). Weinheim, Germany: Wiley-VCH.
Potter, N. N., & Hotchkiss, J. H., (2012). *Food Science*. Springer Science and Business Media.
Qin, J., Li, R., Raes, J., et al., (2010). A human gut microbial gene catalogue established by metagenomic sequencing. *Nature, 464*, 59–65.
Quail, M. A., Smith, M., Coupland, P., et al., (2012). A tale of three next generation sequencing platforms: Comparison of Ion Torrent, Pacific Biosciences and Illumine MiSeq sequencers. *BMC Genomics, 13*, 341.
Quast, C., Pruesse, E., Yilmaz, P., et al., (2013). The SILVA ribosomal RNA gene database project: Improved data processing and web-based tools. *Nucleic Acids Research, 41*, 590–596.
Rachid, C. T., Santos, A. L., Piccolo, M. C., et al., (2013). Effect of sugarcane burning or green harvest methods on the Brazilian Cerrado soil bacterial community structure. *PLoS One, 8*, e59342.
Robitaille, G., Tremblay, A., Moineau, S., St-Gelais, D., Vadeboncoeur, C., & Britten, M., (2009). Fat-free yogurt made using a galactose-positive exopolysaccharide producing recombinant strain of *Streptococcus thermophilus*. *Journal of Dairy Science, 92*, 477–482.
Rodriguez-Valera, F., Martin-Cuadrado, A. B., Rodriguez-Brito, B., et al., (2009). Explaining microbial population genomics through phage predation. *Nature Reviews Microbiology, 7*, 828–836.
Rothberg, J. M., Hinz, W., Rearick, T. M., et al., (2011). An integrated semiconductor device enabling non-optical genome sequencing. *Nature, 475*(7356), 348–352.
Sanger, F., Nicklen, S., & Coulson, A. R., (1977). DNA sequencing with chain-terminating inhibitors. *Proc Natl Acad Sci USA, 74*(12), 5463–5467.
Schadt, E. E., Turner, S., & Kasarskis, A., (2010). A window into third generation sequencing. *Human Molecular Genetics, 19*: R227–240.

Seemann, T., (2014). Prokka: Rapid prokaryotic genome annotation. *Bioinformatics, 30*, 2068–2069.

Sogin, M. L., Morrison, H. G., Huber, J. A., et al., (2006). Microbial diversity in the deep sea and the underexplored "rare biosphere." *Proc Natl Acad Sci USA, 103*(32), 12115–12120.

Solieri, L., Dakal, T. C., & Giudici, P., (2013). Next-generation sequencing and its potential impact on food microbial genomics. *Annals of Microbiology, 63*, 21–37.

Sul, W. J., Cole, J. R., Jesus, E. D. C., et al., (2011). Bacterial community comparisons by taxonomy supervised analysis independent of sequence alignment and clustering. *Proc. Natl. Acad. Sci. USA, 108*, 14637–14642.

Talevi, A., Enrique, A. V., & Bruno-Blanch, L. E., (2012). Anticonvulsant activity of artificial sweeteners: A structural link between sweet-taste receptor T1R3 and brain glutamate receptors. *Bioorganic and Medicinal Chemistry Letters, 22*, 4072–4074.

Travers, K. J., Chin, C. S., Rank, D. R., Eid, J. S., & Turner, S. W., (2010). A flexible and efficient template format for circular consensus sequencing and SNP detection. *Nucleic Acids Research, 38*(15), e159.

Van, H. S. A., Zomer, A. L., Kuipers, O. P., & Kok, J., (2005). Projector 2: Contig mapping for efficient gap-closure of prokaryotic genome sequence assemblies. *Nucleic Acids Research, 33*, W560–566.

Wegkamp, A., Tensink, B., De Vos, W. M., & Smid, E. J., (2010). Development of the minimal growth medium for *Lactobacillus plantarum*. *Letters in Applied Microbiology, 50*, 57–69.

Weirick, T., Sahu, S. S., Mahalingam, R., et al., (2014). LacSubPred: Predicting subtypes of laccases, an important lignin metabolism-related enzyme class using in silico approaches. *BMC Bioinformatics, 15*, S15.

Wilmes, P., Simmons, S. L., Denef, V. J., & Banfield, J. F., (2009). The dynamic genetic repertoire of microbial communities. *FEMS Microbiology Reviews, 33*, 109–132.

Wolfe, B. E., Button, J. E., Santarelli, M., et al., (2014). Cheese rind communities provide tractable systems for in situ and in vitro studies of microbial diversity. *Cell, 158*, 422–433.

Yamada, T., Letumic, I., Okuda, S., Kanehisa, M., & Bork, P., (2011). Ipath 2.0: Interactive pathway explorer. *Nucleic Acids Research, 39*, W412–415.

Yilmaz, M. T., Dertli, E., Toker, O. S., et al., (2015). Effect of in situ exopolysaccharide production on physicochemical, rheological, sensory, and microstructural properties of the yogurt drink ayran: An optimization study based on fermentation kinetics. *Journal of Dairy Science, 98*, 1604–1624.

CHAPTER 10

Whole Genome Sequencing for Food Safety and Quality

CHARLES OLUWASEUN ADETUNJI, MUHAMMAD AKRAM, OLUGBEMI TOPE OLANIYAN, EUGENE AYODELE AYENI, WADZANI DAUDA PALNAM, DEVARAJAN THANGADURAI, AJAYI KOLAWOLE TEMIDAYO, OLUGBENGA SAMUEL MICHAEL, FAHAD SAID KHAN, SAHER ISLAM, JULIANA BUNMI ADETUNJI, and BENJAMIN EWA UBI

ABSTRACT

Whole-genome sequencing (WGS) has been established as a novel biotechnological tool that could be applied for effective and quick investigation in order to determine the food pathogens that might be responsible for foodborne disease outbreaks. These techniques have been identified as novel techniques for the maintenance of quality assurance of food safety and quality. This involves full genome shotgun sequencing, high throughput NGS sequencing, and third era of sequencing (single particles) site read sequencing. This chapter provides comprehensive information on the application of WGS for the food safety and quality. Specific highlights were also provided on different case studies that have utilized WGS for the identification of foodborne pathogens, which are responsible for several types of food-related illness.

Global Food Safety: Microbial Interventions and Molecular Advancements. Saher Islam, PhD, Devarajan Thangadurai, PhD, Jeyabalan Sangeetha, PhD, & Natália Cruz-Martins, PhD (Editors)

10.1 INTRODUCTION

Foodborne microorganisms are one of the foremost causes of mortality worldwide. The capability to perform epidemiological studies and tackle foodborne diseases has been recognized as a fundamental element of the general framework of well-being. It has been established that the second and third-generation sequencing tools are able to sequence the whole genome of microorganisms, which has allowed them to be used routinely in several research centers and in the sequencing of these pathogenic microorganisms. Over the past decade, whole-genome sequencing (WGS) has emerged as one of the most encouraging strategies in food and clinical microbiology. Moreover, a significant progress and gene sequencing has undergone several advancements most especially in several science laboratories as an outcome of high-level sequence growth from a sequence of major bacterial genes in 1995 (Fleischmann et al., 1995; Fraser et al., 1995; Metzker et al., 2010).

Additionally, it has been recognized that food poisoning can lead to significant weakness and death, especially in immunocompromised individuals (Stein et al., 2007; Tox et al., 2010). Many foodborne diseases produced by microorganisms are infectious diseases (Scallen et al., 2011). In addition, due to increasing globalization, foodborne pathogens have spread around the world, which has a significant impact on metabolism and food security. As a result, central governments and international institutions have imposed radiation restrictions to improve hygiene (Frank et al., 2011; Bernard et al., 2014).

The most common methods utilized for the detection of foodborne pathogens include traditional gram staining with unique biochemical properties, which are commonly used for microbial isolation and successful identification of pathogenic microorganisms from clinical, food, or natural sources. But these techniques have several limitations which have necessitated the application of more accurate techniques such as advanced sequencing techniques (Bokulich et al., 2012; Ahn et al., 2014; Lessard et al., 2014; De Philippis et al., 2017). This chapter provides detailed information on the application of WGS for food safety and quality as well as provides detailed information on DNA sequencing methods and how some of these methods represent an important objective to understand the food-related microbes.

10.2 WHOLE GENOME SEQUENCING (WGS)

The application of genome sequencing for food safety and quality involves whole-genome shotgun sequencing, high throughput NGS sequencing and the third-generation sequencing (single molecule) (Loman and Pallan, 2015).

10.3 SEROTYPING

Serotyping is a strategy for collecting microorganisms based on the response between a given antiserum and antigens present on the cell surface, allowing order to be maintained at lower levels. Serotyping of most bacterial species depends on the location of flagellar and important antigens (Grimont et al., 2007), although capsule antigens can also be used (McCallum et al., 1991). Serotyping of *Clostridium botulinum* depends on the identification of different types of neurotoxins (Aoki et al., 2001).

10.4 PHAGE TYPING

The phage used in the epidemiological techniques is applied in the identification of some microorganisms such as *Salmonella* spp. and *Escherichia coli* O157: H7. Although the excretion is limited, the phage composition contains virotoxin to distinguish *E. coli* O 157: H7 contamination cases.

10.5 AMPLIFICATION-BASED PROCEDURES

Improvement-based strategies entail the following: variable number tandem repeats (VNTR), amplified fragment length polymorphism (AFLP), and PCR enhancement techniques. PCR can be used to determine the harmful variables or antimicrobial resistance (AMR) properties (Benzon et al., 1999). Multi-Locus VNTR probes (MLVA) must be demonstrated with different locks to achieve extended resolution (Lindsted et al., 2005).

10.6 RESTRICTION DIGESTION-BASED TECHNIQUES

The constraint-based procedure includes regularly used strategies based on restriction fragment length polymorphism (RFLP), ribotyping, and PFGE.

In the RFLP subtype, bacterial genomic DNA is treated with restriction endonucleases and DNA segments are separated by gel electrophoresis. There are two different approaches: use a few fragment compounds and concentrated electrophoresis to separate large fragments (e.g., PFGE), or hybridize DNA sections layer by layer to important parts (e.g., ribosomes) (Tenover et al., 1995).

10.7 PULSENET

There are two prerequisites for providing resources; the first is a subtype, and the second is a framework that allows successful correspondence to collect data so that experts can use current information from national (or global) case studies. This could lead to an improvement in PulseNet. BioNumerics is a product package that is reliably used to beat PFGE models, and these examples are then transferred to PulseNet. The strength of the PulseNet framework lies in the fact that each research center researches similar atomic algorithms/standards and practices so that the study of local genetic traits can be conducted manually (CDCAP, 2004).

10.8 PREPARATION OF GENOMIC DNA AND SEQUENCING

The foodborne microorganism will be purified on the media to obtain the pure colony. The genomic DNA present could then be isolated using any available DNA isolation kits by following the manufacturer's instructions. Thereafter, this will be followed by amplification of the purified DNA and sequencing of the amplified genes. The result will then be subjected to alignment in NCBI (Wheeler et al., 2008).

10.9 ADVANTAGES OF USING WGS FOR SCREENING OF FOOD SAMPLES

The introduction of NGS in the middle of the 21st Century has reformed the way of DNA sequencing that can be applied for well-being (Sanger et al., 1977; Fleischmann et al., 1995; Swaminathan et al., 2001; Venter et al., 2001; Pridmore et al., 2004; Aarestrup et al., 2012). With the NGS approach, the whole bacterial genome can be cut into small random portions (<100 to a few 1,000 base pairs) several times in one response (a method called "peer-to-peer massive sequencing"), after which the complete DNA aggregation is solved by linking pieces electronically in sequence with overlays (Margulies et al., 2005; Vincent et al., 2017). With the NGS approach, the entire genome can be sequenced very quickly at the expense of $100. WGS offers information about phylogenetic history while it could also contribute to overall well-being by improving food safety, quality, and availability over time. For example, WGS can provide information on disease, adaptation, persistence of biocides, minerals, antimicrobials, and genomic diversity. Metagenomic methods help to directly understand the microbial networks and properties from a dietary experience without distinguishing between pathogens or other microorganisms.

The food industry has recently started using WGS to identify transient microorganisms (new or deleted WGS profiles) and track riders (closely related WGS profiles) as amazing tracking tools for pollution sources. Some organizations use WGS to identify complex areas of specialization within their line of creation, which prevents end products from being regularly contaminated. Usage of WGS can have the greatest influence on redevelopment and overall well-being.

10.10 RELEVANCE OF WHOLE GENOME SEQUENCING (WGS) MAINTENANCE OF FOOD SAFETY AND QUALITY

WGS analysis has provided a better insight during retrospective investigations in order to link food pathogens back to environmental or polluted food sources which have played a significant role in the detection of foodborne disease outbreaks (Brown et al., 2019). WGS analysis was carried out by Nethery et al. (2019) for comparison of eight strains of *Lactobacillus buchneri* with two close reference genomes *L. buchneri* CD034 and NRRL B-30929. Seven out of the eight strains were isolated from spoiled fermented cucumber while the remaining one was a type strain ATCC 4005. Draft genomes obtained after sequencing and assembly were of sizes ranging from 2.49 Mb with contigs > 1 kbp of 20 to 128. Upon comparison, draft genomes of the eight strains revealed a higher percentage in identity (>50%) with the two complete reference genomes. Various mobile genetic elements (transposons), genome metabolic islands and two putative prophages were identified in the incomplete draft genomes having four primary gaps in coverage. It was established in the study that the information provided by the identified prophages and CRISPR-Cas systems (Clustered regularly interspaced short palindromic repeat (CRISPR) and related genes (cas) systems) on the *L. buchneri* genomes through WGS could be further explored for treatment and minimization of food contamination and spoilage by *L. buchneri*, consequently preventing industrial loss.

Taylor et al. (2015) used single nucleotide polymorphism (SNP)-based WGS method for surveillance and outbreak investigation studies of *Salmonella enterica* serovar *enteritidis*. Sequencing of 55 *Salmonella* isolates selected epidemiologically from clinical patients and one environmental site over a period of 14 years in Minnesota and Ohio was carried out followed by pairwise comparison of the genomes to the reference genome of *S. enteritidis* strain P125109. The result obtained revealed 2,580 genome positions with SNPs; about 99% average reference genome coverage and very close relatedness of all the outbreak isolates with other isolates within the outbreak evident by distance of at most 3 SNPs.

Furthermore, it was observed that with a small sample size, little genetic diversity was shown by *S. enteritidis* by reason of small SNP differences observed between the outbreak isolates. This suggested genetic stability in outbreak clades of *S. enteritidis* isolates. In this study, WGS proved to be of greater resolution between isolates within a time frame suitable for outbreak detection and surveillance than the PFGE subtyping method but still requires the support of epidemiological information for appropriate outbreak source traceback (Taylor et al., 2015).

The virulence and genome of *Mucor circinelloides* strain isolated from yogurt was investigated to be incriminated as being a major causative factor responsible for the foodborne fungal infection outbreak of mucormycosis (Lee et al., 2014). In their study, multilocus sequence typing (MLST) was utilized for WGS analysis to identify the strain *Mucor circinelloides* f. *circinelloides* after isolating the fungus from a yogurt container. The increasing incidence of mucormycosis, a foodborne-fungal infection with symptoms of diarrhea, nausea, and vomiting was traced in 300 consumers of yogurts in Texas. The virulence for gastrointestinal illness in animals and humans and the capacity of food spoilage have been demonstrated with the help of WGS. The yogurt isolate was found to belong to *M. circinelloides* f. *circinelloides* subgroup distinct from the other less virulent subspecies of *M. circinelloides* complex which are *M. circinelloides* f.

lustinacus and *M. circinelloides* f. *griceocyanus*. The study opened up the need to pay a closer attention to the prevention of fungal contamination in daily essentials (Lee et al., 2014).

Another genomic analysis of a major disease-causing fungus in edible mushrooms was conducted by Xu et al. (2020). The investigation was carried out to investigate genes associated with its pathogenicity and mycoparasitism. Cobweb disease caused by *Cladobotryum dendroides* was identified and characterized by whole-genome sequencing to put forward a high-quality reference genome for the fungus and to perform a comparative genome analysis with another member of the genus *Cladobotryum* and also gain insights into the fungi-fungi interactions in the mycoparasitic *Cladobotryum*. Using the PacBio Sequel sequencing platform and SMART de novo for the genome sequencing and assembly respectively, 36.69 Mb of assembled *C. dendroides* genome was obtained and found to be very close to the size of *C. prostrusum*. A substantial number of the gene predictions conducted on the genome sequence of *C. dendroides* also revealed strong association with virulence, pathogenicity, resistance, and adaptation. The information provided in this research can enhance further functional analysis towards prevention and treatment of Cobweb disease in edible mushrooms (Xu et al., 2020).

For an accurate discrimination of *Shigella* at the species and serotype levels, WGS was used to carry out genoserotyping of *Shigella* by Wu et al. (2019). The importance of *in silico* serotyping based on WGS became vibrant because it normally points out the level of accuracy during characterization and differentiating different serotypes of *Shigella* species and enteroinvasive *E. coli* (EIEC) which other molecular typing methods that employ multilocus virulence gene *ipaH* have not been able to achieve. From the study, an automated WGS based *in silico* pipeline called ShigaTyper was developed which has been affirmed to be rapid. It also could provide a high level of accuracy during determination of 59 *Shigella* serotypes using Illumina paired-end WGS. This also helps in the differentiation of 221 *Shigella* isolates after examining genetic determinants from EIEC. A direct *Shigella* serotype prediction without complexity in bioinformatic interpretation was made possible by the designed automated ShigaTyper and it was reported to have the potential to detect more novel serotypes and virulence genes (Wu et al., 2019).

Whole-genome MLST based examination of *Salmonella* species isolated from ready-to-eat (RTE) food and wild birds was conducted by

Aung et al. (2019). They analyzed 21,428 samples of cooked and retail foods in Singapore from the period of 2010 to 2015 and obtained 32 *Salmonella* isolates which were lay open to multi-locus sequence typing (MLST) analysis. Majority of the specific sequence types discovered in food and wild birds through MLST WGS method did not overlap and were traced to importation and not from the local environment. However, it was pointed out that limitations in discriminatory power in *Salmonella* exist with the use of MLST when applied on small and less diverse sample size of isolates (Aung et al., 2019).

Painset et al. (2019) presented the strain diversity of *Listeria monocytogenes* in RTE foods and clinical cases of humans based on the virulence and resistance gene profiles using whole-genome sequencing analysis. The study selected 1,142 isolates of *L. monocytogenes* covering RTE foods and sporadic human cases across different regions in the European Union. After sequencing and bioinformatic analysis, 'paired-end' reads were mapped against the reference gene sequences for detection of gene presence, while MLST and assigned clonal complexes (CCs) were obtained from sequences. Genome assembly was performed and core genome SNP phylogenies were also constructed. This study yielded two lineages I and II based on the core SNP phylogenetic analysis, which revealed a consistency in the diversity and distribution observed in previous global studies of *L. monocytogenes*. When assessed for the presence and absence of 115 putative virulence markers across the strain collection, WGS data revealed the presence of over 80% markers in more than 95% of the isolates implying abundance in both lineages I and II of *L. monocytogenes*. The WGS data generated from their study was reported to provide an opportunity for high resolution definition of the population of *L. monocytogenes* as well as to resolve many questions on the strain diversity of *L. monocytogenes* towards food safety for humans (Painset et al., 2019).

Furtherance to the studies on *L. monocytogenes* was conducted by Elson et al. (2019) who employed WGS to investigate the determination and evolution of *L. monocytogenes* strains during the examination of two outbreaks of listeriosis linked to crabmeat as infection vehicle. WGS analysis were performed on 50 isolates which were previously typed by fluorescent amplified fragment length polymorphism (FAFLP) into two as *L. monocytogenes* type V.3 and V.6 strains in order to unveil their phylogenetic relationships in humans and crabmeat for the two outbreaks. The retrospective WGS analysis of the sequenced isolates of outbreak 1 revealed a clear link of the foodborne pathogen in the cases to the crabmeat from producer A and the environment of producer A with similarity of SNP differences ranging from zero to two. The minute diversity in the sequences also pointed to the persistence of the strain over a long time. Analysis of sequenced isolates of outbreak 2 revealed the occurrence and perseverance of multiple strains in the same production environment of producer B as there was a substantial degree of genetic diversity between the typed strains. The results of their study established the value of WGS analysis in high resolution identification and discrimination of *L. monocytogenes* in illness cases, seafoods, and environments with history of seafood consumption for tracing to the food supply chain (Elson et al., 2019).

Another investigation of listeriosis outbreak in the United States linked to milkshakes was conducted by Li et al. (2017) using multiple WGS analyzes and PFGE analysis. This study revealed the differentiation of isolates with no epidemiological relatedness and also showed evolution and persistence of isolates within the period of the outbreak. All the WGS analyzes performed, established the fact that *L. monocytogenes* strain persevered in the ice cream (milkshake) machine but has evolved into a unique clade in the outbreak period of one full year. Their result was congruent with previous findings that linked the outbreak strain to a transmission between two milkshake production facilities. The information provided by the results of core genome MLST scheme helped to elucidate more potential genetic markers related to virulence, AMR, and microevolution of *L. monocytogenes* isolates (Li et al., 2017).

A comparative study of discriminatory power between WGS-based analysis and other multi virulence-locus sequence typing (MVLST) method on 34 isolates of *L. monocytogenes* from rabbit meat was conducted by Palma et al. (2017). The WGS-based analysis used was core-genome multi-locus sequence typing (cgMLST) and able to reveal two more clusters belonging to ST14 isolates in the rabbit meat processing plant which other molecular typing methods (ribotyping and MVLST) could not determine. It was however noted that good knowledge of bioinformatics is required to obtain results that will beat other methods hands down (Palma et al., 2017).

In the same bid to distinguish sporadic strains of *L. monocytogenes* in food-related environments, Stasiewicz et al. (2015) performed WGS analysis of 188 isolates from a previously reported study of *L.*

monocytogenes in food-associated environments. SNP-based phylogenetic analysis revealed two clades of isolates which agreed with the previously discovered lineages I and II by other typing methods. It however provided more information to better resolve the epidemiological relatedness, thus improving on molecular subtyping as well as distinguishing persistent strains from mere sporadic ones. Their study further strengthened the fact that WGS analysis cannot be sufficient on its own but helps to accurately achieve the source tracing and control of foodborne illnesses when in conjunction with epidemiological data (Stasiewicz et al., 2015).

Lakicevic et al. (2017) described WGS as a reliable, fast, and cheap tool utilized in a diverse array of health-promoting and food safety benefits which allow all genomes to be evaluated as compared to other forms of traditional techniques in subtyping. Study revealed that this technique helps in monitoring the distribution, understanding across various production line for rapid prevention and control strategy. It is then suggested that WGS is the most suitable technique for the analysis of pathogen genomes such as antimicrobial resistant organisms.

Apruzzese et al. (2019) described how several countries in the world have incorporated whole-genome sequencing as part of the nation foodborne disease outbreaks monitoring and control systems. The authors revealed that the wide gap between the developed and developing nations in terms of knowledge capacity for data interpretation and analysis for whole-genome sequencing create a level of disparity in food trade at the international level. FAO (2016) revealed that WGS is a very powerful tool and important in the area of food biosafety. With high precision, it is able to detect and categorize different microbes.

Brown et al. (2019) revealed the use of WGS in food biosafety and regulation among different health regulatory agencies across the globe due to the ability of rapidly detecting with high level of accuracy outbreaks in foodborne pathogens. The authors further explained the advantage of WGS technique as the ability to analyze the entire genome of bacteria better and faster with precision compared with any known technique till date. WGS is also allowed at international level for effective monitoring and mapping of contaminated food to specific illness if the genome of pathogens is effectively shared.

Ronholm et al. (2016) demonstrated that modern strategy for categorizing and mapping pathogenic agents during epidemiological evaluations of foodborne infections involves high resolution technique such as WGS based on SNP. The authors showed that pathogen lineages plus phylogenetic relationships can be analyzed utilizing this technique. Most of the data generated can be utilized for antibiotic resistance gene mapping, virulence gene monitoring, synteny characterization, environmental attribution, and mobile genetic component identification. Recently, application of many software packages has increased the efficiency of whole-genome sequencing in producing *in silico* data. The authors further revealed that the success in the categorization of etiological agents during epidemiological outbreaks relies solely on the ability to evaluate related and non-related cases through high resolution tool. Many latest technologies are now being used as subtyping method such as whole-genome sequencing involving the utilization of SNP techniques. Also, several advancements in software packages have allowed the generation of *in silico* data. For non-culturable foodborne agents, metagenomics technology offers the opportunity to detect microbes and investigate food safety utilizing next-generation sequencing.

Foodborne pathogens have been the main cause of mortality across the globe which poses a major public health concern. Many molecular biology tools are available to evaluate diverse strains of these pathogens such as subtyping approaches. Studies have proved that the distribution and sources of food supply chain span across different geographical locations, hence pathogenic outbreaks may sometimes overlap. Studies have indicated that advancement in second and third generation approaches led to a more robust, fast, and accurate way of analyzing genomes of pathogenic organisms in real-time. WGS has been adduced to be cheap, fast, and can easily generate highly revolutionized phylogenetic subtyping in managing foodborne microbial biohazards (Tyson et al., 2015).

Quainoo et al. (2017) revealed that nosocomial infections because of multidrug-resistant bacteria affect a lot of population across the globe. More recently, many scientists have engaged with whole-genome sequencing for the examination and evaluation of pathogen genome. Ruppitsch et al. (2019) showed that the recent advancements witnessed that WGS has changed the field of food microbiology, physiology, and public health with high level and sophisticated analysis of pathogenic bacteria and another deleterious microorganism that pose danger to the survival and health of humans. Many scientists have routinely engaged with the use of WGS for monitoring and investigation of foodborne pathogens, which is current adduced to be the most efficient molecular biology tool for phylogenetic relatedness, virulence-traits, antibiotic resistance, small genetic variations analysis plus multi-locus sequence typing thereby ultimately making transmission of pathogens traceable.

10.11 NGS HIGH THROUGHPUT SEQUENCING

The benefits of NGS sequencing on Sanger can be concise as follows: (1) *In vitro* development of sequence library, (2) *In vitro* clonal enrichment of part of the DNA, (3) allowing sections of DNA to be multiplexed based on sequence, and (4) strong immobilization of DNA. Given the changed strategies used to restrain DNA on solid substrates, three innovations are pervasive: (a) pyrosequencing, (b) sequence by ligation in buffer, and (c) sequence by synthesis on glass substrate.

10.11.1 PYROSEQUENCING

The first advanced high-speed sequencer to become economically accessible was the 454 GS20 Pyrosequencing stage (Roche) (Margulies, 2006). DNA is first sheared by enzyme-based absorption or sonication and ligated with an oligonucleotide adapter. Next, individually ligated segments undergo for emulsion PCR (Ronaghi et al., 1996). Then for sequencing reaction, amplicon-carrying spheres are taken into picotiter plate. In each pyrosequencing cycle, nucleotides cause the release of catalytically mediated particles of inorganic pyrophosphate (PPi) identified by the detector.

10.11.2 SEQUENCING BY LIGATION IN BUFFER

Powerful innovation relies on novelties in multipolar sequencers (Shendure et al., 2005). The DNA piece on the adapter side will be introduced to 1 μm paramagnetic buffer and the PCR reaction is normally enhanced with an oil-water suspension. The buffer with the PCR amplicons was restrained on a solid flat substrate and hybridized to each other at the adapter with a preliminary PCR amplification. Every sequencing reaction generally continues the ligation of fluorescently categorized DNA octamers until the positional identity of the nucleotide is revealed. The resulting cleavage of the substance leaves the pentamer in the DNA layout. The dynamic focus of this procedure reveals DNA placement. This step uses a 2-based coding framework, so errors are identified more quickly and provide 99.94% correctness.

10.11.3 SEQUENCING BY SYNTHESIS ON A GLASS SURFACE

The Illumina genome analyzer (SOLEXA) was developed in 2006 and 2008 (Fedurco et al., 2006; Turcatti et al., 2008). This includes DNA deposition in a DNA library and degradation of oligonucleotide compounds. The method of DNA purification involved is called Bridge PCR (Adessi et al., 2000). Both the front and rear accessories have a clamp attached to the glass surface for proper leveling. The DNA section adjacent to the connector further hybridizes, directing the glass surface away from the original orientation. Scaffold PCR uses formamide-based denaturation and Bst DNA polymerase to improve DNA sections, resulting in "groups" of cloned amplicons. Amplicons delivered from isolated DNA segments accumulate in isolated physical regions of the cluster. After the age of the group, the sequence reserves hybridize to the comprehensive legacy surrounding the conspiracy area. The sequence continues with a cycle containing the modified DNA polymerase and 4 nucleotides.

10.11.4 THIRD GENERATION SEQUENCING (SINGLE-MOLECULE REAL-TIME: SMRT)

Third-generation sequencing involves individual DNA molecules that are sequenced in a real-time manner, which reduces low failure rates. First time, the Pacific Biosciences platform provided the single molecular real time sequencing service (Eid et al., 2009). In this situation, the phase of preparing the library traps the closed DNA particle by linking the connector atom of the DNA atom of interest to be sequenced to two closures. The DNA atoms in the ring are stacked into SMRT R cells containing 150,000 zeptolliter wells. At the bottom of each well is an isolated, restrained DNA polymerase. At this point, the DNA polymerase binds to the clip of the round DNA atom of interest and initiates replication. Four nucleotides with a fluorescent name are then introduced into the reaction wells. Enzymatic fusion of each base creates a light heartbeat that recognizes the base and iteratively studies to obtain a DNA sequence (Rhoads and Au, 2015). The main priority of SMRT R sequences is the distance found. The reading length of the first C1 age sequencer was about 1,500 base pairs. Subsequent C4 scientific conventions contained a normal reading length of 10 kbp. The capacity of the PacBio RS II framework is 1 to 5 billion bases per SMRT R cell. In any case, the stage miscalculation rate is generally high (around 11–15%).

10.11.5 SEQUENCING METHODS UTILIZED FOR EFFECTIVE CHARACTERIZATION OF FOOD-RELATED MICROBIOMES

10.11.5.1 16S rDNA SEQUENCING

It is one of the most significant culturally autonomous technologies used in regular microbiome research. Most microscopic organisms contain 16S rDNA-grade quality consisting of nine hypervariable loci contiguous according to conserved configurations (Neefs et al., 1993). It is an exceptional opportunity to design a non-exclusive PCR pool that could enhance and inherit these hypervariable loci to recognize the comparative scientific classification of bacteria in food web-related species. Therefore, the quality of 16S rDNA can also be used to differentiate organisms. These groups are classified into operational taxonomic units (OTUs), taking into account the similarity of nucleotide inheritance. The OTU database is then compared to distinguish the microbes present in the sample from other microorganisms. The main concept involved in the detection of the microflora using this methodology was to identify the microbial population in a particular environment or food samples (Giovannoni et al., 1990).

With Sanger sequencing, you can only sequence a few amplicons. This result is not taken into account for the smaller numbers of individuals lost from the vegetation population, thus replacing the broader representation of the microbial network. Subsequent NGS studies in the 16S rDNA sequencing significantly extended the limit of evidence for progressive exhaustive discrimination of the bacterial and human network. In all cases, greater involvement was derived due to the fact that only a short amplicon was sequenced (Claesson et al., 2009). Furthermore, pyrosequencing was able to store one example and facilitate multiplexing during each instrument pass (Hamady et al., 2008). The latest advance has paid off in a way that will destroy natural prokaryotes. Since then, 16S rDNA sequencing has become one of the best-known strategies for recognizing individuals of food web-associated microbial communities. One of the main concerns in using the 16S rDNA sequencing method is access to a large number of bioinformatics devices designed to study sequencing information. Programming commonly used to extract 16S rDNA information through food/wild-type examples include QIIME (quantitative insights into microbial ecology) (Caporaso et al., 2010), motor (Schloss et al., 2009) and USEARCH (ultrafast grouping test) (Edgar, 2010).

10.11.5.2 FIRST GENERATION: THE SANGER SHOTGUN APPROACH

Studies have shown that during the early period, Sanger shotgun approach of WGS was developed and adopted for the determination of the sequence through a combination of chain-terminating radiolabeled or fluorescent base labeling dideoxynucleosides by DNA polymerase during DNA replication (Smith et al., 1986). In the early 1990s, the first successful bacterial genome sequence using *Haemophilus influenzae* was launched by the Sanger method. Subsequently, advances in the methodology of Sanger sequencing gave birth to bioinformatics mechanism utilized to accumulate shotgun sequences converted into genomes (Fleischmann et al., 1995).

10.11.5.3 SECOND GENERATION: MASSIVELY PARALLEL SEQUENCING

Studies have shown that NGS was launched by Roche in 2005 and later Illumina's genetic analyzer (GA) was introduced into the market in 2007. Earlier, NGS was not very efficient for short sequencing analysis, hence many other better platforms were launched in 2010, resulting in microbial sequencing with different method of sequencing. In 2016, Ion Torrent S5 platform has been seen to be the most popular due to its ability to generate much data per run. The second-generation sequencing platforms are also very poor in read length plus genome assembly algorithms but most of the available draft genomes generated were used downstream analysis (ECDPC, 2015).

10.12 CONCLUSION

This chapter has provided detailed information on the application of WGS for maintenance of food safety and quality. Detailed information was also provided on some techniques that could lead to monitoring and investigation of foodborne pathogens. It was also recognized that WGS has been identified as a cheap, fast, and easy technique that could be applied for identification of various forms of foodborne microbial infections. Furthermore, it plays a crucial role in the examination and evaluation of genome of foodborne pathogens while it also represents comprehensive information on phylogenetic relatedness, virulence-traits, antibiotic resistance, small genetic variations analysis and multi-locus sequence typing thereby ultimately making transmission of pathogens traceable.

KEYWORDS

- **foodborne disease**
- **food pathogens**
- **food quality**
- **food safety**
- **whole genome sequencing**

REFERENCES

Aarestrup, F. M., Brown, E. W., Detter, C., et al., (2012). Integrating genome-based informatics to modernize global disease monitoring, information sharing, and response. *Emerg. Infect. Dis., 18*(11), e1. doi: 10.3201/eid/1811.120453.

Adessi, C., Matton, G., Ayala, G., et al., (2000). Solid-phase DNA amplification: Characterization of primer attachment and amplification mechanisms. *Nucleic Acids Res., 28*(20), E87. doi: 10.1093/nar/28.20.e87.

Ahn, J. Y., Min, J., Lee, S. H., et al., (2014). Metagenomic analysis for identifying *Kimchi* sp. during the industrial-scale batch fermentation. *Toxicol. Environ. Health Sci., 6*, 8–15. https://doi.org/10.1007/s13530-014-0182-0.

Allard, M. W., Bell, R., Ferreira, C. M., et al., (2018). Genomics of foodborne pathogens for microbial food safety. *Curr. Opin. Biotechnol., 49*, 224–229.

Allard, M. W., Luo, Y., Strain, E., et al., (2012). High resolution clustering of *Salmonella enterica* serovar *Montevideo* strains using a next-generation sequencing approach. *BMC Genomics, 13*, 32. http://dx.doi.org/10.1186/1471-2164-13-32.

Aoki, K. R., (2001). Pharmacology and immunology of botulinum toxin serotypes. *J. Neurol., 248*(1), 3–10. doi: 10.1007/pl00007816.

Apruzzese, I., Song, E., Bonah, E., et al., (2019). Investing in food safety for developing countries: Opportunities and challenges in applying whole-genome sequencing for food safety management. *Foodborne Pathog. Dis., 16*(7), 463–473. doi: 10.1089/fpd.2018.2599.

Aung, K. T., Chen, H. J., Chau, M. L., et al., (2019). *Salmonella* in retail food and wild birds in Singapore prevalence, antimicrobial resistance, and sequence types. *Int. J. Environ. Res. Public Health, 16*(21), 4235. doi: 10.3390/ijerph16214235.

Bernard, H., Faber, M., Wilking, H., et al., (2014). Large multistate outbreak of norovirus gastroenteritis associated with frozen strawberries, Germany, 2012. *Euro Surveill., 19*(8), 20719. doi: 10.2807/1560-7917.ES2014.19.8.20719.

Bokulich, N. A., Joseph, C. M. L., Allen, G., Benson, A. K., & Mills, D. A., (2012). Next-generation sequencing reveals significant bacterial diversity of botrytized wine. *PLoS One, 7*(5), e36357. doi: 10.1371/journal.pone.0036357.

Brown, E., Dessai, U., McGarry, S., & Gerner-Smidt, P., (2019). Use of whole-genome sequencing for food safety and public health in the United States. *Foodborne Pathog. Dis., 16*(7), 441–450. doi: 10.1089/fpd.2019.2662.

Caporaso, J. G., Kuczynski, J., Stombaugh, J., et al., (2010). QIIME allows analysis of high throughput community sequencing data. *Nat. Methods, 7*(5), 335, 336. doi: 10.1038/nmeth.f.303.

Centers for Disease Control and Prevention, (2004). *Standardized Molecular Subtyping of Foodborne Bacterial Pathogens by Pulsed-Field Gel Electrophoresis Training*. Centers for Disease Control and Prevention, Atlanta, GA. http://www.cdc.gov/pulsenet/resources/training-and-outreach.html (accessed on 20 December 2021).

Claesson, M. J., O'Sullivan, O., Wang, Q., et al., (2009). Comparative analysis of pyrosequencing and a phylogenetic microarray for exploring microbial community structures in the human distal intestine. *PLoS One, 4*(8), e6669. doi: 10.1371/journal.pone.0006669.

De Filippis, F., Genovese, A., Ferranti, P., Gilbert, J. A., & Ercolini, D., (2016). Metatranscriptomics reveals temperature-driven functional changes in microbiome impacting cheese maturation rate. *Sci. Rep., 6*, 1–12. doi: 10.1038/srep21871.

Edgar, R. C., (2010). Search and clustering orders of magnitude faster than BLAST. *Bioinformatics, 26*(19), 2460–2461. doi: 10.1093/bioinformatics/btq461.

Eid, J., Fehr, A., Gray, J., et al., (2009). Real-time DNA sequencing from single polymerase molecules. *Science, 323*(5910), 133–138. doi: 10.1126/science.1162986.

Elson, R., Awofisayo-Okuyelu, A., Greener, T., et al., (2019). Utility of whole-genome sequencing to describe the persistence and evolution of *Listeria monocytogenes* strains within crabmeat processing environments linked to two outbreaks of listeriosis. *Journal of Food Protection, 82*(1), 30–38. doi: 10.4315/0362-028X.JFP-18-206.

Fedurco, M., Romieu, A., Williams, S., Lawrence, I., & Turcatti, G., (2006). BTA, a novel reagent for DNA attachment on glass and efficient generation of solid-phase amplified DNA colonies. *Nucleic Acids Res., 34*(3), e22. doi: 10.1093/nar/gnj023.

Fleischmann, R. D., Adams, M. D., White, O., et al., (1995). Whole-genome random sequencing and assembly of *Haemophilus influenzae* Rd. *Science, 269*(5223), 496–512.

Food and Agricultural Organization of the United Nations, (2016). *Applications of Whole Genome Sequencing in Food Safety Management.* http://www.fao.org/3/a-i5619e.pdf (accessed on 20 December 2021).

Frank, C., Werber, D., Cramer, J. P., et al., (2011). Epidemic profile of Shiga-toxin-producing *Escherichia coli* O104: H4 outbreak in Germany. *N. Engl. J. Med., 365*(19), 1771–1780. doi: 10.1056/NEJMoa1106483.

Fraser, C. M., Gocayne, J. D., White, O., et al., (1995). The minimal gene complement of *Mycoplasma genitalium. Science, 270*(5235), 397–404. doi: 10.1126/science.270.5235.397.

Giovannoni, S. J., Britschgi, T. B., Moyer, C. L., & Field, K. G., (1990). Genetic diversity in Sargasso sea bacterioplankton. *Nature, 345*, 60–63. doi: 10.1038/345060a0.

Grimont, P. A. D., & Weill, F. X., (2007). *Antigenic Formulae of the Salmonella Serovars.* WHO Collaborating Centre for Reference and Research on *Salmonella*, Paris, France. https://www.pasteur.fr/sites/default/files/veng_0.pdf (accessed on 20 December 2021).

Hamady, M., Walker, J. J., Harris, J. K., Gold, N. J., & Knight, R., (2008). Error-correcting barcoded primers for pyrosequencing hundreds of samples in multiplex. *Nat. Methods, 5*(3), 235–237. doi: 10.1038/nmeth.1184.

Lakicevic, B., Nastasijevic, I., & Dimitrijevic, M., (2017). Whole-genome sequencing: An efficient approach to ensuring food safety. *IOP Conf. Ser.: Earth Environ. Sci.*, 85, 012052. doi: 10.1088/1755-1315/85/1/012052.

Lee, S. C., Billmyre, R. B., Li, A., et al., (2014). Analysis of a foodborne fungal pathogen outbreak: Virulence and genome of a *Mucor circinelloides* isolate from yogurt. *mBio, 5*(4), e01390-14. doi: 10.1128/mBio.01390-14.

Lessard, M. H., Viel, C., Boyle, B., St-Gelais, D., & Labrie, S., (2014). Metatranscriptome analysis of fungal strains *Penicillium camemberti* and *Geotrichum candidum* reveal cheese matrix breakdown and potential development of sensory properties of ripened Camembert-type cheese. *BMC Genomics, 15*, 235. doi: 10.1186/1471-2164-15-235.

Li, Z., Pérez-Osorio, A., Wang, Y., et al., (2017). Whole-genome sequencing analyses of *Listeria monocytogenes* that persisted in a milkshake machine for a year and caused illnesses in Washington State. *BMC Microbiology, 17*(1), 134. doi: 10.1186/s12866-017-1043-1.

Liu, B., & Pop, M., (2009). ARDB - antibiotic resistance genes database. *Nucleic Acids Res., 37*, D443–447.

Loman, N. J., & Pallen, M. J., (2015). Twenty years of bacterial genome sequencing. *Nat. Rev. Microbiol., 13*(12), 1–9. doi: 10.1038/nrmicro3565.

Margulies, M., Egholm, M., Altman, W. E., et al., (2005). Genome sequencing in microfabricated high-density picolitre reactors. *Nature, 437*(7057), 376–380.

McCallum, K. L., & Whitfield, C., (1991). The rcsA gene of *Klebsiella pneumoniae* O1:K20 is involved in expression of the serotype-specific K (capsular) antigen. *Infect Immun., 59*(2), 494–502.

Metzker, M. L., (2010). Sequencing technologies - the next generation. *Nat Rev Genet., 11*(1), 31–46. doi: 10.1038/nrg2626.

Neefs, J. M., Van De, P. Y., De Rijk, P., Chapelle, S., & De Wachter, R., (1993). Compilation of small ribosomal subunit RNA structures. *Nucleic Acids Res., 21*(13), 3025–3049. doi: 10.1093/nar/21.13.3025.

Nethery, M. A., Henriksen, E. D., Daughtry, K. V., Johanningsmeier, S. D., & Barrangou, R., (2019). Comparative genomics of eight *Lactobacillus buchneri* strains isolated from food spoilage. *BMC Genomics, 20*, 902.

Painset, A., Björkman, J. T., Kiil, K., et al., (2019). LiSEQ – whole-genome sequencing of a cross-sectional survey of *Listeria monocytogenes* in ready-to-eat foods and human clinical cases in Europe. *Microbial Genomics, 5*(2), e000257. doi: 10.1099/mgen.0.000257.

Palma, F., Pasquali, F., Lucchi, A., De Cesare, A., & Manfreda, G., (2017). Whole-genome sequencing for typing and characterization of *Listeria monocytogenes* isolated in a rabbit meat processing plant. *Italian Journal of Food Safety, 6*(3), 6879. doi: 10.4081/ijfs.2017.6879.

Pridmore, R., Berger, B., Desiere, F., et al., (2004). The genome sequence of the probiotic intestinal bacterium *Lactobacillus johnsonii* NCC 533. *Proc. Natl. Acad. Sci. USA, 101*(8), 2512–2517.

Quainoo, S., Coolen, J. P. M., Van, H. S. A. F. T., et al., (2017). Whole-genome sequencing of bacterial pathogens: The future of nosocomial outbreak analysis. *Clin Microbiol Rev., 30*(4), 1015–1063. https://doi.org/10.1128/ CMR.00016-17.

Quast, C., Pruesse, E., Yilmaz, P., et al., (2013). The SILVA ribosomal RNA gene database project: Improved data processing and web-based tools. *Nucleic Acids Res., 41*, 590–596. doi: 10.1093/nar/gks1219.

Rhoads, A., & Au, K. F., (2015). PacBio sequencing and its applications. *Genomics, Proteomics and Bioinformatics, 13*(5), 278–289. doi: 10.1016/j.gpb.2015.08.002.

Ronaghi, M., Karamohamed, S., Pettersson, B., Uhlén, M., & Nyrén, P. L., (1996). Real-time DNA sequencing using detection of pyrophosphate release. *Anal. Biochem. 242*(1), 84–89. doi: 10.1006/abio.1996.0432.

Ronholm, J., Nasheri, N., Petronella, N., & Pagotto, F., (2016). Navigating microbiological food safety in the era of whole-genome sequencing. *Clin Microbiol. Rev., 29*(4), 837–857. doi: 10.1128/CMR.00056-16.

Ruppitsch, W., Pietzka, A., Cabal, A., et al., (2019). Advances in foodborne outbreak investigation and source tracking using whole-genome sequencing. *IOP Conf. Series: Earth Environ. Sci., 333*, 012010. doi: 10.1088/1755-1315/333/1/012010.

Sanger, F., Nicklen, S., & Coulson, A. R., (1977). DNA sequencing with chain-terminating inhibitors. *Proc Natl Acad Sci USA, 74*(12), 5463–5467.

Scallan, E., Hoekstra, R. M., Angulo, F. J., et al., (2011). Foodborne illness acquired in the United States - major pathogens. *Emerg. Infect. Dis., 17*(1), 7–15. doi: 10.3201/eid1701.p11101.

Schloss, P. D., Westcott, S. L., Ryabin, T., et al., (2009). Introducing mothur: Open-source, platform-independent, community-supported software for describing and comparing microbial communities. *Appl. Environ. Microbiol., 75*(23), 7537–7541. doi: 10.1128/AEM.01541-09.

Shendure, J., Porreca, G. J., Reppas, N. B., et al., (2005). Accurate multiplex polony sequencing of an evolved bacterial genome. *Science, 309*(5741), 1728–1732. doi: 10.1126/science.1117389.

Smith, L. M., Sanders, J. Z., Kaiser, R. J., et al., (1986). Fluorescence detection in automated DNA sequence analysis. *Nature, 321*(6071), 674–679. http://dx.doi.org/10.1038/321674a0.

Stasiewicz, M. J., Oliver, H. F., Wiedmann, M., & Den, B. H. C., (2015). Whole-genome sequencing allows for improved identification of persistent *Listeria monocytogenes* in food-associated environments. *Appl Environ Microbiol., 81*(17), 6024–6037.

Stein, C., Kuchenmüller, T., Henrickx, S., et al., (2007). The global burden of disease assessments - WHO is responsible? *PLoS Negl. Trop. Dis., 1*(3), e161. doi: 10.1371/journal.pntd.0000161.

Swaminathan, B., Barrett, T. J., Hunter, S. B., Tauxe, R. V., & CDC PulseNet Task Force, (2001). PulseNet: The molecular subtyping network for foodborne bacterial disease surveillance, United States. *Emerg Infect Dis., 7*(3), 382–389. doi: 10.3201/eid0703.010303.

Taylor, A. J., Lappi, V., Wolfgang, W. J., et al., (2015). Characterization of foodborne outbreaks of *Salmonella enterica* serovar *enteritidis* with whole-genome sequencing single-nucleotide polymorphism-based analysis for surveillance and outbreak detection. *J Clin Microbiol., 53*(10), 3334–3340. doi: 10.1128/JCM.01280-15.

Tenover, F. C., Arbeit, R. D., Goering, R. V., et al., (1995). Interpreting chromosomal DNA restriction patterns produced by pulsed-field gel electrophoresis: Criteria for bacterial strain typing. *J. Clin. Microbiol., 33*(9), 2233–2239.

Turcatti, G., Romieu, A., Fedurco, M., & Tairi, A. P., (2008). A new class of cleavable fluorescent nucleotides: Synthesis and optimization as reversible terminators for DNA sequencing by synthesis. *Nucleic Acids Res., 36*(4), e25. doi: 10.1093/nar/gkn021.

Tyson, G. H., McDermott, P. F., Li, C., et al., (2015). WGS accurately predicts antimicrobial resistance in *Escherichia coli. J. Antimicrob. Chemother., 70*(10), 2763–2769. http://dx.doi.org/10.1093/jac/dkv186.

Venter, J. C., Adams, M. D., Myers, E. W., et al., (2001). The sequence of the human genome. *Science, 291*(5507), 1304–1351.

Vincent, A. T., Derome, N., Boyle, B., Culley, A. I., & Charette, S. J., (2017). Next-generation sequencing (NGS) in the microbiological world: How to make the most of your money. *J. Microbiol Methods, 138*, 60–71.

Wheeler, D., Barrette, T., Benson, D. A., et al., (2008). Database Resources of the National Center for Biotechnology Information. *Nucleic Acids Res., 36*, D13–21.

Wu, Y., Lau, H. K., Lee, T., Lau, D. K., & Payne, J., (2019). *In silico* serotyping based on whole-genome sequencing improves the accuracy of *Shigella* identification. *Appl. Environ. Microbiol., 85*(7), e00165-19. https://doi.org/10.1128/AEM.00165-19.

Xu, R., Liu, X., Peng, B., et al., (2020). Genomic features of *Cladobotryum dendroides*, which causes cobweb disease in edible mushrooms, and identification of genes related to pathogenicity and mycoparasitism. *Pathogens, 9*(3), 232. doi: 10.3390/pathogens9030232.

Part IV

Controlling Foodborne Illness for Public Health

CHAPTER 11

Application of Microorganism in Food Production and Control of Foodborne Illness

DHRUV THAKUR, ANURAG SINGH, RAJAT SUHAG, and ATUL DHIMAN

ABSTRACT

Microorganisms associated with food not only help in the production process of certain food products but also cause a great number of diseases. This chapter discusses the characteristics of the most common microorganisms like bacteria, yeasts, molds, viruses, and parasites that are important to the food industry. In addition, foodborne outbreaks associated with these microorganisms, along with methods for controlling them, which take into account the risk-based approach and good manufacturing practices, are discussed. Some important applications of food-related microorganisms are also elaborated in this chapter. Knowledge of the control and harmful effects of foodborne pathogens will further help to ensure safe and microbes' free food.

11.1 INTRODUCTION

Microorganisms have been in this world well before we humans came into existence. Without even knowing about what complex changes they

Global Food Safety: Microbial Interventions and Molecular Advancements. Saher Islam, PhD, Devarajan Thangadurai, PhD, Jeyabalan Sangeetha, PhD, & Natália Cruz-Martins, PhD (Editors)

carry out in food; humans are still unknowingly getting the benefits out of these microbes. Microorganisms benefit society in many ways. They are necessary for the production of different products like bread, beer, cheese, and many more which are not limited to food. They also help to maintain the balance by acting as decomposers in the food chain, which are essential for regulating the nutrient cycle, and by keeping the waste in check. Beneficial microorganisms can be utilized for the production of healthy products. An example of beneficial microorganisms is the gut flora, which has various benefits, such as inducing intestinal development, enhance nutrition and also help in immunity (Sears, 2005). There is no doubt that such small creatures, invisible to the naked eye, are important for not only humans, but also to the planet as a whole. But not all of them are beneficial and some cause harm, such as the latest pandemic we have all witnessed with the novel Corona Virus (COVID-19).

The correlation between food consumption and disease was recognized at an early stage and it was Hippocrates (460 BC) who reported a strong link between food consumption and human disease (Bintsis, 2017). Foodborne diseases are one of the biggest public health issues in the world. Pathogens such as bacteria, fungi, viruses, etc., are biological agents that cause a foodborne illness event (Bintsis, 2017). These foodborne pathogens are likely to cause illness or in some extreme cases, may even cause death. Hence, it is important to know the possibility of such occurrences and to be prepared to mitigate such events. This is why these organisms have to be studied well and methods need to be developed to prevent them from causing hazards.

Generally, foodborne illness is categorized into: (a) Infection and (b) Intoxication. When a pathogen is ingested while consuming food during which the pathogen multiplies in the human host, it is known as an infection whereas; intoxication occurs when the pathogen produces a toxin in food consumed by the human host. Time varies between ingestion and onset of symptoms with infection symptoms taking longer to develop. Foodborne outbreak is defined as a similar illness in $\geq$2 people after consuming a common food and epidemiological and/or microbiological evidence implicates that the source of illness is food (Thirkell et al., 2019). These details will be covered in the further sections of the chapter, with the focus on the classification and types of microorganisms in food, their applications and the different diseases caused by them, and various methods or techniques used to control them.

11.2 TYPES OF MICROORGANISMS IN FOOD

The human food supply consists mainly of plants and animals and products thereof. Naturally, microorganisms are present in all our surroundings, so plants and animals' outer surfaces are infected by different microorganisms. There are various classifications of microorganisms. The most commonly used classification divides microorganisms majorly into bacteria, viruses, fungi, and parasites. Based on their function or effect, microorganisms that are related to foods may be categorized as "spoilage," "pathogenic" or "useful."

Any change in food quality due to which it becomes unfit for human consumption is known as food spoilage. The most prominent type of food spoilage is because of microorganisms that produce enzymes resulting in the formation of certain by-products that are undesirable in food leading to spoilage (Benner, 2014). Such microorganisms are known as spoilage microorganisms and cause huge economic losses in the supply chain. According to Benner (2014), microbiological food spoilage depends on both intrinsic and extrinsic characteristics such as moisture content, storage temperature, oxidation-reduction potential and many more. Apart from economic loss and food insecurity, spoiled food also contributes to food waste, which causes various other problems (Odeyemi et al., 2020).

Microbes that lead to desirable changes in food are the useful organisms, e.g., converting milk to curd, sugar to alcohol, and cabbage to sauerkraut. The common name for these changes is fermentation (Banwart, 1989). Both spoilage and useful organisms result in modification of food, but one is desirable and the other is not. The pathogen, as already discussed, is an organism that causes illnesses in the host (Pigłowski, 2019). According to the World Health Organization (WHO), one out of 10 people become sick and death of 420,000 people occurs annually because of eating contaminated food (WHO, 2018). Pathogens are majorly mesophilic, having an ideal growth temperature range between 20–45°C (Bintsis, 2017), which is why the human body is the preferred host for various pathogens because of the uniform temperature and nutrient rich environment (Sarmah et al., 2018). Various types of microorganisms in food are discussed in subsections.

11.2.1 BACTERIA IN FOODS

Though bacteria are important for various aspects of food production (fermentation), but they lead the charts in causing diseases related to food and are found in different shapes, sizes, types, and properties. They are unicellular prokaryotes and generally found in five forms: cocci (spherical), bacilli (rod-shaped), spirillum (spiral), spirochaete (corkscrew) and vibrio (comma). They may also form associations like chains, clusters or tetrads and may be motile or non-motile (NM) (Bibek, 2004). Bacterial cells are either Gram-positive or Gram-negative, based on the behavior of Gram-staining. In general, bacteria are composed of single membrane surrounded by a thick peptidoglycan are Gram-positive (e.g., *Lactobacillus*, *Bacillus*, *Listeria*, *Staphylococcus*, *Streptococcus*, etc.), whereas Gram-negative bacteria's thinner peptidoglycan layer is surrounded by two membranes (e.g., *E. coli*, *Salmonella*, *Shigella*, etc.). The leading pathogens are Gram-positive cocci, which causes approximately one-third of all human bacterial infections (Todar, 2012). Bacteria generally multiply by binary fission in which two cells are produced asexually from one cell, which are replicas of the original cell. Various applications of bacteria as well as pathogenic bacteria are discussed in the later section.

11.2.2 YEAST IN FOODS

Yeasts are single-celled, heterotrophic eukaryotes that belong to the kingdom fungi. In context of food sector, the yeast is generally referred to *Saccharomyces cerevisiae*, due to its immense diversity, yeasts are categorized under two distinct phyla, Ascomycota and Basidiomycota. Yeast produces a wide variety of fermented foods, including alcoholic drinks made from various substrates, fermented milk products, leavened cereal foods, and condiments (Rai and Jeyaram, 2017). For the production of traditional foods and beverages that involve fermentation, yeasts are the most commonly used microbe all over the globe.

In bakery and pastry items, yeast is widely used as a leavening agent. It produces alcohol and carbon dioxide from sugar during dough fermentation cycle, which controls the texture of baked products. *Saccharomyces cerevisiae* is the mostly used leavening agent in bakery products thus it is popularly known as baker's yeast (Değirmencioğlu et al., 2016). Apart from the generation of gas, the yeasts also generate succinic acid, which develops dough rheology as well as flavor in bread (Jayaram et al., 2014). Organic acids (acetic acid, butyric acid, etc.), are also produced using yeast as starter culture during fermenting milk (Álvarez-Martín et al., 2008). Therefore, a suitable co-starter must be chosen to limit the production of organic acid necessary for a product.

There are various forms in which baker's yeast is found, such as dried pellet, compressed, cream, granular, frozen, or encapsulated as well as in instant form (Cauvain, 2015). Its fundamental desirable qualities are the rapid use of maltose, high sugar resistance, lasting freezing tension, and the development of high CO_2 gas. The high fermentation rate is the highest sought characteristic of baker's yeast strains. Two of the most significant factors for bread volume are yeast fermentation (production of CO_2) and gas retention. The second is achieved by a network of gluten. Therefore, the components mentioned above must be balanced in order to achieve a quality of end product (Bell et al., 2018; Farid et al., 2019). Yeast helps bakers not only to increase the dough volume, but also produce compounds that develop the aroma. The most commonly developed aroma compounds in bread crumbs are alcohol, aldehydes, 3-hydroxy-2-butanone (acetoin), 2,3-butanedione (diacetyl), and esters (Hazelwood et al., 2008; Birch et al., 2013).

Alcoholic beverages, particularly beer, wine, and distilled spirit are produced by fermentation using yeast. The most important wine and brewing species is *Saccharomyces cerevisiae* or brewer's yeast (Yu et al., 2018). In alcoholic beverage processing, fermentation is a key microbial activity that affects the quality of the product and regulating these effects is important for assuring the quality of alcoholic beverages. Depending on the flocculation behavior, brewer's yeast is classified into two categories, ale, and lager type, which represent the top and bottom-fermenting, respectively. In the beverage industry, *Saccharomyces cerevisiae* is called ale type and *Saccharomyces uvarum* (*carlsbergensis*) is called lager type (Ferreira et al., 2010). Ale beer is produced by fermentation carried above 15°C temperature by top-fermenting yeast. In this process, after completion of fermentation, yeast is carried to the top surface of the wort. Bottom fermenting yeast produces lager beer by fermentation at below 15°C temperature. Lager beer is consumed worldwide as compared to ale type beer (Gonçalves et al., 2016).

The major fermenting yeast for the production of wine is also *Saccharomyces cerevisiae* (Blanco et al., 2012). Some non-*Saccharomyces* yeast species have been isolated from wine such as *Kloeckera* spp. and *Candida* spp. *Kloeckera apiculata* dominates the non-*Saccharomyces* yeast found in grape (Azhar et al., 2017).

In dairy products, the most common yeast species are: *Candida lusitaniae, Saccharomyces cerevisiae, Candida krusei, Galactomyces geotrichum, Kluveromyces lactis, Yarrowia lipolytica, Debaryomyces hansenii, Kluyveromyces marxianus, Candida zeylanoides* and several *Pichia* species. The contribution of these species of yeast is to add flavor and color, which plays a vital role for manufacturing fermented dairy products (Fleet, 2007; Jacques and Casaregola, 2008; Ballin, 2010).

Reports suggest that meat and meat products are also the ideal media for the growth of yeast. The frequently isolated species from fermented meat particularly sausages are *Candida, Debaryomyces, Yarrowia, Trichosporon, Pichia, Cryptococcus*, and *Rhodotorula* (Fleet, 2011; Steensels et al., 2015). Because of its positive contribution to the final product, *Debaryomyces hansenii* is often used in fermented meat products as a starter culture (Parente et al., 2017; Rai et al., 2019).

11.2.3 MOLD IN FOODS

Acidified and high acid foods are susceptible to the growth of mold spores. Mold growth may lead to an increase in the pH of food above 4.6, which enhance the chances of *C. botulinum* growth (Awuah et al., 2007). Commonly occurring spore-forming molds linked to spoilage in such foods are *Byssochlamys nivea, Eupenicillium javanicum, Byssochlamys fulva, Talaromyces avellanus, Penicillium expansum, Eurotium repens, Neosartorya fischeri* and *Talaromyces flavus* (Daher et al., 2017; Xiao et al., 2017). Though there is a less chance of low-acid foods being spoiled by fungi as compared to high acidified ones because they are usually held at lower temperatures. Spoilage is assumed to be caused by contamination during bottling or processing, or because of heat-resistant nature of mold spores (Garnier et al., 2017).

Mold spore-formers commonly occurring in foods are *B. nivea, Eupenicillium brefeldianum, N. fischeri* and *Hamigera avellanea* which are usually found in cream cheese (Pitt and Hocking, 2009). Most commonly found molds in dairy products are *Penicillium* and *Aspergillus* (Garnier et al., 2017). Some *Aspergillus* species have also been reported in peanuts, tree nuts, other oilseeds (cottonseed), and milk. Molds like *Aspergillus*, *Penicillium*, and *Eurotium* were also obtained from spoiled refrigerated as well as fresh poultry and meat products (Dave and Ghaly, 2011).

El-Badry and Raslan (2016) studied mold contamination in different types of cheese. The prevalent mold genera were *Penicillium*, *Aspergillus*, *Alternaria*, and *Cladosporium*. Detailed study revealed specific species like *A. niger*, *A. parasiticus*, *A. flavus* and *A. ochraceous*. Highest mold content was reported in Roumy cheese preceded by cheddar and white cheese.

Several issues have been reported by different researchers which are associated with mold spoilage in food products. *Pichia burtonii* causes surface spoilage of bread, popularly called chalk mold which results in pink or white patches (Deschuyffeleer et al., 2011; Saranraj and Geetha, 2012). Thread mold, commonly associated with cheese is caused due to the growth of mycelium of several molds such as *Phoma*, *Aspergillus*, and *Penicillium* (Hocking and Faedo, 1992). Mold growth in beverage containers at the bottom; often described as a "mouse in my product" due to the dense filamentous appearance is called puffballs which is commonly due to the growth of *Paecilomyces (Byssochlamys)* in heat-treated juices (Snyder and Worobo, 2018).

11.2.4 VIRUS IN FOODS

Viruses are intercellular obligate parasites that need host cells to propagate and infect. From a structural standpoint, the infectious extracellular particle (virion) consists of either single stranded (ss) or double stranded (ds) DNA or RNA surrounded by a protein shell (International Committee on Taxonomy of Viruses, 2012).

In the human gastrointestinal tract, a large number of different viruses, which cause a wide range of diseases, can be found. While any virus capable of causing diseases after consumption can probably be considered foodborne and/or transmitted by water, most of the viral foodborne diseases identified are gastroenteritis or hepatitis, which are caused respectively by Norovirus (NoV) and Hepatitis A virus (HAV). In addition to this, enteroviruses, rotavirus, sapoviruses, adenoviruses, astrovirus, and hepatitis E virus (HEV) are also associated. Patients with gastroenteritis (gastrointestinal inflammation) can excrete around 10^{13} whereas those with hepatitis excrete around 10^{10} virus particles per gram of stool (Costafreda et al., 2006; Ozawa et al., 2007; Caballero et al., 2013). Nausea, diarrhea, and stomach pain and sometimes fever and headache are symptoms for viral gastroenteritis. While the most severe cases are usually caused by viral gastroenteritis agents, viruses like NoV are responsible for most cases (Hall et al., 2014; Bosch et al., 2018).

The most common outbreaks which are foodborne are related to bivalve molluscan shellfish and fruits which may be contaminated prior to harvest. According to European Commission RASFF (2019), 85% of the foodborne risks involved bivalve mollusks, while 15% of the risks involved fruits. Clams normally obtained from bivalves, which were frozen, triggered 57% of all foodborne risks, preceded by oysters (15%) and mussels (11%). Frozen strawberries as well as raspberries participated in 5% of foodborne risks and frozen berry mixes in 3% of risks. Improper hygiene practices lead to contamination and thus uncooked or partially cooked food products are the most at risk. Surfaces and other types of fomites used in food preparation can serve as vehicles for transmission of foodborne viruses (Bosch et al., 2016).

11.2.5 PARASITES IN FOODS

Parasites are eukaryotic organisms that are divided between single-cell protozoa and complex hermaphrodite helminths. Including the flagellate *Giardia*, *Cryptosporidium*, and *Toxoplasma*, cestodes *Echinococcus* and *Taenia*, *Anisakis*, and *Trichinella* roundworms and trematodes *Clonorchis*, *Opisthorchis*, and *Fasciola* are prevalent worldwide (FAO/WHO, 2014).

Many sources have triggered human parasite infections and various transmission modes are obtained. Most commonly oral transmission through food and water takes place. Because of different environments in which food is produced globally, different survival, multiplication, and transmission strategies are available for parasites. While several parasites which are waterborne (*Giardia* and *Entamoeba*) have simple life cycles, the foodborne parasites contain multiple life cycles and host species which harness the exogenous, endogenous environment and food habits of consumers. Meat-borne parasites typically reproduce and achieve infectiousness in protective structures which are formed during animal development such as muscle cysts. Some examples are *Toxoplasma gondii*, *Trichinella spiralis* and *Sarcocystis hominis*. Others like *Giardia* and *Cryptosporidium* contaminate the surfaces of fruits and vegetables, thereby infecting consumers (Robertson et al., 2015; Bintsis, 2017).

11.3 APPLICATIONS OF MICROORGANISMS IN FOOD PRODUCTION

Microbes play an essential role in producing a variety of products with flavor characteristics and also provide technological advantages (Wood, 2013). In many food preparations, microorganisms and their enzymes are widely used to enhance the tastes and textures of products and give industry a significant economic gain (Raveendran et al., 2018). Microorganisms application in food includes organic acid production like citric, lactic, gluconic, and ascorbic acid; manufacturing enzymes for food industries like rennet, protease, glucoamylase, α-amylase, glucose isomerise and pullulanase; production of amino acids like L-glutamate, lysine, methionine, etc., by fermentation; producing yogurt, cheese, probiotics, etc., by lactic acid fermentation; producing beer by alcoholic fermentation (Pai, 2003); producing flavors which are used as a food additive and producing xanthan gum (Couto and Sanromán, 2006) and also preparing food colorants from microbial pigments (Nigam and Luke, 2016). Fermentation technology used for food production is a fairly cost-effective, low-energy operation, important for the shelf life of product and microbiological health (Liu et al., 2011). Fermentation is an anaerobic process involving oxidation of carbohydrates generating different metabolic products and end products generally includes acids, alcohols, and carbon dioxide. Preservation is achieved by the production of metabolites that inhibits pathogens and spoilage microbes. It also helps in removing antinutritional factors and toxins from food. Metabolites like ethanol, carbon dioxide, antimicrobials, bacteriocins, and hydrogen peroxides are produced during fermentation. Apart from this, fermentation improves texture and nutritional value along with the production of flavors and aromatic compounds (Caplice and Fitzgerald, 1999; Terefe, 2016b). Lactic acid bacteria (LAB) cause the fermentation of sugars via homo-, hetero-, or mixed acid fermentation (Sauer et al., 2008). Moreover, LAB has potential importance for food preservation and fermentation. They produce proteinases, bacteriocins, and exopolysaccharides which are responsible for its preservation action (Topisirovic et al., 2006). Bacteriocins produced by these groups show inhibiting activity against pathogenic organisms such as *Clostridium botulinum* and *Listeria monocytogenes*. Fermentation by LAB also produces flavoring compounds like acetaldehyde and diacetyl, and increases nutritional values by producing vitamins and antioxidants. *Lactobacillus*, *Lactococcus*, *Enterococcus*, *Streptococcus*, *Pediococcus*, and *Leuconostoc* genus are included in it (Paul et al., 2002). Fungal, lactic acid and alkaline fermentation can be categorized widely in a typical food fermentation process. For example, natto is a product prepared by alkaline fermentation, soy sauce prepared by fungal fermentation and yogurt prepared by lactic acid fermentation (Terefe, 2016b). In industries, solid-state fermenter and stirred tank reactors are used for fermentation (Wood, 2013). Similarly, acetic acid bacteria (AAB) fermentation is important in production of vinegar which produces high concentration of acetic acid from ethanol. Two main genera of AAB are *Acetobacter* and *Gluconobacter* (Sengun and Karabiyikli, 2011). *Aspergillus niger*, a filamentous fungus and *Yarrowia lipolytica* a yeast is used industrially for manufacturing citric acid. LAB and filamentous fungus *Rhizopus oryzae* produce lactic acid. In food industries, these organic acids are used as food additive ingredient (Sauer et al., 2008). *Xanthomonas campestris* bacteria produce xanthan gum which is a commonly used as an emulsifier, stabilizer, and texture enhancer in food industry (Couto and Sanromán, 2006). Microbial enzymes are used in brewing, starch liquefaction, beer production, meat tenderization, clarification of juice, cheddar cheese production, etc. (Raveendran et al., 2018). Thus, microorganisms have huge application in food production as well as preservation. More applications have been shown in Table 11.1.

TABLE 11.1 Applications of Microbes in Food Industry

Product	Microorganism
Fermented Food	
Bread	*Saccharomyces cerevisiae*, other yeasts, LAB
Idli	LAB (*Leuconostoc mesenteroides, Candida, Trichosproron pullulans, Torulopsis, Enterococcus faecalis*)
Soy sauce	*Aspergillus oryzae* or *A. soyae, Lactobacillus, Zygosaccharomyces rouxii*
Nan	*Saccharomyces cerevisiae*, LAB

TABLE 11.1 *(Continued)*

Product	Microorganism
Sauerkraut	LAB, *Ln. mesenteroides, Lb. brevis, Lb. plantarum, Lb. curvatus, Lb. sake*
Pickles	*Pediococcus cerevisiae, Lb. plantarum*
Yogurt	*Streptococcus thermophilus, Lactobacillus delbrueckii* subsp. *bulgaricus*
Lassi	LAB and yeast
Butter	LAB
Dosa	*Ln. mesenteroides, Streptococcus faecalis, Torulopsis candida, Torulopsis pullulans*
Natto	*Bacillus subtilis natto*
Enzymes	
α-Amylase	*Bacillus amyloliquefaciens, Bacillus stearothermophilus* or *Bacillus licheniformis*
Glucoamylase	*Aspergillus niger* and *Aspergillus awamori*
Protease	*Aspergillus usamii*
Lactase (β-galactosidase)	*Kluyveromyces lactis* and *Kluyveromyces fragilis*
Lipase	*Penicillium camemberti, Rhizopus miehei, Candida antarctica* (CALB) and *Pseudomonas* spp.
Phospholipase	*Fusarium oxysporum*
Food Additives	
Citric acid	*Aspergillus niger* and *Candida guilliermondii*
Lactic acid	LAB, *Lactobacillus delbruckii, L. casei, L. acidophilus*
Gluconic acid	*Aspergillus niger* and some *Penicillium* species
Riboflavin	*Bacillus subtilis*
β-carotene	*Blakeslea trispora* and *Dunaliella salina*
Astaxanthin	*Haematococcus Pluvialis, Haematococcus lacustris* and *Xanthophylomyes dendrorhous*
Canthaxanthin	*Haematococcus lacustris* and *Bradyrhizobium* spp.
Carrageenan	*Chondrus crispus, Chondrus ocellatus, Eucheuma cottonii, Eucheuma spinosum*
Furcelleran	*Furcellaria fastigiata*
L-glutamate, lysine, threonine	*Corynebacterium glutamicum*
Xanthan gum	*Xanthomonas campestris*

Source: Adapted from: Caplice and Fitzgerald (1999); Pai (2003); Couto and Sanromán (2006); Tamang et al. (2015); Terefe (2016a); Raveendran et al. (2018).

11.4 FOODBORNE ILLNESS

Microorganisms use food as a source of nutrition for their growth. They metabolize food and secrete by-products which make the food inedible, also result in health-related problems when consumed (Addis and Sisay, 2015). Foodborne pathogens cause many sporadic cases of disease and clinical complications and massive challenging spreads in many countries. *Escherichia coli* O157:H7, found in hamburgers from a fast-food chain was pathogen identified in 1982 in an eruption of bloody diarrhea. *Listeria monocytogenes* cause meningitis and other infections (Tauxe, 1997). These foodborne illnesses can also be life threatening (Linscott, 2011). Contamination in food can be due to bacteria, viruses, parasites, toxins, or chemicals which cause foodborne diseases or foods poisoning, or food-borne illness. It is estimated that foodborne illnesses are caused by major pathogens which not only affects human health but economy also. Pathogenic bacteria (*Escherichia coli*, *Campylobacter jejuni*, *Bacillus cereus*, *Clostridium botulinum*, etc.), parasites (*Toxoplasma gondii*, *Cyclospora cayetanensis*, and *Trichinella spiralis*) as well as viruses (Noroviruses) are responsible for different foodborne diseases (Bintsis, 2017). The two most common foodborne illnesses are food infections and food intoxications.

11.4.1 FOODBORNE INFECTIONS AND FOOD INTOXICATIONS

Ingesting contaminated food and water with pathogens, their toxins and chemicals cause food poisoning. The term "food poisoning" is generally used for both food infections and food intoxication (Addis and Sisay, 2015). Ingestion of viable pathogens along with food leads to their establishment in consumers and hence is termed as food infection. Disease causing microorganisms like bacteria, viruses, and parasites release toxins and infect the gastrointestinal tract, damages intestinal epithelium and cause gastroenteritis. Whereas in food intoxication, either microbially produced toxins or biotoxicants or any poisonous substance after getting ingested with food infects the intestine. Toxicants can come from plants, animals, bacteria, fungus, and chemicals (Al-Mamun et al., 2018). In the majority of food infections and intoxications symptoms like diarrhea, vomiting, nausea, abdominal cramps, fever, chills, headaches, dehydration, etc., arise (Addis and Sisay, 2015). Various foodborne infections caused due to different microorganisms are shown in Table 11.2.

TABLE 11.2 Foodborne Infections Caused Due to Different Microorganisms

Organism	Major Infections/ Toxin Produced	Symptoms, Clinical, and Other Issues
Bacteria		
Aeromonas	Septicemia Gastroenteritis	Fever, vomiting Summer diarrhea Meningitis Chronic enterocolitis Pulmonary and wound infections
Bacillus cereus	Diarrheal food poisoning Emetic intoxication	Vomiting, respiratory distress diffuse bleeding, muscle cramps
Brucella	Brucellosis	Sweating, chills, weakness, migraine, weight loss and fever
Arcobacter	Enteritis Septicemia Bacteremia	Mastitis, Diarrhea, Gastritis, Reproductive abnormalities
Campylobacter	Dysentery	Diarrhea, fever, abdominal cramps, headache, asthenia, and anorexia
Clostridium botulinum	Botulism	Blurred vision, dropping eyelids, slurred speech, dry mouth, muscle weakness and difficulty swallowing
Clostridium perfringens	Food poisoning Necrotic enteritis	Abdominal pains, nausea, and diarrhea
Escherichia coli	EPEC EHEC/STEC EIEC ETEC	Diarrhea, fever, vomiting Diarrhea, hemorrhagic colitis Dysentery and fever Dysentery
Listeria monocytogenes	Hemolysin Listeriosis	Prenatal infection, meningitis, gastroenteritis, septicemia and cephalitis
Plesiomonas shigelloides	Gastroenteritis	Dysentery, nausea, vomiting, fever, chills, abdominal pain, and migraine headache
Salmonella enterica	Enterocolitis Enteric fever	Abdominal pain, diarrhea, vomiting, fever, nausea
Shigella	Acute inflammatory colitis	Headache, dysentery, intestinal cramps
Staphylococcus aureus	Food poisoning	Nausea, diarrhea, sweating, headache, dehydration, drop in blood pressure

TABLE 11.2 *(Continued)*

Organism	Major Infections/ Toxin Produced	Symptoms, Clinical, and Other Issues
Vibrio cholerae	Cholera	Watery diarrhea, vomiting, abdominal pain, nausea, headache, fever, and chills
Yersinia	Gastroenteritis Septicemia	Nausea, headache, fever, vomiting
Fungal Mycotoxins		
Aspergillus flavus, A. parasiticus, A. nominus, A. pseudotamarii	Aflatoxins	Aflatoxicosis Toxins are carcinogenic, mutagenic, hepatotoxic, teratogenicity
Fusarium graminearum *F. culmorum*	Deoxynivalenol (DON or vomitoxin)	Adverse effects immunity
F. verticillioides *F. proliferatum*	Fumonisins	Carcinogenic and neural tube defects
Aspergillus alutaceus *A. carbonarius* *A. niger*	Ochratoxin A	Carcinogenic, Balkan nephropathy, immunosuppressive, teratogen, and genotoxic
Penicillium spp. *Aspergillus clavatus*	Patulin	–
Fusarium graminearum *F. crookwellese*	Zearalenone	Premature thelarche, pubarche
Protozoa		
Cryptosporidium	Gastrointestinal problems	Watery diarrhea, vomiting, nausea, fever, abdominal pain, anorexia
Cyclospora	Cyclosporiasis	Watery stools, burping, and flatulence
Entamoeba histolytica	Amebiasis	Dysentery with blood and mucoid diarrhea
Giardia lamblia	–	Diarrhea, vomiting, nausea, abdominal pain, flatulence, fatigue, and weight loss
Parasites		
Anisakis simplex	–	Diarrhea, vomiting, nausea, abdominal pain
Ascaris	–	Pneumonia, coughing, wheezing, vomiting, nausea, abdominal pain

TABLE 11.2 (Continued)

Organism	Major Infections/ Toxin Produced	Symptoms, Clinical, and Other Issues
Viruses		
Hepatitis A Hepatitis E	–	Infection in liver, vomiting, nausea, abdominal discomfort, fever, headache
Norovirus	–	Gastroenteritis, fever, headache, vomiting, nausea
Rotavirus	–	Vomiting, diarrhea, fever, dehydration

Source: Adapted from: Jay et al. (2008); Heredia et al. (2009); Linscott (2011); Addis and Sisay (2015); Bintsis (2017); Al-Mamun et al. (2018).

11.5 IMPORTANT FOODBORNE PATHOGENS

11.5.1 BACTERIA

As already discussed, the most common cause of foodborne diseases are bacteria. Some of these are capable of forming spores which are highly resistant, dormant structures that help the organisms to survive in harsh environmental conditions (e.g., *Clostridium botulinum*). Some produce toxins which are heat-resistant (e.g., *Staphylococcus aureus*). Mostly they are mesophilic, but psychrotrophs are also common (Bintsis, 2017). Some of the major pathogenic bacteria are discussed in this section.

11.5.1.1 BACILLUS CEREUS

Bacillus cereus, a member of the Bacillaceae family, is facultative anaerobic, Gram-positive, toxin, and spore producing bacteria (Granum, 1994; Griffiths and Schraft, 2017). Like other Bacilli, it is commonly found in soil (Kotiranta et al., 2000), but is also found in air, water, plant, and animal material as well (Griffiths and Schraft, 2017). It multiplies quickly at room temperature. It is a mesophilic species which grows in the temperature range of 10–50°C, a pH of 4.9–9.3 and a minimum water activity of 0.91 (Bintsis, 2017).

Among different species of *Bacillus* group, the most common species associated with foodborne diseases is *B. cereus* (Griffiths and Schraft, 2017). Among the two main types of illness caused by *B. cereus*, one is diarrheal, which is caused by complex enterotoxins which form in the small intestine during their vegetative growth while the other is the emetic (vomiting) toxin which forms in food itself by the growing cells (Granum and Lund, 1997; Tewari and Abdullah, 2014). The quick onset of the emetic form includes nausea and vomiting, while diarrhea and abdominal pain characterize the late start of the diarrheal type. Nevertheless, these occur as a result of the survival of its endospores after cooking, which germinate at a favorable condition during storage and result in the growth of vegetative cells. Different foods which are associated with diarrheal type includes vegetables, meat, milk, and fish, whereas the vomiting-type outbreaks are generally associated with starchy foods such as rice, potato, pasta, and cheese products (Dilbaghi and Sharma, 2007).

Because of the symptoms being mild and of short duration (Griffiths and Schraft, 2017), it is not a notifiable disease and hence the data related to their incidence is limited. 58 outbreaks which resulted in 1,123 cases between 1973 and 1987 were recorded (Bibek, 2004) and other outbreaks were also mentioned (Granum and Lund, 1997; Bintsis, 2017; Griffiths and Schraft, 2017).

Thirkell et al. (2019) reported an outbreak of *B. cereus* at a restaurant in Canberra, Australia. Four people reported gastrointestinal illness and a high level of *Bacillus cereus* in beef was reported (19,000 colony forming units/gram). Another food poisoning outbreak of diarrheal *B. cereus* was reported in the literature by Son et al. (2019) which was associated with contaminated chicken satay after attending mass *iftar* in a mosque in Tegal Kenongo village in Indonesia. Out of the 303 villagers involved in the event, 188 villagers had symptoms such as diarrhea, nausea as well as abdominal cramps. In other research carried out by Chen et al. (2019), 209 cases of *B. cereus* related outbreak were confirmed due to consumption at high-risk stalls at a university in the Shunyi District of Beijing, China in 2018. The factors responsible for making *B. cereus* a potential threat in foods is endospore producing ability, capability of growing and surviving at sub-zero temperatures and production of toxin with rice and milk being the most commonly contaminated food items (Tewari and Abdullah, 2014).

11.5.1.2 CAMPYLOBACTER JEJUNI

It is a member of family Campylobacteriaceae which results in one of the most commonly identified bacterial causes of acute gastroenteritis (AGE) in the world (Acheson and Allos, 2001; Nyati and Nyati, 2013; Wieczorek et al., 2018). They are Gram-negative, spiral, rod-shaped, or curved non-spore-forming bacteria and maybe motile or NM depending on the species (Kaakoush et al., 2015). Contrary to other organisms, their main sources of energy are amino acids or tricarboxylic acid intermediates rather than carbohydrates (Banwart, 1989). They require microaerophilic environment having 5% oxygen, 8% CO_2, and 87% N_2 with the temperature between 32–45°C. They are also sensitive to various other extrinsic parameters, like pH<5.0, NaCl>2.5%, pasteurization temperatures as well as drying (Bibek, 2004).

Raw chicken and raw milk are frequently contaminated by *C. jejuni* and it is usually carried through healthy cattle and flies. Another potential source is non-chlorinated water (Bibek, 2004; Dilbaghi and Sharma, 2007; Kaakoush et al., 2015). Mainly they are transmitted to people during handling, preparation, and consumption of contaminated food (Wieczorek et al., 2018). But practicing proper sanitation as well as some mild treatments such as pasteurization can easily control such infections.

Human campylobacteriosis varies significantly and affects all ages which result in mild to serious injuries to permanent and neurological symptoms. Infections are generally self-limited and can be treated in a short time without antibiotic treatment. However, persistent infections in young, elderly, or immune-compromised individuals may occur in particular (Ammar et al., 2019). It causes around 166 million diarrheal cases and 37,600 deaths globally per year, and it may also result in the development of Guillain-Barré syndrome (GBS) which is an acute flaccid paralysis (Oh et al., 2018).

Various outbreaks related to *C. jejuni* have been reported (Kaakoush et al., 2015; Bintsis, 2017). Some recent outbreaks were covered in the literature by Glashower et al. (2017) in which 5 people who consumed undercooked chicken liver mousse in Washington were identified in 2016. In Seoul, Korea, out of the 257 policemen who consumed lunch, 55 were infected. It was found out that the infection was a result of the cross-contamination of *C. jejuni* from raw chicken via environmental sources (Kang et al., 2019). According to the Ministry of Health, Labor, and Welfare (MHLW), in Japan *Campylobacter* overtook *Salmonella* and *Vibrio parahaemolyticus* as the leading cause of food poisoning in 2003 but since some revision in regulations related to the consumption of chicken, the incidents have decreased (Vetchapitak and Misawa, 2019).

Controlling its access to uncooked foods especially which are of animal origin is quite difficult. However, an easy and simple way for reducing its load in raw foods is properly following sanitation practices. Avoiding consuming raw foods, heating it, and preventing the contamination after heating are crucial to control campylobacteriosis in foods of animal origin.

11.5.1.3 CLOSTRIDIUM BOTULINUM

The genus *Clostridium* is one of the largest bacterial genera which comprises of about 180 species (Dürre, 2014). *Clostridium botulinum* is a Gram-positive, anaerobic, spore-forming, rod-shaped bacterium that produces a neurotoxic protein. Its spores are distributed in different places such as soil, mud, sewage, fruits, and vegetables and in the intestines of animals and fishes. They are susceptible to low water activity (0.93), low pH (<4.6) and moderately high salt (5.5%). But because of the highly heat resistant spores, it can survive in minimally or insufficiently processed foods (Dilbaghi and Sharma, 2007).

It produces botulinum neurotoxin, the deadliest poisonous substance (LD_{50} of 1 ng/kg) which in fact is also viewed as a potential bioweapon (Shukla and Sharma, 2005). On consuming this toxin-containing food, a severe form of food poisoning occurs which is known as Botulism, but it is quite rare. 962 outbreaks related to botulism occurred in the United States from 1899 to 1990 in which 1,036 deaths were reported out of 2,320 cases, which show its high mortality rate (Solomon and Lilly, 2017). The botulinum neurotoxins are of seven types, ranging from A to G, which is based on the specificity of the antigen of the toxin produced by each strain. The type A, B, E, and F cause botulism in humans; types C and D in birds and mammals; and type G, is not clearly implicated in a botulism case yet (Bintsis, 2017). Recently, a species producing type H toxin has also been identified by Barash and Arnon (2014), which needs to be studied more. Mostly contaminated type of foods by types A and B are home-canned vegetables (Solomon and Lilly, 2017). At initial stages, some gastrointestinal symptoms (nausea, vomiting, diarrhea, and constipation) can be seen. Nevertheless, if the amount of toxin ingested is higher, neurological symptoms that include blurry vision, trouble speaking, swallowing, breathing, dry mouth and paralysis of involuntary muscles of lung and heart occur after a short period which results in respiratory failure and ultimately death (Dilbaghi and Sharma, 2007).

A failure in applying the *C. botulinum* cook to foods (121°C for 3 min) led to many outbreaks associated with *C. botulinum* as covered by Bintsis (2017). There were 197 outbreaks reported between 1920 and 2014 out of which majority (55%) occurred in the United States with type A toxin responsible for almost half of them (Fleck-Derderian et al., 2018). Mazuet et al. (2015) also reported human botulism outbreak, which was caused by ham consumption containing botulinum neurotoxins B and E. Another case was registered in New York in 2018 where 3 women were subjected to botulinum neurotoxin from home-canned peas (Bergeron et al., 2019). Such outbreaks show the importance of processing of foods suitably as well as storing and consuming safe food free of such diseases. Continued surveillance and prompt investigation is the key to reducing the impact of possible botulism outbreaks on public health.

11.5.1.4 CLOSTRIDIUM PERFRINGENS

The anaerobic, Gram-positive spore-former *Clostridium perfringens* is present in soil, foods, sewage, intestines, and feces of animals and humans (Li et al., 2013; Bhattacharya et al., 2020). The cells can form small chains and differ in size. Optimum growth conditions are a temperature between 10 to 52°C. They do not prosper at a low pH (<5.0), in high NaCl concentration ($>5\%$), and water activity less than 0.93. Similar to *C. botulinum*, its vegetative cells are also prone to mild-heat treatment such as pasteurization, but because of spores, it can withstand several hours of boiling (Bibek, 2004). It causes about a million illnesses each year, which makes it a common bacterial cause of foodborne illness which is generally related to the consumption of contaminated meat and poultry (Grass et al., 2013).

There are five groups (A–E) of *C. perfringens* strains based on the production of four major toxins namely alpha-, beta-, epsilon-, and iota-toxins (Shimizu et al., 2002). Almost all cases of foodborne gastroenteritis which involve *C. perfringens* are because of type A infection after the ingestion of highly contaminated foods with greater than 10^6–10^7 viable vegetative cells, which produce enterotoxin following sporulation in the small intestine. During cell lysis, enterotoxin produced during sporulation is released along with the spores. The enterotoxin binds to epithelial cells after release, causing cytotoxic damage to the cell membrane and subsequent permeability changes leading to diarrhea and abdominal cramping (Bibek, 2004; Bintsis, 2017). Its single strain can carry about three different toxin plasmids, with up to three distinct toxin genes in one plasmid (Li et al., 2013).

Several outbreaks have been reported by *C. perfringens* from contaminated food. One such outbreak took place in 2018 when 34 reports of diarrhea with abdominal cramps were reported in the West Midlands, England. On investigating it was found that cheese sauce containing leeks which was leftover and reheated was associated with illness (Bhattacharya et al., 2020). In 2019, 4 athletes of 47 team members who took part at the Panhellenic Handball Championship for children in Greece reported gastroenteritis. On analyzing it was found that minced beef was positive for *Clostridium perfringens* because of improper storage and it was also not re-heated properly. It was the first instance of a *Clostridium perfringens* foodborne outbreak in Greece (Mellou et al., 2019). Packer et al. (2020) also reported about an outbreak related to *Clostridium perfringens* in England in 2018. Out of 293 people, 81 were affected because of the meat buffet.

11.5.1.5 CRONOBACTER SAKAZAKII

It belongs to the Enterobacteriaceae family and was known as *Enterobacter sakazakii* from 1980 to 2007 (Mardaneh and Dallal, 2017; Henry and Fouladkhah, 2019). It is a Gram-negative, motile, facultative anaerobic, mesophilic, rod shaped and opportunistic organism which does not form spores (Fakruddin et al., 2013). It is highly aggressive in hypersensitive people, such as children and the elderly people (Parra-Flores et al., 2018). It is present in a number of foods as well as beverages which are not subjected to inactivation of pathogens. The appropriate nature of infant cereals and some fruit and vegetables for its growth is especially concerning (Beuchat et al., 2009). It grows optimally at 37–43°C and under low moisture conditions with a water activity of 0.30–0.83; it can survive for about a year (Henry and Fouladkhah, 2019). Thus, it is highly found in powdered infant formula and powdered milk because of lack of processing before consumption and low moisture levels (Beuchat et al., 2009; Mardaneh and Dallal, 2017).

C. sakazakii causes life-threatening meningitis (swelling of protective membrane which covers the spinal cord and brain), septicemia (systemic spread of bacteria and toxins in the blood), and necrotizing enterocolitis in infants (Drudy et al., 2006), but diarrhea and urinary tract infection are also observed (Parra-Flores et al., 2018) with a mortality rate of 10–80% (Mardaneh and Dallal, 2017). Other potential sources of contamination are retail foods such as dried meats, legumes, nuts, dried flours, and spices (Bintsis, 2017).

Various outbreaks from 1958–2016 have been covered in the article by Henry and Fouladkhah (2019) which have been majorly associated with infant mortality and morbidity. One in 100,000 infants in the USA has been reported for *Cronobacter* infection. In very low birth weight infants, this incidence rate rises to 9.4 per 100,000 (Holý and Forsythe, 2014). In October 2016, foodborne AGE resulted in 156 cases in a school in Jiangning District, China (Yong et al., 2018). In this study, a noteworthy feature was that *C. sakazakii* caused foodborne AGE in healthy adults, despite being an opportunistic pathogen which affects hypersensitive people. The species shows great genetic diversity, e.g., in 1994 an outbreak of *C. sakazakii* occurred in a French newborn intensive care unit, the entire genome phylogeny of 26 isolates revealed four distinct clusters, each associated with a distinct sequence type (Masood et al., 2015).

The crucial recommendation given in the risk assessments done by the Food and Agriculture Organization-World Health Organization (FAO-WHO) was to reconstitute powdered infant formula with hot water (>70°C), minimizing storage time, and storage at <5°C if necessary (Holý and Forsythe, 2014).

11.5.1.6 ESCHERICHIA COLI

E. coli is a Gram-negative, oxidase-negative, rod-shaped bacterium which belongs to the family Enterobacteriaceae. In 1885, Theodor Escherich described this microorganism for the first time. It grows at 37°C, both aerobic and anaerobic environments are suitable, and it may be NM or motile, with peritrichous flagella and is easily extracted from the fecal samples (Croxen et al., 2013). The required pH for optimum growth is 6.0–8.0 (Bintsis, 2017). Its carriers include animals, people, food crops and meat can also be contaminated as a result. Contamination generally occurs when food or water come in contact with feces.

Most *E. coli* are harmless and live in human's and animal's gastrointestinal tract, thus contributing to intestinal health. Usually, *E. coli* and its human host have a symbiotic relationship; thereby coexist in good health (Kaper et al., 2004). However, consuming food or water which is contaminated with *E. coli* may cause moderate to AGE. In 1997 the estimated number of cases each year in the USA caused due to *E. coli* was 115,000 (Tauxe, 2002).

Some of the strains acquired virulence factors through plasmids, transposons, bacteriophages, and/or pathogenicity islands (PAIs), thereby evolving into pathogenic *E. coli*. These can be grouped on the basis of serogroups, clinical symptoms, pathogenicity mechanisms or virulence factors (Lim et al., 2010). There are six well-described groups amongst the intestinal pathogens which include: enteropathogenic *E. coli* (EPEC), enterohemorrhagic *E. coli* (EHEC) also known as Shiga toxin-producing *E. coli* (STEC), enterotoxigenic *E. coli* (ETEC), enteroaggregative *E. coli* (EAEC), enteroinvasive *E. coli* (EIEC) and diffusely adherent *E. coli* (DAEC) (Kaper et al., 2004).

Among these, EHEC produces Shiga toxins (Stxs) in humans and cause hemorrhagic colitis (HC) and sequelae hemolytic uremic (HU) syndrome which are life-threatening (Lim et al., 2010). While colonizing the intestinal tract, Stx is produced by EHEC, which enters the host through epithelial cells and acts on submucosal immune cells that release cytokines. The cytokines further result in inflammation and increase the expression of the Stx receptor globotriaosylceramide (Gb3), thereby targeting the intestine, kidneys, and brain (García et al., 2010). Research indicated that 85–95% of cases of HU syndrome in North America are caused because of the STEC strain O157:H7 and the remaining 5–15% are because of non-O157 Shiga toxin-producing *E. coli* (Armstrong et al., 1996) which shows the higher pathogenicity of STEC. As STEC infections result in various outcomes from asymptomatic infection to death, this is a significant public health problem and is therefore listed in many countries as a notifiable infectious disease (Terajima et al., 2017).

Recently in Gyeongsangbuk-Do province of South Korea, an outbreak occurred in elementary school children as reported by Lim et al. (2020). In this out of 106 students, 13 developed food poisoning due to diarrheagenic enteropathogenic *Escherichia coli* infection because they consumed cucumber chili which had contaminated water along with other food items. In another outbreak in 2013, 7 people were diagnosed with the same outbreak strain of *E. coli* O157:H7 in three different Canadian provinces. They consumed various forms of raw (tartare) products made of beef, salmon, veal, duck, or tuna, between 3 and 6 December at two different restaurants in the same city (Gaulin et al., 2015). The second-largest outbreak related to *E. coli* occurred in 2011 which was caused by a rare strain of *E. coli* O104:H4. More than 4,000 people in 16 countries became ill between 21 May and 22 July 2011 and 50 people died (Bintsis, 2017).

11.5.1.7 LISTERIA MONOCYTOGENES

It is a psychotropic, Gram-positive, motile, and non-sporulating facultative anaerobe which causes invasive human and animal diseases, especially central nervous system (CNS) infections (Drevets and Bronze, 2008). It can be obtained from various sources, which include food, soil, water, plants, and human and animal feces. Among healthy individuals, the infection may result in feverish gastroenteritis, whereas in immunocompromised individuals, it may result in meningitis, septicemia, and encephalitis (Dussurget et al., 2014). Growth may occur anywhere between 1–44°C, but the preferred temperature is 35–37°C. The cells are susceptible to pasteurization but are resilient to conditions such as freezing, high salt, drying, and a pH > 5.0 (Bibek, 2004).

L. monocytogenes enters the host cells by using surface proteins known as internalins, especially InlA and InlB. But in the post-intestinal stages of the infection, InlC, and InlJ are also involved (Jamshidi and Zeinali, 2019). Listeriosis is relatively rare, but it is one of the deadliest foodborne illnesses with up to 50% mortality in some cases (Dussurget et al., 2014). It occupies the fourth place among zoonotic disease in Europe with 0.41 cases per 100,000 people annually (Jamshidi and Zeinali, 2019). While most human listeriosis occurs at irregular intervals, but some of these cases may be a result of common source which was previously unrecognized (Farber and Peterkin, 1991). When healthy people consume food in which *L. monocytogenes* is present, symptoms may or may not be produced. Mostly it takes about 1 to 7 days for symptoms to show up. But in sensitive people (e.g., pregnant women, unborn fetuses, infants, elderly, etc.), symptoms are related to the nausea, vomiting, abdominal cramps, and diarrhea, along with fever and headache. Various vital organs, like the

CNS, are then invaded via the bloodstream. In pregnant women, it can also invade the fetal organs via the placenta (Bibek, 2004).

Buchanan et al. (2017) reported various outbreaks of listeriosis in which 1,763 cases were reported in 2013 in the European Union from food sources such as cheese, shellfish, crustaceans, molluscs, meat, etc. Other outbreaks in the same article covered different foods such as ice cream, pre-cut celery, and mung bean sprouts, etc., which shows that it is very common in agriculture, aquaculture, and food processing environments. In Spain, from the 35 people infected by listeriosis since 2013, 3 adults died, 2 miscarriages, and 1 stillbirth occurred. These infections may have been more lethal without early diagnosis and treatment (Pérez-Trallero et al., 2014). According to the Centers for Disease Control and Prevention (CDCP), in USA 24 people infected by *Listeria monocytogenes* were reported as of November 7, 2019. Out of these 24 people, 22 hospitalizations and 2 deaths were reported, but the source of infection was not identified (Centers for Disease Control and Prevention, 2019b).

Because of its ubiquitous presence, foods free of this pathogen are nearly impossible. Apart from control measures in processing facilities, consumer education is also important to reduce foodborne listeriosis. This involves cooking raw food items of animal origin; washing raw vegetables thoroughly before consuming; avoiding cross-contamination; avoiding consumption of raw milk as well as sanitizing hands and equipment after handling uncooked foods (Bibek, 2004).

11.5.1.8 SALMONELLA SPP.

In 1855, Theobald Smith discovered *Salmonella* by isolating it from the intestines of pigs which had swine fever. The bacteria got its name after Dr Daniel Elmer Salmon who accompanied Smith in his work (Eng et al., 2015). It is a rod-shaped, Gram-negative facultative anaerobe, may be motile, non-sporulating bacteria that belongs to the family Enterobacteriaceae and is the most frequently isolated foodborne pathogen (Percival and Williams, 2014; Eng et al., 2015). The optimum growth temperature is 35–37°C, but they survive in a range of 5–46°C which indicates their mesophilic nature. Sensitivity towards low water activity and pH does not allow them to grow below 0.94 and 4.5, respectively (Dilbaghi and Sharma, 2007).

Being a natural inhabitant of the digestive system of animals, birds, and insects they can induce salmonellosis and continue being in a carrier state. After being infected, people also become carriers of the infection and shed the bacteria for a long time through feces. These were also isolated from fecal pollution in soil, water, and sewage (Bibek, 2004; Alberto et al., 2012; Eng et al., 2015).

Two major groups are: typhoid salmonellosis (*S. typhi* and *S. paratyphi*) and non-typhoid salmonellosis (other *Salmonella* serovars) out of which *S. enteritidis* and *S. typhimurium* contribute to 75% of salmonellosis in humans (Elafify et al., 2019). Mostly its strains are pathogenic due to its ability to enter, replicate, and flourish in the cells of the human host, which contributes to life-threatening disorders (Eng et al., 2015). After ingestion, it invades the mucous membrane of the small intestine, grows rapidly, and produces a toxin, leading to inflammation and fluid accumulation in the intestine. The bacteria multiply in the epithelial cells and develop a thermolabile enterotoxin that is related to fluid and electrolyte secretions (Bibek, 2004).

A military establishment reported an outbreak of food poisoning on 29 May 2011, where 43 cases in a matter of hours of food toxicity occurred. Analyzing the results, it was found out that the cause was *Salmonella* spp. present in potato-bitter gourd vegetable (Kunwar et al., 2013). During 2017, Qatar's public health authorities documented an increase in infant *Salmonella* cases where the attack rate of infection among infants was calculated at 3.23 cases per 1,000 infants and children aged less than three years comprised almost half of the reported cases. The majority (69.5%) of infants were formula-fed, and bottled water was used for preparing the formula in 85.2% cases and boiled municipal tap water in 14.8% cases (Al-Dahshan et al., 2019). In 2019 Tailor Cut Produce of New Jersey, recalled its Fruit Luau mix, because it may be contaminated with *Salmonella* (FDA, 2020). As of February 14, 2020, 165 illnesses were reported with 73 people hospitalized and no deaths occurred.

Salmonella infection is a distressing problem for public health around the world. Its strains adapt easily to different environments of hosts because of their genetic make-up due to which its elimination becomes difficult (Eng et al., 2015). To mitigate this food risk, regular cleaning and sanitizing of contact surfaces of food and utensils as well as hands is necessary.

11.5.1.9 SHIGELLA SPP.

This Gram-negative species belongs to the family Enterobacteriaceae, which also contains other enteric pathogens such as ETEC (The et al., 2016). They are non-spore-forming, NM prokaryotic rods. It is considered to be one of the oldest human pathogens for which evolutionary origins are known (Mumy, 2014). The strains are able to grow between 7–46°C, with an optimum temperature 37°C. Under refrigeration, freezing, 5% NaCl and pH 4.5, they can survive for days but are sensitive to pasteurization (Dilbaghi and Sharma, 2007).

Within the genus *Shigella*, there are four species, namely *Shigella dysenteriae*, *Shigella flexneri*, *Shigella boydii*, and *Shigella sonnei* which are also known as A, B, C, and D, respectively (Mumy, 2014). *S. sonnei* appears to be more likely than other serogroups to withstand lower temperatures (Bintsis, 2017). Due to its close resemblance to *E. coli*, various researchers believe that they both should be considered one genus (Mumy, 2014) but because of some differences, they are separate.

It cannot naturally infect other host species because it is very well adapted to the human host, particularly the gut; animals therefore rarely present a problem as infection sources and reservoirs (Mumy, 2014). They are intracellular pathogens which are transmitted through the fecal-oral route (Mattock and Blocker, 2017), and can trigger an infection via an exceptionally low infectious dose (<10 bacteria) (The et al., 2016). Generally, salads, raw vegetables, milk, and dairy products, and poultry are associated with its outbreak. After recovery, a person can still remain a carrier for some time (Bibek, 2004). It thrives in the small intestine after ingestion and thereby enters the colon where shigella enterotoxins and serotype toxin are produced which induces symptoms like high fever, vomiting, diffuse colicky abdominal pain (pain that starts and stops abruptly) followed by bloody diarrhea with mucous and tenesmus. Usually within 5–7 days it is self-resolved, but individuals with high risk may end up with complications (Aslam and Okafor, 2020).

Although there is a significant decrease in mortality, the incidence of shigellosis is reported to be 188 million cases with approximately 1 million deaths per year (Kotloff et al., 2017). One of the largest foodborne outbreaks of shigellosis in US history occurred in a wedding in Oregon in 2018 (Rekant et al., 2019). Of the 263 wedding attendees, a staggering 107 (54%) cases were identified. Nearly 10 people were hospitalized and none of them died. The investigators found that only asparagus consumption was associated with illness and the organism responsible was *S. flexneri* which is uncommon in the region. Another outbreak occurred in West Bengal, India in a housewarming party in November 2016 (Debnath et al., 2018). Out of the 34 people attending the party, 73% (25/34) were attacked and 76% (19/25) of them required hospitalization and the food-related to the outbreak was tomato salad.

11.5.1.10 STAPHYLOCOCCUS AUREUS

Staphylococcus aureus is pathogenic as well as commensal as approximately 30% of the humans are colonized by it (Tong et al., 2015). The cells are Gram-positive, facultative anaerobic and have a spherical shape. Generally, they are found in clusters, and when observed under a light microscope, it resembles a bunch of grapes (Gnanamani et al., 2017). They are mesophilic (7–48°C) and grow at a low water activity of 0.86, pH about 5, high concentrations of sugar and salt (15%) (Dilbaghi and Sharma, 2007). *S. aureus* can be naturally present in healthy humans, animals, and birds' nose, throat, skin, and hair (feathers). It can be found in human, animal, and bird infections and cuts in hands as well as facial-erupted acne in humans which contribute to the contamination of food (Bibek, 2004). It ranks third among the most common cause of foodborne disease and is also found in soil, water, and air (Pal et al., 2020).

S. aureus is a versatile pathogen that can cause various diseases. Mostly, it causes infections of the skin and respiratory tract. However, it also causes a range of other diseases which may be life-threatening, e.g., toxic shock syndrome, infective endocarditis, scalded skin syndrome and many more (Otto, 2014). Moreover, it can acquire new genetic elements when stressed due to external conditions through regulatory mechanisms, thus becoming an extremely dangerous organism (Ondusko and Nolt, 2018). The toxins secreted by *S. aureus* are of three different categories: (a) membrane-damaging toxins, which may work with or without the help of a receptor; (b) receptor function interfering toxins which are not membrane-damaging, and (c) secreted enzymes that disrupt or damage essential defense mechanisms of host (Otto, 2014). According to Gnanamani et al. (2017) the process of infections involves five stages: (1) colonization, (2) local infection, (3) systemic dissemination and/or sepsis, (4) metastatic infections and (5) toxinosis. Primary symptoms include salivation, nausea, vomiting, abdominal cramps, and diarrhea and secondary symptoms include sweating, chills, headache, and dehydration (Bibek, 2004).

As estimated by CDC approximately 241,188 illnesses, 1,064 hospitalizations, and 6 deaths annually in the US are caused due to staphylococcal food poisoning. Although there are nine identified staphylococcal enterotoxins (SEs), designated as A, B, C1, C2, C3, D, E, F, and G; the majority of outbreaks are caused by types A and D (Bintsis, 2017). Several food items are involved in foodborne outbreaks of staphylococcal origin such as raw meat, dairy products and pancakes were also involved once (Pal et al., 2020). Despite the production of toxin, it does not affect the quality of food (Bibek, 2004); thus, its detection becomes difficult. An incidence involved 26 workers working for post-earthquake reconstruction which occurred in Central Italy in 2017. Gastrointestinal symptoms were observed after eating a meal provided by a catering service, and 23 people among them were hospitalized (Guidi et al., 2018). On investigating it was found that outbreak occurred because of ingesting SEs and the most likely source of poisoning was pasta salad. Another outbreak occurred in Umbria, Italy, 2015 (Ercoli et al., 2017). Out of 42 customers, 24 were affected after having food in a local restaurant. After preparation the Chantilly cream dessert instead of refrigeration, was kept at room temperature for about 5 hours, which resulted in the growth of microbes and production of SE. For decreasing the incidents related to staphylococcal food poisoning, the initial load of *S. aureus* should be reduced by properly selecting ingredients and cleaning the surrounding environments as well as maintaining adequate personal hygiene (Bibek, 2004).

11.5.1.11 VIBRIO SPP.

Vibrios are Gram-negative, mesophilic, and chemoorganotrophic, usually motile rods which have a facultatively fermentative metabolism (Thompson et al., 2005). The genus *Vibrio*, belongs to the Vibrionaceae family, which contains about 35 species, of which more than one-third are pathogenic to humans (Bintsis, 2017). The growth temperature lies between 5–42°C and in the presence of 3–5% NaCl also the cells multiply rapidly but are sensitive to 10% salt (Bibek, 2004).

Despite their genomic diversity, all of them are inhabitants of marine environments and they prefer warm, brackish water, and their prevalence is directly related with the environmental temperatures (Baker-Austin et al., 2018). Environmental factors which affect the concentration and distribution of *Vibrio* in aquatic ecosystems include temperature, exposure to UV light, concentration of organic and inorganic chemicals, pH, oxygen tension and salinity which is the most important factor (Farmer et al., 2015). When displaying virulence traits, even single strains can be pathogenic to humans. Generally, the outbreaks which are related to *Vibrio* occur in tropical or subtropical regions or coincide with seawater warming conditions (Hackbusch et al., 2020).

Vibrio spp. is involved in most human diseases which occur because of the natural microbiota of aquatic environment and seafood; and the most common species include *Vibrio cholerae*, *Vibrio parahaemolyticus* and *Vibrio vulnificus* (Bonnin-Jusserand et al., 2019). The two major infections caused in humans by the *Vibrio* genus are: cholera and non-cholera infections. Cholera is caused by *V. cholerae*, which results in watery diarrhea because of contaminated food or water. Whereas non-cholera *Vibrio* spp., e.g., *V. vulnificus*, are a causative agent of vibriosis, a category of infections with multiple symptoms (Baker-Austin et al., 2018).

V. cholerae usually enter the human body because of consuming contaminated seafood like mollusks (raw oysters) and raw crustaceans or by exposure of an open wound to a contaminated water source (Bintsis, 2017). The leading cause of bacterial diarrhea is *V. parahaemolyticus*, which is associated with consuming seafood in the US and occasionally causes septicemia; whereas *V. vulnificus* is the leading cause of death associated with the consumption of seafood in the US (Kaysner et al., 2004).

Generally, a *Vibrio* related outbreak in aquafarming impacts the economy directly and also threatens public health. The density of *Vibrio* species increases as a result of increased temperature of freshwater; thus, the European Food Safety Authority (EFSA) has alerted of its growth in oceans as a potential emerging issue (Bonnin-Jusserand et al., 2019). One indication of this was the cholera epidemic in Yemen, which took over 1,500 lives and now affects 21 of 22 Yemeni provinces with more than 246,000 cases (Osunla and Okoh, 2017). In June 2019, CDC investigated an outbreak of gastrointestinal illnesses which was related to the harvesting of raw oysters from Estero El Cardon estuary in Baja California Sur, Mexico. About 16 ill people were reported from five states (Centers for Disease Control and Prevention, 2019a). A total of 82 laboratory-confirmed cases of *Vibrio parahaemolyticus* infection with raw shellfish consumption in Canada occurred between January and October, 2015, which was the largest outbreak related to this organism in the country (Taylor et al., 2018).

11.5.1.12 YERSINIA ENTEROCOLITICA

These Gram-negative bacteria in the family Enterobacteriaceae are generally associated with gastrointestinal infections, such as pseudo-appendicular syndrome and enteritis (Cantwell et al., 2020). They are non-spore, rod shaped, motile at temperatures less than 37°C, and facultative anaerobic. The optimum temperature for growth is 25–29°C but they can also survive at a range of 0–44°C. Cells can grow in 5% NaCl and at a pH above 4.6 (Bibek, 2004).

Major transmission occurs by consuming raw or poorly cooked pork, although food products such as pasteurized milk and tofu have also been involved (Schaake et al., 2013; Longenberger et al., 2014). Transmission methods may be foodborne, animal to human, waterborne or in some rare cases human-to-human transmission also occurs (Sabina et al., 2011). The disease caused by *Y. enterocolitica* is known as yersiniosis and young children are more affected by it (Dilbaghi and Sharma, 2007). Every serotype of *Y. enterocolitica* is not enteropathogenic, and those involved in causing yersiniosis in humans are prevalent primarily in swine (Bintsis, 2017). Cases of *Y. enterocolitica* mostly self-resolve without the need for antibiotics, either from transmission through animals or by consuming contaminated foods (Cantwell et al., 2020). Incidences relating to the foodborne infection from *Y. enterocolitica* vary according to the variation of climate and geography (Sabina et al., 2011). Immunocompromised people as well as children under the age of 15 are most commonly infected. Yersiniosis-related extraintestinal infections include meningitis, Reiter syndrome, myocarditis, septicemia, etc. (Bintsis, 2017).

A series of enterocolitical infections of *Y. enterocolitica* were reported in May 2014 from a military base in Northern Norway (Macdonald et al., 2016). A total of 133 cases were reported with salad mix being the likely source of the outbreak. It was the largest outbreak of *Y. enterocolitica* infection in Norway as of 2014 and the first to be reported from a military context. Another outbreak occurred in July 2011, in south-western Pennsylvania, USA (Longenberger et al., 2014). The outbreak was investigated for the source, in order to prevent further transmission. It was connected to milk products obtained from a family run dairy which carried out on-site pasteurization and a total of 22 people were diagnosed. Refrigeration cannot control its growth due to the psychotropic nature of strains. Proper sanitation and adequate heat treatment are essential in controlling the occurrence of yersiniosis. Moreover, one should avoid consuming raw milk and meat which is cooked at low temperatures (Bibek, 2004).

11.5.2 VIRUSES

Actually, 22 families are divided into recognized viruses that can infect human beings. Furthermore, recent molecular technological developments have contributed to the discovery of a variety of new viruses, most of which remain entirely characterized. For at least 10 taxonomic families of viruses, foodborne transmission has been reported and the associated diseases may be anywhere between mild diarrhea to extreme encephalitis. The transmission from food can occur through:

- Contaminated food by infected food producers (common);
- Food contamination in the manufacturing process (often-bivalve shellfish molluscans or berry fruit);
- Through ingestion of animal food products carrying virus (very rare).

The first two methods of transmission occur via fecal-oral route, thereby infecting the person following ingestion and then entering cells in the gut epithelium and then replicating at or elsewhere in the body (Velebit et al., 2019). WHO and FAO of the United Nations (2008) have also related the risk of the virus to a particular commodity and its significance in causing viral foodborne diseases and have identified combinations of viral commodity that should take careful note of the measures of prevention and control:

- Bivalve molluscan shellfish related to NoV and HAV
- Fresh produce related to NoV and HAV
- Prepared foods related to NoV and HAV
- Water for food preparation related to rotavirus

11.5.2.1 HEPATITIS A VIRUS (HAV)

HAV that is of the *Hepatovirus* genus within the *Picornaviridae* family causes hepatitis A. Only human and primate Hepatoviruses were identified, indicating that there is no other reservoir introduction (Cella et al., 2018). HAV accounts for nearly half of the globally diagnosed human hepatitis infections. At least a million hepatitis A cases are detected annually around the globe according to the WHO (WHO Position Paper on Hepatitis A Vaccines: June 2012-Recommendations 2013). HAV can spread from person to person, food, water, and virus-infected surfaces. HAV is extremely stable virus. Thus it can remain in foodstuffs for long periods in the right environment (Sánchez, 2013). It is possibly because of the unique use of codon that prevents direct competition with the cell system of the host.

Outbreaks of HAV-borne foods are mainly linked to foods such as fruits, shellfish, leafy greens and ready to eat meals. If water contaminated with human excreta comes in contact with shellfish, it can accumulate HAV. It is often high risk because it is either consumed as raw or may be cooked slightly or steamed. The contamination of fresh produce by water irrigation or by individuals infected by viruses such as workers who harvest berries is possible (Boxstael et al., 2013). Contamination in ready to eat foods is because of contact with fecal infected hands or surfaces (fomites) during preparation.

11.5.2.2 HEPATITIS E VIRUS (HEV)

It is a member of the Hepeviridae family those are positive-stranded RNA viruses and infect various mammals. Hepeviridae is composed of two genera: *Orthohepadnavirus* (mammals and avian) and *Piscihepevirus* (infecting trout). The most commonly found HEV strains are of the genus *Orthohepadnavirus* which are classified as A, B, C, and D (Smith et al., 2014). HEV epidemiology is very complex, and its transmission to humans via food is an emergent danger, especially in the developed European countries. Several studies have shown that pork pies, shellfish, wild boar, unpasteurized milk, undercooked or raw pork, homemade sausages, and ethnic foods pose the risk of HEV infection (Nan et al., 2017).

Pigs constitute a significant source of zoonotic HEV. The genotype 3 (gt3) of HEV is endemic to the swine population in European countries (Aprea et al., 2018). In the last decade, reports of human accidents involving foodborne routes have increased (Ricci et al., 2017). Despite being susceptible to infection, pigs do not suffer clinical diseases. Hence, inspecting it visually is not ideal for its detection (Oliveira-Filho et al., 2014). Laboratory tests are therefore needed to ascertain whether the tissues, organs, muscles, and fluids which have a possibility of entering the food chain can be contaminated. In order to establish specific strategies for HEV infection, prevention, and monitoring; improvement in the awareness of the source, epidemiology, and control procedures is needed. Due to the high popularity of pork and its products in Europe, the possible risks to HEV infection of human beings must be assessed (García et al., 2020).

11.5.2.3 NOROVIRUS

NoV belongs to the Caliciviridae family which is classified into five genera. The two genera of the Caliciviridae family, NoV, and Sapovirus show viruses that infect people. Pigs, goats, rodents, cats, dogs, and sheep, along with Sapoviruses have also been identified with NoV (Andreoletti et al., 2011). The infection of NoV induces AGE, with nausea, vomiting, diarrhea, and stomach pain being the most common signs. In general, symptoms develop in 12–48 hours following infection and sickness usually remains for 1 to 3 days (Velebit et al., 2019).

In young children (<5 years) and elders, NoV disease prevalence is the highest. A broad human reservoir, environmental responsiveness, low infection dose (10–100 viral particles), short staying immunity (at most 18 months) and the ability to transfer by different pathways are all important factors that contribute to the significant impact of the Noroviruses. Most of the foods in question involve shellfish that feed on water by filtration, berry fruits as well as green vegetables polluted by fertilizers before collection or irrigated by polluted municipal water (Koopmans and Duizer, 2004; Karp et al., 2017).

11.5.3 PARASITES

11.5.3.1 TAENIA SAGINATA

The tapeworm *Taenia saginata* develops taeniosis upon infection. Its definitive hosts are humans, and it exists in the intestine or release eggs into the fecal matter. It is often known as beef tapeworm, because, as the name suggests the interim hosts are cattle. Cattle ingest *Taenia saginata* eggs, the larvae of the oncosphere, which migrate to muscle tissues such as tongue, heart, esophagus, striated muscles, and diaphragm. A cysticercus forms in the tissues of the larva. Intake of active cysticerci in poorly cooked beef infects humans; infectious dose is not established, in principle, an active cysticercus. Mostly it results in asymptomatic infections in humans and yet moderate abdominal health problems can occur, which include discomfort, lack of appetite, weight loss, nausea, and proglottides that may cause pruritis (Trevisan et al., 2018).

11.5.3.2 TRICHINELLA SPP.

The ingestion of unsuitable meat made up of the *Trichinella* spp. larvae (nematode) leads to trichinellosis, a world-wide illness. The most prevalent route of infection in humans is pork, meat from other sources like horses, bears, wild boars, badgers, and walruses have also caused the outbreak (Bai et al., 2017). Larvae are released when striated muscle infected with *Trichinella* is consumed which enters the enterocytes located in the small intestine and reaches the adult stage through 4 molts. Larvae are produced upon mating of adult worms that migrates through the bloodstream to tissues of newly infected host (Gottstein et al., 2009).

A serious disease characterized by diarrhea, fever, periorbital edema and myalgia and associated risks like thromboembolic, encephalitis, and myocarditis disease can be caused by the infection in humans. Most human infections throughout the world are due to *Trichinella spiralis*, preceded by *Trichinella britovi* (Bruschi and Dupouy-Camet, 2014). There is substantial infection risk in wildlife and in some areas the concern is growing about potential outdoor freelance pigs (Murrell, 2016). In the EU, the incidence of human trichinellosis has steadily declined in the last few decades, but 168 confirmed humans were reported in 2017, an upsurge of about 65% as compared to 101 confirmed cases of human infection in 2016 (EFSA and ECDC, 2018).

11.5.4 PROTOZOA

11.5.4.1 TOXOPLASMA GONDII

This parasitic protozoon employs household cats as the definitive hosts, whereas other animals such as cattle and humans are the intermediate hosts. The oocysts are released by infected cats into feces that may infect intermediate hosts if ingested following sporulation which contributes to the rapid growth of tachyzoites across the body. Tachyzoites may pass the placenta and infect the fetus in pregnant women. Tachyzoites become tissue cysts (bradyzoites) following localization in muscles and the CNS. The infection of foodborne *Toxoplasma gondii* may therefore either be acquired by consuming raw or undercooked meat which contains tissue cysts or by drinking oocysts from infected fruits, water, or shellfish (Koutsoumanis et al., 2018). Dairy products that are not pasteurized also contain tachyzoites. There is a significant lack of awareness of tachyzoites, tissue cysts or oocyst mediated infections; however, intake of meats is considered as a risk factor for bradyzoites (Belluco et al., 2018). *Toxoplasma gondii* infections during pregnancies can trigger abortion or childbirth with serious health problems in a congenital infected infant. In people with weakened immune systems, acute or chronic *Toxoplasma gondii* infections can be particularly severe and life-threatening complications may occur, e.g., encephalitis (Chalmers et al., 2020).

11.5.4.2 SARCOCYSTIS SPP.

The seldom occurring parasite in humans was sarcocystosis. Two clinical presentations, bowel, and muscle sarcocystosis, can be presented to this parasite infection (Fayer et al., 2015). *Sarcocystis hominis* and *Sarcocystis suihominis* infections can lead to intestinal sarcocystosis, while infections can be caused by muscular sarcocystosis with various *Sarcocystis* species specially *Sarcocystis nesbitii*. More cases were identified among returning travelers with muscular sarcocystosis and the medical community has gained interest (Ortega and Cama, 2018). *Sarcocystis* is a parasite that is able to complete its life cycle by a definite and intermediate host. This parasite affects many species that have a significant impact on the market, including those that affect poultry, cattle, and other mammals, insects, reptiles, and fish (Fayer et al., 2015).

11.6 CONTROL OF FOODBORNE PATHOGENS

Preventing and controlling foodborne pathogens at different levels does require proper microbial risk assessment, good manufacturing practices, implementing hazard analysis and critical control points (HACCP), using proper packaging, and using various thermal and non-thermal technologies to destroy it (Heredia et al., 2009). Industries should apply proper risk-based approach which includes assessment of risk by using a science-based approach, policy-based risk management and proper communication and exchange of information at each level (Schlundt, 2002). By using proper food handling precautions at every stage, right from handling of raw materials to preparation of finished foods, can mitigate the risk of foodborne pathogens. Cross-contamination should be avoided by storing raw materials and cooked food separately. Also, food workers should wash their hands, equipment, cutters, and contaminated surfaces properly to avoid cross-contamination. Proper technologies are required to prevent any kind of foodborne pathogen to become a hazard (Altekruse et al., 1997). With time many technologies have come for controlling foodborne pathogens. Food preservation and control of pathogens is possible by application of heat, storing at low temperatures, reducing water activity, chemical preservation, radiation application, hurdle technology and many more. Traditionally thermal technologies like pasteurization have been used for destruction of pathogenic microbes. In thermal treatment the product is exposed to a particular temperature for a specific time. Sterilization is more intense thermal treatment using temperature up to 135–150°C that leads to destruction of all microbes along with their spores present in food and leading to increase shelf stability of product. Commercially, sterilization is done using batch systems, continuous retort systems, pouch processing system and aseptic processing systems (Toledo, 2007; Singh and Heldman, 2008). *Clostridium botulinum* is one of the most used indicator organisms which is the most potent source of food poisoning. By using correct temperature and time combination, the probability of survival of *C. botulinum* can be reduced to 10^{-12} (Earle and Earle, 1983). Low temperature storage of product is also one of the mostly used and effective methods for retarding the growth of bacterial foodborne pathogens. Most of these organisms have a minimum growth temperature of 7°C except psychrophiles which can even grow at –5°C. For low temperature storage method like chilling (0–15°C), refrigeration (0–7°C) and freezing (<–18°C) (Erkmen and Bozoglu, 2016). Controlling pathogens is more effectively achieved by using hurdle technology. In hurdle technology, hurdles like physical hurdles (aseptic packaging, ionizing radiation, ultrasounds, high temperature treatment, low temperature storage, etc.), physicochemical hurdles (lactic acid, carbon dioxide, low pH, ethanol, low water activity, chemical preservatives, etc.), and microbial hurdles (antibiotics, bacteriocins, competitive flora, etc.), are used for inhibiting the growth of foodborne pathogens (Fellows, 2000). Although heating food effectively minimizes microorganisms but they do have detrimental effects on nutritional composition, natural taste, and flavor of product. Therefore, novel technologies like high pressure processing, pulsed electric field, osmotic dehydration, membrane processing, high intensity pulsed light, radiofrequency electric fields, ultrasounds, irradiation, microwaves, radiofrequency processing, vacuum cooling and ohmic heating, etc., are being explored more for their effect on foodborne pathogens. Organizations should apply proper food safety management system by using a proper risk-based food safety approach and also conduct food safety training programs for people involved in production (Bintsis, 2017).

11.7 CONCLUSION

Microbes are essential part of the food industry as they are used for producing a variety of products since ages. Microbes such as yeast and molds are widely used for producing fermented food products offering a wide range of taste and health benefits. With these advantages, microbes are also pathogenic, thus they can affect human health on consumption. Food safety has always been of great concern to the food sector and consumers to eliminate food spoilage microbes during and after processing. Knowledge about the control of foodborne pathogens and their harmful effects will further help to ensure safe and pathogens-free food.

KEYWORDS

- **aerobe**
- **anaerobe**
- **cross-contamination**
- **enterotoxin**
- **fermentation**
- **foodborne infection**
- **foodborne intoxication**
- **foodborne outbreak**
- **foodborne pathogens**
- **gastroenteritis**
- **good manufacturing practices**
- **hurdle technology**
- **risk assessment**
- **spore**

REFERENCES

Acheson, D., & Allos, B. M., (2001). *Campylobacter jejuni* infections: Update on emerging issues and trends. *Clinical Infectious Diseases, 32*(8), 1201–1206.

Addis, M., & Sisay, D., (2015). A review on major foodborne bacterial illnesses. *Journal of Tropical Diseases, 3*(4), 1–7.

Al-Dahshan, A., Elyamani, R., Naja, S., et al., (2019). Epidemiological characteristics of a salmonella outbreak among infants in Qatar, 2017. *Qatar Medical Journal, 2019*(3), 1–8.

Al-Mamun, M., Chowdhury, T., Biswas, B., & Absar, N., (2018). Food poisoning and intoxication: A global leading concern for human health. In: Grumezescu, A. M., & Holban, A. M., (eds.), *Food Safety and Preservation* (pp. 307–352). London: Academic Press.

Alberto, C., Torres-Vitela, M. R., Villarruel-Lpez, A., & Castro-Rosas, J., (2012). The role of foods in *Salmonella* infections. In: Mahmoud, B. S. M., (ed.), *Salmonella - A Dangerous Foodborne Pathogen*. London: IntechOpen. doi: 10.5772/28316.

Altekruse, S. F., Cohen, M. L., & Swerdlow, D. L., (1997). Emerging foodborne diseases. *Emerging Infectious Diseases, 3*(3), 285–293.

Álvarez-Martín, P., Flórez, A. B., Hernández-Barranco, A., & Mayo, B., (2008). Interaction between dairy yeasts and lactic acid bacteria strains during milk fermentation. *Food Control, 19*(1), 62–70.

Ammar, A. M., El-Aziz, N. K., Elgdawy, A. A., Emara, M. S., & Hamdy, M. M., (2019). Genotyping and antimicrobial resistance of *Campylobacter jejuni*: A review. *Advances in Animal and Veterinary Sciences, 7*(2), 129–136.

Andreoletti, O., Budka, H., Buncic, S., et al., (2011). Scientific opinion on an update on the present knowledge on the occurrence and control of foodborne viruses. *EFSA Journal, 9*(7), 2190.

Aprea, G., Amoroso, M. G., Di Bartolo, I., et al., (2018). Molecular detection and phylogenetic analysis of hepatitis E virus strains circulating in wild boars in south-central Italy. *Transboundary and Emerging Diseases, 65*(1), e25–e31.

Armstrong, G. L., Hollingsworth, J., & Morris, J. G., (1996). Emerging foodborne pathogens: *Escherichia coli* 0157:H7 as a model of entry of a new pathogen into the food supply of the developed world. *Epidemiologic Reviews, 18*(1), 29–51.

Aslam, A., Okafor, C. N., (2020). *Shigella.* In: StatPearls [Internet]. Treasure Island (FL): StatPearls Publishing.

Awuah, G. B., Ramaswamy, H. S., & Economides, A., (2007). Thermal processing and quality: Principles and overview. *Chemical Engineering and Processing: Process Intensification, 46*(6), 584–602.

Azhar, S. H. M., Abdulla, R., Jambo, S. A., et al., (2017). Yeasts in sustainable bioethanol production: A review. *Biochemistry and Biophysics Reports, 10*, 52–61.

Bai, X., Hu, X., Liu, X., Tang, B., & Liu, M., (2017). Current research of trichinellosis in China. *Frontiers in Microbiology, 8*, 1472.

Baker-Austin, C., Oliver, J. D., Alam, M., et al., (2018). *Vibrio* spp. infections. *Nature Reviews Disease Primers, 4*(1), 8.

Ballin, N. Z., (2010). Authentication of meat and meat products. *Meat Science, 86*(3), 577–587.

Banwart, G. J., (1989). Microorganisms associated with food. In: Banwart, G. J., (ed.), *Basic Food Microbiology* (pp. 49–100). Boston, MA: Springer.

Barash, J. R., & Arnon, S. S., (2014). A novel strain of *Clostridium botulinum* that produces Type B and type H botulinum toxins. *Journal of Infectious Diseases, 209*(2), 183–191.

Bell, V., Ferrão, J., Pimentel, L., Pintado, M., & Fernandes, T., (2018). One health, fermented foods, and gut microbiota. *Foods, 7*(12), 195.

Belluco, S., Simonato, G., Mancin, M., Pietrobelli, M., & Ricci, A., (2018). *Toxoplasma gondii* infection and food consumption: A systematic review and meta-analysis of case-controlled studies. *Critical Reviews in Food Science and Nutrition, 58*(18), 3085–3096.

Benner, R. A., (2014). Organisms of concern but not foodborne or confirmed foodborne: Spoilage microorganisms. In: Motarjem, Y., Moy, G., & Todd, E. C., (eds.), *Encyclopedia of Food Safety* (Vol. 2, pp. 245–250). San Diego: Academic Press.

Bergeron, G., Latash, J., Costa-Carter, C. D., et al., (2019). Notes from the field: Botulism outbreak associated with home-canned peas, New York City. *Morbidity and Mortality Weekly Report, 68*(10), 251–252.

Beuchat, L. R., Kim, H., Gurtler, J. B., Lin, L. C., Ryu, J. H., & Richards, G. M., (2009). *Cronobacter sakazakii* in foods and factors affecting its survival, growth, and inactivation. *International Journal of Food Microbiology, 136*(2), 204–213.

Bhattacharya, A., Shantikumar, S., Beaufoy, D., et al., (2020). Outbreak of *Clostridium perfringens* food poisoning linked to leeks in cheese sauce: An unusual source. *Epidemiology and Infection, 148*(43), 1–7.

Bibek, R., (2004). *Fundamental Food Microbiology*, 624. Boca Raton: CRC Press.

Bintsis, T., (2017). Foodborne pathogens. *AIMS Microbiology, 3*(3), 529–563.

Birch, A. N., Petersen, M. A., & Hansen, A. S., (2013). The aroma profile of wheat bread crumb influenced by yeast concentration and fermentation temperature. *LWT - Food Science and Technology, 50*(2), 480–488.

Blanco, P., Mirás-Avalos, J. M., & Orriols, I., (2012). Effect of must characteristics on the diversity of *Saccharomyces* strains and their prevalence in spontaneous fermentations. *Journal of Applied Microbiology, 112*(5), 936–944.

Bonnin-Jusserand, M., Copin, S., Le Bris, C., et al., (2019). *Vibrio* species involved in sea foodborne outbreaks (*Vibrio cholerae, V. parahaemolyticus* and *V. vulnificus*): Review of microbiological versus recent molecular detection methods in seafood products. *Critical Reviews in Food Science and Nutrition, 59*(4), 597–610.

Bosch, A., Gkogka, E., Le Guyader, F. S., et al., (2018). Foodborne viruses: Detection, risk assessment, and control options in food processing. *International Journal of Food Microbiology, 285*, 110–128.

Bosch, A., Pintó, R. M., & Guix, S., (2016). Foodborne Viruses. *Current Opinion in Food Science, 8*, 110–119.

Boxstael, V. S., Habib, I., Jacxsens, L., et al., (2013). Food safety issues in fresh produce: Bacterial pathogens, viruses and pesticide residues indicated as major concerns by stakeholders in the fresh produce chain. *Food Control, 32*(1), 190–197.

Bruschi, F., & Dupouy-Camet, J., (2014). Trichinellosis. In: Bruschi, F., (ed.), *Helminth Infections and Their Impact on Global Public Health* (pp. 229–273). Vienna: Springer.

Buchanan, R. L., Gorris, L. G. M., Hayman, M. M., Jackson, T. C., & Whiting, R. C., (2017). A review of *Listeria monocytogenes*: An update on outbreaks, virulence, dose-response, ecology, and risk assessments. *Food Control, 75*, 1–13.

Caballero, P., Tuells, J., Duro-Torrijos, J. L., & Nolasco, A., (2013). Acceptability of pandemic A(H1N1) influenza vaccination by essential community workers in 2010 Alicante (Spain), perceived seriousness and sources of information. *Preventive Medicine, 57*(5), 725–728.

Cantwell, P., Zaman, R., & Rawlins, J., (2020). Rare case of *Yersinia enterocolitica* causing infection of a temporal venolymphatic malformation. *Annals of Case Reports and Reviews*, 3. doi: 10.39127/2574-5747/ACRR.

Caplice, E., & Fitzgerald, G. F., (1999). Food fermentations: Role of microorganisms in food production and preservation. *International Journal of Food Microbiology, 50*, 131–149.

Cauvain, S., (2015). *Technology of Breadmaking*, 408. Cham: Springer International Publishing. doi: 10.1007/978-3-319-14687-4.

Cella, E., Golkocheva-Markova, E. N., Trandeva-Bankova, D., et al., (2018). The genetic diversity of hepatitis A genotype I in Bulgaria. *Medicine, 97*(3), e9632.

Centers for Disease Control and Prevention, (2019a). *Multistate Outbreak of Gastrointestinal Illnesses Linked to Oysters Imported from Mexico*. https://www.cdc.gov/vibrio/investigations/rawoysters-05-19/index.html (accessed on 20 December 2021).

Centers for Disease Control and Prevention, (2019b). *Outbreak of Listeria Infections.* https://www.cdc.gov/listeria/outbreaks/monocytogenes-08-19/index.html (accessed on 20 December 2021).

Chalmers, R. M., Robertson, L. J., Dorny, P., et al., (2020). Parasite detection in food: Current status and future needs for validation. *Trends in Food Science and Technology, 99*, 337–350.

Chen, D., Li, Y., Lv, J., et al., (2019). A foodborne outbreak of gastroenteritis caused by Norovirus and *Bacillus cereus* at a university in the Shunyi district of Beijing, China 2018: A retrospective cohort study. *BMC Infectious Diseases, 19*(1), 910.

Costafreda, M. I., Bosch, A., & Pintó, R. M., (2006). Development, evaluation, and standardization of a real-time TaqMan reverse transcription-PCR assay for quantification of hepatitis A virus in clinical and shellfish samples. *Applied and Environmental Microbiology, 72*(6), 3846–3855.

Couto, S. R., & Sanromán, M. A., (2006). Application of solid-state fermentation to food industry - A review. *Journal of Food Engineering, 76*(3), 291–302.

Croxen, M. A., Law, R. J., Scholz, R., Keeney, K. M., Wlodarska, M., & Finlay, B. B., (2013). Recent advances in understanding enteric pathogenic *Escherichia coli*. *Clinical Microbiology Reviews, 26*(4), 822–880.

Daher, D., Le Gourrierec, S., & Pérez-Lamela, C., (2017). Effect of high pressure processing on the microbial inactivation in fruit preparations and other vegetable based beverages. *Agriculture, 7*(9), 72.

Dave, D., & Ghaly, A. E., (2011). Meat spoilage mechanisms and preservation techniques: A critical review. *American Journal of Agricultural and Biological Science, 6*(4), 486–510.

Debnath, F., Mukhopadhyay, A. K., Chowdhury, G., Saha, R. N., & Dutta, S., (2018). An outbreak of foodborne infection caused by *Shigella sonnei* in West Bengal, India. *Japanese Journal of Infectious Diseases, 71*(2), 162–166.

Değirmencioğlu, N., Gurbuz, O., & Şahan, Y., (2016). The monitoring, via an in vitro digestion system, of the bioactive content of vegetable juice fermented with *Saccharomyces cerevisiae* and *Saccharomyces boulardii*. *Journal of Food Processing and Preservation, 40*(4), 798–811.

Deschuyffeleer, N., Audenaert, K., Samapundo, S., Ameye, S., Eeckhout, M., & Devlieghere, F., (2011). Identification and characterization of yeasts causing chalk mold defects on par-baked bread. *Food Microbiology, 28*(5), 1019–1027.

Dilbaghi, N., & Sharma, S., (2007). *Food and Industrial Microbiology: Food Spoilage, Food Infections and Intoxications Caused by Microorganisms and Methods for Their Detection.* https://www.yumpu.com/en/document/read/21116825/food-and-industrial-microbiology (accessed on 20 December 2021).

Drevets, D. A., & Bronze, M. S., (2008). *Listeria monocytogenes*: Epidemiology, human disease and mechanisms of brain invasion. *FEMS Immunology and Medical Microbiology, 53*(2), 151–165.

Drudy, D., Mullane, N. R., Quinn, T., Wall, P. G., & Fanning, S., (2006). *Enterobacter sakazakii*: An emerging pathogen in powdered infant formula. *Clinical Infectious Diseases, 42*(7), 996–1002.

Dürre, P., (2014). Physiology and sporulation in *Clostridium*. *Microbiology Spectrum, 2*(4), TBS-0010–2012.

Dussurget, O., Bierne, H., & Cossart, P., (2014). The bacterial pathogen *Listeria monocytogenes* and the interferon family: Type I, type II and type III interferons. *Frontiers in Cellular and Infection Microbiology*, 4. doi:10.3389/fcimb.2014.00050.

Earle, R. L., & Earle, M. D., (1983). *Unit Operations in Food Processing*, 1–314. New York: Pergamon Press.

EFSA and ECDC (European Food Safety Authority and European Centre for Disease Prevention and Control), (2018). The European Union summary report on trends and sources of zoonoses, zoonotic agents and foodborne outbreaks in 2017. *EFSA Journal, 16*(12), 5500. doi: 10.2903/j.efsa.2018.5500.

El-Badry, S., & Raslan, A., (2016). Mould contamination of some Egyptian cheese. *Benha Veterinary Medical Journal, 30*(2), 28–33. doi:10.21608/bvmj.2016.31325.

Elafify, M., Darwish, W. S., Al-Ashmawy, M., et al., (2019). Prevalence of *Salmonella* spp. in Egyptian dairy products: Molecular, antimicrobial profiles and a reduction trial using D-Tryptophan. *Journal of Consumer Protection and Food Safety, 14*(4), 399–407.

Eng, S., Pusparajah, P., Syakima, N., et al., (2015). *Salmonella*: A review on pathogenesis, epidemiology and antibiotic resistance. *Frontiers in Life Science, 8*(3), 284–293.

Ercoli, L., Gallina, S., Nia, Y., et al., (2017). Investigation of a staphylococcal food poisoning outbreak from a Chantilly cream dessert, in Umbria (Italy). *Foodborne Pathogens and Disease, 14*(7), 407–413.

Erkmen, O., & Bozoglu, T. F., (2016). Food preservation by low temperatures. In: Erkmen, O., & Bozoglu, T. F., (eds.), *Food Microbiology: Principles into Practice* (pp. 34–43). John Wiley and Sons Ltd. doi:10.1002/9781119237860.

European Commission RASFF, (2019). *RASFF — The Rapid Alert System for Food and Feed — 2018 Annual Report.* doi: 10.2875/914558.

Fakruddin, M., Rahaman, M. M., Ahmed, M. M., & Hoque, M. M., (2013). *Cronobacter sakazakii* (*Enterobacter sakazakii*): An emerging foodborne pathogen. *International Journal of Biomedical and Advance Research, 4*(6), 349–359.

FAO/WHO (Food and Agriculture Organization of the United Nations/World Health Organization), (2014). *Multicriteria-Based Ranking for Risk Management of Foodborne Parasites* (p. 302). Microbiological Risk Assessment Series No. 23. Rome.

Farber, J. M., & Peterkin, P. I., (1991). *Listeria monocytogenes*, a foodborne pathogen. *Microbiological Reviews, 55*(3), 476–511.

Farid, F., Sideeq, O., Khan, F., & Niaz, K., (2019). *Saccharomyces cerevisiae*. In: Nabavi, S. M., & Silva, A. S., (eds.), *Nonvitamin and Nonmineral Nutritional Supplements* (pp. 501–508). London, United Kingdom: Elsevier BV. doi: 10.1016/B978-0-12-812491-8.00066-7.

Farmer, J. J., Janda, J. M., Brenner, F. W., Cameron, D. N., & Birkhead, K. M., (2015). *Vibrio*. In: Whitman, W. B., Rainey, F., Kämpfer, P., Trujillo, M., Chun, J., DeVos, P., Hedlund, B., &. Dedysh, S., (eds.), *Bergey's Manual of Systematics of Archaea and Bacteria* (pp. 1–79). John Wiley and Sons, Ltd. doi: 10.1002/9781118960608.gbm01078.

Fayer, R., Esposito, D. H., & Dubey, J. P., (2015). Human infections with *Sarcocystis* species. *Clinical Microbiology Reviews, 28*(2), 295–311. doi: 10.1128/CMR.00113-14.

FDA, (2020). *Outbreak Investigation of Salmonella Javiana: Fruit Mix*. https://www.fda.gov/food/outbreaks-foodborne-illness/outbreak-investigation-salmonella-javiana-fruit-mix-december-2019 (accessed on 20 December 2021).

Fellows, P. J., (2000). *Food Processing Technology: Principles and Practice*, 591. Cambridge: Woodhead Publishing Limited.

Ferreira, I. M. P. L. V. O., Pinho, O., Vieira, E., & Tavarela, J. G., (2010). Brewer's *Saccharomyces* yeast biomass: Characteristics and potential applications. *Trends in Food Science and Technology, 21*(2), 77–84. doi: 10.1016/j.tifs.2009.10.008.

Fleck-Derderian, S., Shankar, M., Rao, A. K., et al., (2018). The epidemiology of foodborne botulism outbreaks: A systematic review. *Clinical Infectious Diseases, 66*(1), S73–81.

Fleet, G. H., (2007). Yeasts in foods and beverages: Impact on product quality and safety. *Current Opinion in Biotechnology, 18*(20), 170–175.

Fleet, G. H., (2011). Yeast spoilage of foods and beverages. In: Kurtzman, C. P., Fell, J. W., & Boekhout, T., (eds.), *The Yeasts* (Vol. 1, pp. 53–63). Elsevier. doi: 10.1016/B978-0-444-52149-1.00005-7.

García, A., Fox, J. G., & Besser, T. E., (2010). Zoonotic enterohemorrhagic *Escherichia coli*: A one health perspective. *Institute for Laboratory Animal Research Journal, 51*(3), 221–232.

García, N., Hernández, M., Gutierrez-Boada, M., et al., (2020). Occurrence of Hepatitis E Virus in pigs and pork cuts and organs at the time of slaughter, Spain, 2017. *Frontiers in Microbiology, 10*, 2990. doi:10.3389/fmicb.2019.02990.

Garnier, L., Valence, F., & Mounier, J., (2017). Diversity and control of spoilage fungi in dairy products: An update. *Microorganisms, 5*(3), 42.

Gaulin, C., Ramsay, D., Catford, A., & Bekal, S., (2015). *Escherichia coli* O157:H7 outbreak associated with the consumption of beef and veal tartares in the province of Quebec, Canada, in 2013. *Foodborne Pathogens and Disease, 12*(7), 612–618.

Glashower, D., Snyder, J., Welch, D., & McCarthy, S., (2017). Outbreak of *Campylobacter jejuni* associated with consuming undercooked chicken liver mousse - Clark County, Washington, 2016. *Morbidity and Mortality Weekly Report, 66*(38), 1027.

Gnanamani, A., Periasamy, H., & Paul-Satyaseela, M., (2017). *Staphylococcus aureus*: Overview of bacteriology, clinical diseases, epidemiology, antibiotic resistance and therapeutic approach. In: Enany, S., & Alexander, L. E. C., (eds.), *Frontiers in Staphylococcus aureus*. London: IntechOpen.

Gonçalves, M., Pontes, A., Almeida, P., et al., (2016). Distinct domestication trajectories in top-fermenting beer yeasts and wine yeasts. *Current Biology, 26*(20), 2750–2761.

Gottstein, B., Pozio, E., & Nöckler, K., (2009). Epidemiology, diagnosis, treatment, and control of trichinellosis. *Clinical Microbiology Reviews, 22*(1), 127–145.

Granum, P. E., (1994). *Bacillus cereus* and its toxins. *Journal of Applied Bacteriology Symposium Supplement, 76*, 61S–66S.

Granum, P. E., & Lund, T., (1997). *Bacillus cereus* and its food poisoning toxins. *FEMS Microbiology Letters, 157*(2), 223–228.

Grass, J. E., Gould, L. H., & Mahon, B. E., (2013). Epidemiology of foodborne disease outbreaks caused by *Clostridium perfringens*, United States, 1998–2010. *Foodborne Pathogens and Disease, 10*(2), 131–136.

Griffiths, M. W., & Schraft, H., (2017). *Bacillus cereus* food poisoning. In: Christine E. R. D., Tim, A., Richard, A. S., Dean, O. C., & Hans, P. R., (eds.), *Foodborne Diseases* (pp. 395–405). Elsevier Inc. doi: 10.1016/B978-0-12-385007-2.00020-6.

Guidi, F., Duranti, A., Gallina, S., et al., (2018). Characterization of a staphylococcal food poisoning outbreak in a workplace canteen during the post-earthquake reconstruction of Central Italy. *Toxins, 10*(12), 523.

Hackbusch, S., Wichels, A., Gimenez, L., Döpke, H., & Gerdts, G., (2020). Potentially human pathogenic *Vibrio* spp. in a coastal transect: Occurrence and multiple virulence factors. *Science of the Total Environment, 707*, 136113.

Hall, A. J., Wikswo, M. E., Pringle, K., Gould, L., & Parashar, U. D., (2014). Vital signs: Foodborne norovirus outbreaks-United States, 2009–2012. *Morbidity and Mortality Weekly Report, 63*(22), 491–495.

Hazelwood, L. A., Daran, J. M., Van, M. A. J. A., Pronk, J. T., & Dickinson, J. R., (2008). The Ehrlich pathway for fusel alcohol production: A century of research on *Saccharomyces cerevisiae* metabolism. *Applied and Environmental Microbiology, 74*(8), 2259–2266.

Henry, M., & Fouladkhah, A., (2019). Outbreak history, biofilm formation, and preventive measures for control of *Cronobacter sakazakii* in infant formula and infant care settings. *Microorganisms, 7*(3), 77.

Heredia, N. L., Wesley, I. V., & Garcia, J. S., (2009). *Microbiologically Safe Foods*, 696. Hoboken: John Wiley and Sons.

Hocking, A. D., & Faedo, M., (1992). Fungi causing thread mould spoilage of vacuum packaged cheddar cheese during maturation. *International Journal of Food Microbiology, 16*(2), 123–130.

Holý, O., & Forsythe, S., (2014). *Cronobacter* spp. as emerging causes of healthcare-associated infection. *Journal of Hospital Infection, 86*(3), 169–177.

International Committee on Taxonomy of Viruses, (2012). *Virus Taxonomy: Classification and Nomenclature of Viruses. Ninth Report of the International Committee on Taxonomy of Viruses*. doi: 10.1016/B978-0-12-384684-6.00057-4.

Jacques, N., & Casaregola, S., (2008). Safety assessment of dairy microorganisms: The hemiascomycetous yeasts. *International Journal of Food Microbiology, 126*(3), 321–326.

Jamshidi, A., & Zeinali, T., (2019). Significance and characteristics of *Listeria monocytogenes* in poultry products. *International Journal of Food Science, 2019*, 1–7.

Jay, J. M., Loessner, M. J., & Golden, D. A., (2008). *Modern Food Microbiology*, 790. Boston: Springer.

Jayaram, V. B., Cuyvers, S., Verstrepen, K. J., Delcour, J. A., & Courtin, C. M., (2014). Succinic acid in levels produced by yeast (*Saccharomyces cerevisiae*) during fermentation strongly impacts wheat bread dough properties. *Food Chemistry, 151*, 421–428.

Kaakoush, N. O., Castaño-Rodríguez, N., Mitchell, H. M., & Man, S. M., (2015). Global epidemiology of *Campylobacter* infection. *Clinical Microbiology Reviews, 28*(3), 687–720.

Kang, C. R., Bang, J. H., & Cho, S. I., (2019). *Campylobacter jejuni* foodborne infection associated with cross-contamination: Outbreak in Seoul in 2017. *Infection and Chemotherapy, 51*(1), 21–27.

Kaper, J. B., Nataro, J. P., & Mobley, H. L. T., (2004). Pathogenic *Escherichia coli*. *Nature Reviews Microbiology, 2*(2), 123–140.

Karp, B. E., Tate, H., Plumblee, J. R., et al., (2017). National antimicrobial resistance monitoring system: Two decades of advancing public health through integrated surveillance of antimicrobial resistance. *Foodborne Pathogens and Disease, 14*(10), 545–557.

Kaysner, C. A., DePaola, A., & Jones, J., (2004). *Vibrio. Bacteriological Analytical Manual.* https://www.fda.gov/food/laboratory-methods-food/bam-vibrio (accessed on 20 December 2021).

Koopmans, M., & Duizer, E., (2004). Foodborne viruses: An emerging problem. *International Journal of Food Microbiology, 90*(1), 23–41.

Kotiranta, A., Lounatmaa, K., & Haapasalo, M., (2000). Epidemiology and pathogenesis of *Bacillus cereus* infections. *Microbes and Infection, 2*(2), 189–198.

Kotloff, K. L., Riddle, M. S., Platts-mills, J. A., Pavlinac, P., & Zaidi, A. K. M., (2017). Shigellosis. *Lancet, 391*(10122), 801–812.

Koutsoumanis, K., Allende, A., Alvarez-Ordóñez, A., et al., (2018). Public health risks associated with foodborne parasites. *EFSA Journal, 16*(12). doi: 10.2903/j.efsa.2018.5495.

Kunwar, R., Singh, H., Mangla, V., & Hiremath, R., (2013). Outbreak investigation: *Salmonella* food poisoning. *Medical Journal Armed Forces India, 69*(4), 388–391.

Li, J., Adams, V., Bannam, T. L., et al., (2013). Toxin plasmids of *Clostridium perfringens. Microbiology and Molecular Biology Reviews, 77*(2), 208–233.

Lim, J. Y., Yoon, J. W., & Hovde, C. J., (2010). A brief overview of *Escherichia coli* O157:H7 and its plasmid O157. *Journal of Microbiology and Biotechnology, 20*(1), 5–14.

Lim, M. A., Kim, J. Y., Acharya, D., et al., (2020). A diarrhoeagenic enteropathogenic *Escherichia coli* (EPEC) infection outbreak that occurred among elementary school children in Gyeongsangbuk-Do Province of South Korea was associated with consumption of water-contaminated food items. *International Journal of Environmental Research and Public Health, 17*(9).

Linscott, A. J., (2011). Foodborne illnesses. *Clinical Microbiology Newsletter, 33*(6), 41–45.

Liu, S., Han, Y., & Zhou, Z., (2011). Lactic acid bacteria in traditional fermented Chinese foods. *Food Research International, 44*(3), 643–651.

Longenberger, A. H., Gronostaj, M. P., Yee, G. Y., et al., (2014). *Yersinia enterocolitica* infections associated with improperly pasteurized milk products: Southwest Pennsylvania. *Epidemiology and Infection, 142*(8), 1640–1650.

Macdonald, E., Einöder-Moreno, M., Borgen, K., et al., (2016). National outbreak of *Yersinia enterocolitica* infections in military and civilian populations associated with consumption of mixed salad, Norway. *Euro Surveillance, 21*(34), 1–9.

Mardaneh, J., & Dallal, M. M. S., (2017). Study of *Cronobacter sakazakii* strains isolated from powdered milk infant formula by phenotypic and molecular methods in Iran. *Archives of Pediatric Infectious Diseases, 5*(1).

Masood, N., Moore, K., Farbos, A., et al., (2015). Genomic dissection of the 1994 *Cronobacter sakazakii* outbreak in a French neonatal intensive care unit. *BMC Genomics, 16*(1), 750.

Mattock, E., & Blocker, A. J., (2017). How do the virulence factors of *Shigella* work together to cause disease? *Frontiers in Cellular and Infection Microbiology, 7*, 1–24.

Mazuet, C., Sautereau, J., Legeay, C., Bouchier, C., Bouvet, P., & Popoff, M. R., (2015). An atypical outbreak of foodborne botulism due to *Clostridium botulinum* types B and E from ham. *Journal of Clinical Microbiology, 53*(2), 722–726.

Mellou, K., Kyritsi, M., Chrysostomou, A., Sideroglou, T., Georgakopoulou, T., & Hadjichristodoulou, C., (2019). *Clostridium perfringens* foodborne outbreak during an athletic event in northern Greece. *International Journal of Environmental Research and Public Health, 16*(20).

Mumy, K. L., (2014). *Shigella*. In: Wexler, P., Abdollahi, M., Peyster, A., Gad, S. C., Greim, H., Harper, S., Moser, V. C., et al., (eds.), *Encyclopedia of Toxicology* (pp. 254, 255). Elsevier.

Murrell, K. D., (2016). The dynamics of *Trichinella spiralis* epidemiology: Out to pasture? *Veterinary Parasitology, 231*, 92–96. doi: 10.1016/j.vetpar.2016.03.020.

Nan, Y., Wu, C., Zhao, Q., & Zhou, E. N., (2017). Zoonotic hepatitis E virus: An ignored risk for public health. *Frontiers in Microbiology, 8*, 2396. doi:10.3389/fmicb.2017.02396.

Nigam, P. S., & Luke, J. S., (2016). Food additives: Production of microbial pigments and their antioxidant properties. *Current Opinion in Food Science, 7*, 93–100.

Nyati, K. K., & Nyati, R., (2013). Role of *Campylobacter jejuni* infection in the pathogenesis of Guillain-Barré syndrome: An update. *BioMed Research International, 2013*, 852195. doi:10.1155/2013/852195.

Odeyemi, O. A., Alegbeleye, O. O., Strateva, M., & Stratev, D., (2020). Understanding spoilage microbial community and spoilage mechanisms in foods of animal origin. *Comprehensive Reviews in Food Science and Food Safety, 19*(2), 311–331.

Oh, E., Andrews, K. J., & Jeon, B., (2018). Enhanced biofilm formation by ferrous and ferric iron through oxidative stress in *Campylobacter jejuni*. *Frontiers in Microbiology, 9*, 1204.

Oliveira-Filho, E. F., Bank-Wolf, B. R., Thiel, H. J., & König, M., (2014). Phylogenetic analysis of hepatitis E virus in domestic swine and wild boar in Germany. *Veterinary Microbiology, 174*(1, 2), 233–238.

Ondusko, D. S., & Nolt, D., (2018). *Staphylococcus aureus*. *Pediatrics in Review, 39*(6), 287–298. doi: 10.1542/pir.2017–0224.

Ortega, Y. R., & Cama, V. A., (2018). *Cystoisospora belli* and *Sarcocystis* spp. In: Ortega, Y. R., & Sterling, C. R., (eds.), *Foodborne Parasites* (pp. 57–72). Cham: Springer International Publishing.

Osunla, C. A., & Okoh, A. I., (2017). *Vibrio* pathogens: A public health concern in rural water resources in Sub-Saharan Africa. *International Journal of Environmental Research and Public Health, 14*(10), 1–27.

Otto, M., (2014). *Staphylococcus aureus* toxins. *Current Opinion in Microbiology, 17*(1), 32–37. doi: 10.1016/j.mib.2013.11.004.

Ozawa, K., Oka, T., Takeda, N., & Hansman, G. S., (2007). Norovirus infections in symptomatic and asymptomatic food handlers in Japan. *Journal of Clinical Microbiology, 45*(12), 3996–4005.

Packer, S., Day, J., Hardman, P., et al., (2020). A cohort study investigating a point source outbreak of *Clostridium perfringens* associated with consumption of roasted meat and gravy at a buffet on mothering Sunday 2018, South West, England. *Food Control, 112*, 107097. doi: 10.1016/j.foodcont.2020.107097.

Pai, J. S., (2003). Applications of microorganisms in food biotechnology. *Indian Journal of Biotechnology, 2*, 382–386.
Pal, M., Kerorsa, G. M., Marami, L. M., & Kandi, V., (2020). Epidemiology, pathogenicity, animal infections, antibiotic resistance, public health significance, and economic impact of *Staphylococcus aureus*: A comprehensive review. *American Journal of Public Health Research, 8*(1), 14–21.
Parente, E., Cogan, T. M., & Powell, I. B., (2017). Starter cultures: General aspects. In: McSweeney, P. L. H., Fox, P. F., Cotter, P. D., & Everett, D. W., (eds.), *Cheese: Chemistry, Physics and Microbiology* (pp. 201–226). Academic Press. doi: 10.1016/B978-0-12-417012-4.00008-9.
Parra-Flores, J., Cerda-Leal, F., Contreras, A., Valenzuela-Riffo, N., Rodríguez, A., & Aguirre, J., (2018). *Cronobacter sakazakii* and microbiological parameters in dairy formulas associated with a food alert in Chile. *Frontiers in Microbiology, 9*, 1708.
Paul, R. R., Morgan, S., & Hill, C., (2002). Preservation and fermentation: Past, present and future. *International Journal of Food Microbiology, 79*(1, 2), 3–16.
Percival, S. L., & Williams, D. W., (2014). *Salmonella*. In: Percival, S. L., Yates, M. V., Williams, D. W., Chalmers, R. M., & Gray, N. F., (eds.), *Microbiology of Waterborne Diseases* (pp. 209–222). Elsevier. doi: 10.1016/B978-0-12-415846-7.00010-X.
Pérez-Trallero, E., Zigorraga, C., Artieda, J., Alkorta, M., & Marimón, J. M., (2014). Two outbreaks of *Listeria monocytogenes* infection, Northern Spain. *Emerging Infectious Diseases, 20*(12), 2155–2157.
Pigłowski, M., (2019). Pathogenic and non-pathogenic microorganisms in the rapid alert system for food and feed. *International Journal of Environmental Research and Public Health, 16*(477).
Pitt, J. I., & Hocking, A. D., (2009). *Fungi and Food Spoilage*, 529. Boston MA: Springer. doi: 10.1007/978-0-387-92207-2.
Rai, A. K., & Jeyaram, K., (2017). Role of yeasts in food fermentation. In: Satyanarayana, T., & Kunze, G., (eds.), *Yeast Diversity in Human Welfare*. Singapore: Springer. doi: 10.1007/978-981-10-2621-8.
Rai, A. K., Pandey, A., & Sahoo, D., (2019). Biotechnological potential of yeasts in functional food industry. *Trends in Food Science and Technology, 83*, 129–137. doi: 10.1016/j.tifs.2018.11.016.
Raveendran, S., Parameswaran, B., Ummalyma, S. B., et al., (2018). Applications of microbial enzymes in food industry. *Food Technology and Biotechnology, 56*(1), 16–30.
Rekant, S. I., Poissant, T., Tran, D., et al., (2019). *Shigellosis at a Wedding – Oregon - 2018*. https://www.cdc.gov/eis/downloads/eis-conference-2019-508.pdf (accessed on 20 December 2021).
Ricci, A., Allende, A., Bolton, D., et al., (2017). Public health risks associated with hepatitis E virus (HEV) as a foodborne pathogen EFSA panel on biological hazards (BIOHAZ), panel members. *EFSA Journal, 15*(7), e04886.
Robertson, L. J., Sehgal, R., & Goyal, K., (2015). An Indian multicriteria-based risk ranking of foodborne parasites. *Food Research International, 77*(3), 315–319.
Sabina, Y., Rahman, A., Ray, R. C., & Montet, D., (2011). *Yersinia enterocolitica*: Mode of transmission, molecular insights of virulence, and pathogenesis of infection. *Journal of Pathogens, 2011*, 429069. doi: 10.4061/2011/429069.

Sánchez, G., (2013). *Hepatitis A Virus in Food: Detection and Inactivation Methods*, 50. New York: Springer. doi: 10.1007/978-1-4614-7104-2.
Saranraj, P., & Geetha, M., (2012). Microbial spoilage of bakery products and its control by preservatives. *International Journal of Pharmaceutical and Biological Archives, 3*(1), 38–48.
Sarmah, P., Dan, M. M., Adapa, D., & Tk, S., (2018). A review on common pathogenic microorganisms and their impact on human health. *Electronic Journal of Biology, 14*(1), 50–58.
Sauer, M., Porro, D., Mattanovich, D., & Branduardi, P., (2008). Microbial production of organic acids: Expanding the markets. *Trends in Biotechnology, 26*(2), 100–108.
Schaake, J., Kronshage, M., Uliczka, F., et al., (2013). Human and animal isolates of *Yersinia enterocolitica* show significant serotype-specific colonization and host-specific immune defense properties. *Infection and Immunity, 81*(11), 4013–4025.
Schlundt, J., (2002). New directions in foodborne disease prevention. *International Journal of Food Microbiology, 78*(1, 2), 3–17.
Sears, C. L., (2005). A dynamic partnership: Celebrating our gut flora. *Anaerobe, 11*(5), 247–251. doi: 10.1016/j.anaerobe.2005.05.001.
Sengun, I. Y., & Karabiyikli, S., (2011). Importance of acetic acid bacteria in food industry. *Food Control, 22*(5), 647–656.
Shimizu, T., Ohtani, K., Hirakawa, H., et al., (2002). Complete genome sequence of *Clostridium perfringens*, an anaerobic flesh-eater. *Proceedings of the National Academy of Sciences USA, 99*(2), 996–1001.
Shukla, H. D., & Sharma, S. K., (2005). *Clostridium botulinum*: A bug with beauty and weapon. *Critical Reviews in Microbiology, 31*(1), 11–18.
Singh, R. P., & Heldman, D. R., (2008). *Introduction to Food Engineering, 53*, 864. Tokyo: Elsevier.
Smith, D. B., Simmonds, P., Jameel, S., et al., (2014). Consensus proposals for classification of the family *Hepeviridae*. *Journal of General Virology, 95*(10), 2223–2232.
Snyder, A. B., & Worobo, R. W., (2018). The incidence and impact of microbial spoilage in the production of fruit and vegetable juices as reported by juice manufacturers. *Food Control, 85*, 144–150.
Solomon, H. M., & Lilly, T., (2017). *Clostridium botulinum. Bacteriological Analytical Manual.* https://www.fda.gov/food/laboratory-methods-food/bam-clostridium-botulinum (accessed on 20 December 2021).
Son, K. L., Nugroho, A. S. D., Rahayujati, B., & Gozali, L. K., (2019). Food poisoning outbreak caused by diarrhoeal *Bacillus cereus* in Tegalkenongo village, Bantul, Yogyakarta, Indonesia: A retrospective study. *Journal of Asia Pacific Family Medicine, 18*(1), 1–5.
Steensels, J., Daenen, L., Malcorps, P., Derdelinckx, G., Verachtert, H., & Verstrepen, K. J., (2015). *Brettanomyces* yeasts - from spoilage organisms to valuable contributors to industrial fermentations. *International Journal of Food Microbiology, 206*, 24–38.
Tamang, J. P., Thapa, N., Tamang, B., Rai, A., & Chettri, R., (2015). Microorganisms in fermented foods and beverages. In: Tamang, J. P., (ed.), *Health Benefits of Fermented Foods and Beverages* (pp. 1–110). Boca Raton: CRC Press.
Tauxe, R. V., (1997). Emerging foodborne diseases: An evolving public health challenge. *Emerging Infectious Diseases, 3*(4), 425–434.

Tauxe, R. V., (2002). Emerging foodborne pathogens. *International Journal of Food Microbiology, 78*(1, 2), 31–41. doi: 10.1016/S0168-1605(02)00232-5.

Taylor, M., Cheng, J., Sharma, D., et al., (2018). Outbreak of *Vibrio parahaemolyticus* associated with consumption of raw oysters in Canada, 2015. *Foodborne Pathogens and Disease, 15*(9), 554–559.

Terajima, J., Izumiya, H., Hara-Kudo, Y., & Ohnishi, M., (2017). Shiga toxin (Verotoxin)-producing *Escherichia coli* and foodborne disease: A review. *Food Safety, 5*(2), 35–53.

Terefe, N. S., (2016a). Emerging trends and opportunities in food fermentation. In: Smithers, G., (ed.), *Reference Module in Food Science* (pp. 1–9). Amsterdam: Elsevier. doi: 10.1016/B978-0-08-100596-5.21087-1.

Terefe, N. S., (2016b). Food fermentation. In: Smithers, G., (ed.), *Reference Module in Food Science* (pp. 1–3). Victoria: Elsevier. doi: 10.1016/B978-0-08-100596-5.03420-X.

Tewari, A., & Abdullah, S., (2014). *Bacillus cereus* food poisoning: International and Indian perspective. *Journal of Food Science and Technology, 52*(5), 2500–2511.

The, H. C., Thanh, D. P., Holt, K. E., Thomson, N. R., & Baker, S., (2016). The genomic signatures of *Shigella* evolution, adaptation and geographical spread. *Nature Reviews Microbiology, 14*(4), 235–250.

Thirkell, C. E., Sloan-Gardner, T. S., Kaczmarek, M. C., & Polkinghorne, B. G., (2019). An outbreak of *Bacillus cereus* toxin-mediated emetic and diarrhoeal syndromes at a restaurant in Canberra, Australia 2018. *Communicable Diseases Intelligence, 43*, 1–9.

Thompson, F. L., Gevers, D., Thompson, C. C., et al., (2005). Phylogeny and molecular identification of *Vibrios* on the basis of multilocus sequence analysis. *Applied and Environmental Microbiology, 71*(9), 5107–5115.

Todar, K., (2012). *Bacterial Pathogens of Humans*. http://textbookofbacteriology.net/medical_4.html (accessed on 20 December 2021).

Toledo, R. T., (2007). *Fundamentals of Food Process Engineering*, 570. Boston: Springer. doi: 10.1007/0-387-29241-1.

Tong, S. Y. C., Davis, J. S., Eichenberger, E., Holland, T. L., & Fowler, V. G., (2015). *Staphylococcus aureus* infections: Epidemiology, pathophysiology, clinical manifestations, and management. *Clinical Microbiology Reviews, 28*(3), 603–661.

Topisirovic, L., Kojic, M., Fira, D., Golic, N., Strahinic, I., & Jelena, L., (2006). Potential of lactic acid bacteria isolated from specific natural niches in food production and preservation. *International Journal of Food Microbiology, 112*(3), 230–235.

Trevisan, C., Sotiraki, S., Laranjo-González, M., et al., (2018). Epidemiology of taeniosis/cysticercosis in Europe, a systematic review: Eastern Europe. *Parasites and Vectors, 11*(1), 569.

Velebit, B., Djordjevic, V., Milojevic, L., Babic, M., Grkovic, N., Jankovic, V., & Yushina, Y., (2019). The common foodborne viruses: A review. *IOP Conference Series: Earth and Environmental Science, 333*(1), 012110. doi: 10.1088/1755-1315/333/1/012110.

Vetchapitak, T., & Misawa, N., (2019). Current status of *Campylobacter* food poisoning in Japan. *Food Safety, 7*(3), 61–73.

WHO (2013). WHO position paper on hepatitis A vaccines: June 2012-recommendations. *Vaccine, 31*(2), 285–286. doi: 10.1016/j.vaccine.2012.10.102.

WHO, (2018). In: *Food Safety*. https://www.who.int/news-room/facts-in-pictures/detail/food-safety (accessed on 20 December 2021).

Wieczorek, K., Wolkowicz, T., & Osek, J., (2018). Antimicrobial resistance and virulence-associated traits of *Campylobacter jejuni* isolated from poultry food chain and humans with diarrhea. *Frontiers in Microbiology, 9*, 1508. doi:10.3389/fmicb.2018.01508.

Wood, B. J. B., (2013). Bioprocessing as a route to food ingredients: An introduction. In: McNeil, B., David, A., Ioannis, G., & Linda, H., (eds.), *Microbial Production of Food Ingredients, Enzymes and Nutraceuticals* (pp. 1–15). Cambridge: Woodhead Publishing Limited. doi: 10.1533/9780857093547.1.

World Health Organization and Food and Agriculture Organization of the United Nations, (2008). *Viruses in Food: Scientific Advice to Support Risk Management Activities.* https://apps.who.int/iris/handle/10665/44030 (accessed on 20 December 2021).

Xiao, H. W., Pan, Z., Deng, L., et al., (2017). Recent developments and trends in thermal blanching – A comprehensive review. *Information Processing in Agriculture, 4*(2), 101–127.

Yong, W., Guo, B., Shi, X., et al., (2018). An investigation of an acute gastroenteritis outbreak: *Cronobacter sakazakii*, a potential cause of foodborne illness. *Frontiers in Microbiology, 9*, 2549.

Yu, T., Zhou, Y. J., Huang, M., et al., (2018). Reprogramming yeast metabolism from alcoholic fermentation to lipogenesis. *Cell, 174*(6), 1549–1558.

CHAPTER 12

Food Quality and Food Safety: Management Systems and Analytical Tools for Public Health

SHIKHA PANDHI, ARVIND KUMAR, and SADHNA MISHRA

ABSTRACT

Foodborne pathogens pose a serious threat to public health and the economy due to their alarming association with various foodborne diseases. Hence, food safety and quality service are two essential parameters of concern during the production and processing of food. Food safety refers to the assurance that the food will not pose any harm to the consumer's health whereas, food quality deals with the parameters that govern the product's value to the consumer. Various food safety and quality management systems based on novel approaches have been proved to be efficient over the classical approaches. Pre-requisite programs such as GMP and GHP along with HACCP form an essential part of food safety management systems. This chapter provides an introductory overview of the food quality and safety concept. It also discusses about the different management systems adopted by agri-food sector along with risk analysis approach in brief.

12.1 INTRODUCTION

Food quality and food safety are considered to be the two most critical factors that must be continuously monitored throughout the food chain

Global Food Safety: Microbial Interventions and Molecular Advancements. Saher Islam, PhD, Devarajan Thangadurai, PhD, Jeyabalan Sangeetha, PhD, & Natália Cruz-Martins, PhD (Editors)

starting from the production stage to human consumption (Guérin et al., 2019). Food safety is generally defined as the scientific discipline that deals with handling, preparation, packaging, storage, and distribution of food in such a manner that prevents any chance of occurrence of food-borne illness (Chung et al., 2020). It principally focuses on eliminating all those hazards that make food harmful for human consumption whether exerting short-term and long-term effects on human consumption and are not negotiable whereas, food quality mainly deals with all those attributes that govern the value of a food product for the consumer (Manning and Soon, 2016). Food quality and food safety aspects are generally put on stake when there is a change in any product quality attribute such as food spoilage which affects safety due to presence of any foreign substance, contaminant, toxin, pesticide residue or any other sort of chemical, physical, and microbial hazard that may possess severe health risks to consumers. Safety of food is endangered due to the presence of various pathogens that cause a range of foodborne diseases. Microbial toxins are mostly associated with acute toxicity incidences but may exert certain chronic effects like teratogenic, immunotoxic, and nephrotoxic effects. Apart from this, the other important industrial activity that puts food across in contact with various heavy metals also needs great attention as these toxic metals such as lead, mercury, arsenic, and cadmium display diverse toxicities. The use of certain chemicals has also been banned formerly that persist in the environment and contaminate our food. Hence, they are considered as the two key rudiments that govern the perception of consumers towards the food as well as regulate the decision-making process regarding the preference of food (Van Rijswijk and Frewer, 2008). Also, it exerts a great impact on the reliability of food manufacturers and product cost (Wijtzes et al., 1998).

The risk of foodborne disease is significantly associated with the presence of various biological and chemical contaminants in the food chain starting from production, processing to the ultimate consumption of the product (Käferstein and Abdussalam, 1999). Risk analysis is effectively utilized by various governmental institutions as a tool for foods that are suspiciously associated with any sort of foodborne illness to (1) safeguard public health from potential food hazard, (2) estimate the appropriate level of hazard to protect public health, (3) set up standards to assure the safety of foods (Gorris, 2005). Various agencies are working for establishing acceptable levels of these major contaminants and toxins for effectively monitoring their levels during the product line. In the case of new hazards or hazards with greater health risks are subject to risk analysis procedures (Borchers et al., 2010). These risk assessment procedures consist of stages like hazard identification and characterization, exposure assessments, and subsequent risk characterization. Amongst the major food associated health risks, the risk of occurrence of microbial hazards has been frequently associated with food quality and safety issues as food provides a favorable environment that microorganisms can thrive. An increase in consumer awareness about the relationship of infrequent illness and consumption of certain types of food products has affected consumer choice and ultimately causes changes in economy and food production patterns making food safety a major topic amongst the creators and consumers.

The adoption of various food quality and safety management system has become the need of the hour. These systems are designed to consistently provide the consumer with all the promised product characteristics (Kotsanopoulos and Arvanitoyannis, 2017). Adoption of these systems can either be obligatory or non-obligatory. Pre-requisite programs such as GMP and GHP along with HACCP form an essential part of food safety management systems. HACCP has been at present identified as the best management tool that couples all together all food safety-related management action into a single management system. Still, it cannot stand alone and needs to be supported by a hygienic code of practice like GHP and GMP as pre-requisite requirements for effective management. GMP mainly involves all those precautionary measures that need to be followed to assure adherence to all food quality and food safety requirements, whereas GHP is a part of GMP that deals with all the measures taken to ensure hygienic conditions in the food processing facility. HACCP is a science-based approach that sequentially identifies, evaluates, and controls hazards that may possess a significant risk to food safety and consumer health by reducing risks to satisfactory levels (Heggum, 2001; Trienekens and Zuurbier, 2008). HACCP enables inspection at every step of processing instead of end-product inspection. These three major food quality and safety management systems stand obligatory whereas systems such as ISO 9000 and ISO 22000 are non-obligatory food quality and safety management systems. ISO 9000 is the family of norms that gives guidelines that regulate effective implementation and maintenance of the quality management system within an organization ultimately targeting consumer satisfaction. Whereas, ISO 22000 harmonies the requirements relating to the food safety management system within a food enterprise at a global level (Bilska and Kowalski, 2014).

Foodborne illness has acquired serious concerns and has been described as a global concern which possesses largely avoidable public health challenge. Assuring food safety at each level is necessary to safeguard public health. In recent times, much attention has been acquired by various modern analytical tools used to authenticate the quality and safety of food. This has necessitated the requirement to target resources for efficient recognition and reduction of various foodborne pathogens as well as enhancing our knowledge about their associated concerns (Ellis et al., 2018). There is a steady requirement to extend current technology, as well as refining the present approaches for the detection and analysis of foodborne pathogens and contaminants to mitigate the global threat of foodborne diseases (Ramos et al., 2016). Detection of various chemical and microbial hazards is a cumbersome process due to their different physicochemical behavior, complex nature of food and their presence in low amounts. To ensure the safety of foods, various government institutions have established or determined certain tolerable limits for major contaminants, but lack of synchronized legislation and incidence of new hazards that enter the food chain has made it difficult. This chapter provides a comprehensive introduction about the food quality and food safety concept, its importance as well as exemplifies the role of various modern analytical tools in combating food quality and safety issues as a prospect to safeguard human health.

12.2 FOOD QUALITY AND FOOD SAFETY: A CONCEPT

The concept of food quality generally relates to the product attributes such as freshness, appearance, taste, smell, nutritional value, functional quality, textural, and other physicochemical attributes. Food composition analysis helps in the characterization of food and provides evidence about the presence of desired constituents either being the natural component or in the form of food additives. Also, it helps in making a decision regarding the freshness, informing the presence of microorganisms and toxins as a result of spoilage or deterioration. Whereas, the concept of food safety entails the production and commercialization of food which do not pose any health risk to the consumer, so it should be free from any sort of allergens, pesticides, fertilizer, heavy metals, pathogens, and microbial toxins. The presence of these contaminants may cause harmful effects on human health and well-being, leading to foodborne diseases that directly affect the health care system and economic growth (Scognamiglio et al., 2014).

Failure to meet food safety standards may lead to the chances of the occurrence of foodborne illness. Foodborne illnesses are the result of either presence of any chemical, foreign matter, microorganisms, or their toxin in the food that when ingested along with the food cause public health problem (Giuseppe et al., 2010). The current trend that has been followed during global food production and distribution has aroused the need to enhance or strengthen the area of food safety research to ensure global food safety. It mainly focuses on enhancing current technologies for the detection of foodborne hazards and to understand their possible fundamental mechanism so that accordingly prevented measures can be adopted (Beier and Pillai, 2007). A "farm to fork" approach has been effectively adopted by various food manufacturers to recognize all those points throughout the food chain where the chances of occurrence of any sort of contamination are most likely to occur and are carried out with extra precautionary measures. It necessitates the identification and setting up a series of procedures, inspection, and control systems to lessen the risk associated with unsafe food or bad quality end product.

12.3 FOOD AND PUBLIC HEALTH

The public health sector mainly deals with a set of scientific measures that aims at safeguarding and promoting the good health. Food is essential for the sustenance of life. Safe and wholesome food should be consumed in adequate amounts to meet the requirement to sustain good health. Food being a source of a variety of nutrients offers a tremendous environment to microbes for their growth which sometimes could be beneficial as well as undesirable. When food gets contaminated, it acts as a suitable vector for microorganisms such as bacteria, fungi, viruses, and parasites, etc., causing foodborne illness. Further, the presence of certain chemical contaminants that have been either present naturally or added during processing may exhibit toxic behavior and possess potential harmful effects on public health. Consumption of unsafe food not only affects the health of the people but also leads to monetary consequences for persons, families, communities, and countries. Nevertheless, the most imperative concern is to offer safe food to the customer and necessitates the need to attain a food safety bridge as a requirement to build between the public health sector and other sectors that mainly deals with production and processing of food to make the certain effectual cross-sectoral association. Hence, food safety acts as an essential pre-requisite to assure food security and safety to protect consumers from the risk of any foodborne infection and illness (Chaudhuri, 2015). Advancements made in food safety as one of the preventive approaches to assure good public health has found to be successful in the control and prevention of foodborne illness. But the rising concerns associated with emerging toxicants and foodborne pathogens and the understanding of long-term health effects pose on public health has aroused the need to critically analyze the food chain for food safety ambiguities more vigilantly. The reasons associated with these rising concerns towards food safety may include a combination of factors such as urbanization, large-scale food production, changing lifestyles and diets of people and modern trade practices. Longer distribution chain in terms of time and distance has been the sole reason governing these changes in the food production scenario. Our community is certainly undergoing a thoughtful transition worldwide in the manner food is produced, processed, and distributed. Food safety includes all those necessary conditions and measures that followed during production, processing, storage and distribution of food throughout the chain to assure safety after ingestion. Hence, food safety is undoubtedly an important segment associated with the branch of public health which needs to be prioritized in any public health program (Miyagishima et al., 1995).

12.4 NEED FOR FOOD QUALITY AND FOOD SAFETY

Food is a necessity to sustain life. Safe and quality food is considered as two important factors in product success. The quality of a product needs to remain stable throughout the food chain. A food manufacturer must ensure that the product should fulfill all consumer expectations related to product wholesomeness, freshness, nutritive value as well as sensory attributes during all stages of the food chain, whether during production or post-production. Hence, it is important to estimate the period up to which the product retains its quality attributes. The parameters that change during storage and evaluation of microbial contamination should be continuously monitored to provide safe food as any deterioration of the product may indicate the presence of spoilage or pathogenic microorganism. Therefore, it is necessary to carry out a safety assessment of a food product (Piskuła et al., 2011).

The occurrence of foodborne illness has made food a vehicle for transmitting hazards that may lead to various health implications or in some cases can be lethal. Illness due to the ingestion of contaminated food is the most prevalent health problem in today's world and has eventually affected economic productivity (WHO, 1999). Increased negligence during production and processing of food may lead to changes in chemical and microbiological contamination, thereby increasing the risk of foodborne illness when consumed (Käferstein and Abdussalam, 1999). Hence, the struggle to make sure the production of safe food has predominantly intensified in recent times. Both traditional, as well as modern analytical tools need to be fully exploited and improved to ensure food safety. To strengthen the food safety system, it is important to concentrate not only on the technical development in the production area but also to increase awareness and commitment of all the professionals belonging to the food production chain. Also, there is a need to take appropriate legislative measures where necessary with enhanced dependence on voluntary compliance and awareness among the consumer and food handling professionals to ensure an efficient food safety system aiming at "safe food for good health." The huge importance of these aspects in the production and distribution of food is supported by broad law regulations in the area and by a constant strive to improve food production and distribution processes (Bilska and Kowalski, 2014). The efficient functioning of these standards provides a consumer-friendly market that ultimately results in ensuring the health safety of consumers.

12.5 RISK ANALYSIS

Risk analysis was conducted by various government bodies globally to address the food safety risks linked with biological hazards. It is purposeful, planned, and formalized approach to identify risk and wherever necessary, also reduces risk. It consists of three major components (WHO, 1999) (Figure 12.1): (i) risk assessment, (ii) risk management, and (iii) risk communication.

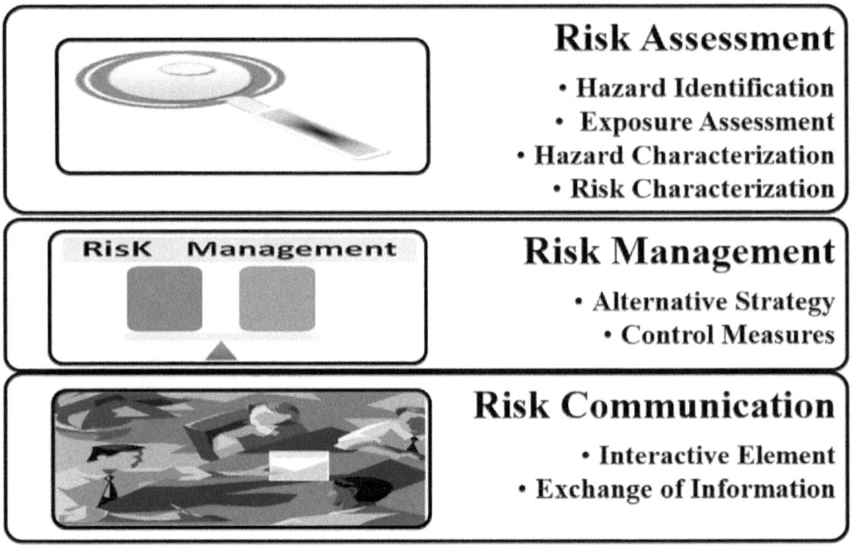

FIGURE 12.1 Elements of risk analysis.

Risk assessment consists of a science-based approach to estimate risks and associated influential factors. Initially, a statement needs to be formulated, and the process includes four components such as hazard identification, exposure assessment, hazard characterization, and risk characterization. Hazard identification consists of the detection of any physical, chemical, and mainly biological agents present in a food chain that are competent to cause any adverse effect. Hazard identification is mainly equipped with public health data, the preliminary estimate of sources, incidence, and quantity of agents under care. Exposure assessment deals with the qualitative or quantitative estimation of probable ingestion of any hazardous agent within the food chain as well as the chances of its exposure. It mainly considers (i) regularity or the tendency of contamination with a pathogenic agent, its chances of occurrence over some time till its consumption, and (ii) dietary information. Hazard characterization involves evaluation of the nature of the physical, chemical, and microbial hazards that may possess adverse health effects. In the case of biological agent determination, parameters include physiology, the pathogenicity of microorganism, severity of infection as well as the susceptibility of the host. After relevant data is obtained, a dose assessment was carried out to evaluate the correlation between the dose of a biological agent, chemical, and physical agent with the severity and/or incidence of occurrence of health hazards. Further, risk characterization involves qualitative or quantitative estimation of the level of uncertainties, probability of incidence and rigorousness of recognized potential hazard in a particular population based on previous steps such as hazard identification, exposure assessment, and hazard characterization.

Risk management is an alternative strategy based on results obtained from risk assessment and involves identification, implementation of suitable control measures, together with regulatory measures. This necessitates the need for an interactive element for an organized risk analysis procedure, there must be a functional link that needs to be present to ensure the integrity of both risk assessment and risk management as two separate aspects as well as maintains transparency in the decisive process for effective management. Risk communication involves the exchange of information and judgment regarding the risk amongst risk assessors, managers, consumers, and other concerned parties.

Risk analysis is mainly carried out by government bodies that have accessibility to all the required research data and findings. As of now, the risk analysis procedure is in the initial phase, which is likely to be strengthening further on a broader scenario to perform an important role in consumer protection against various foodborne hazards. For the successful implementation of this system, government authorities must effectively communicate the probable level of protection into food safety objectives. These objectives set precise targets for food business operators that must be met through suitable interventions. This objective provides a statement about the maximum level of a microbiological hazard that is tolerable for food consumption. These objectives could be quantified and verified at any point (Jouve et al., 1998).

12.6 FOOD QUALITY AND FOOD SAFETY MANAGEMENT SYSTEMS

With rising concerns associated with food quality, food safety aspects of food have necessitated the development of various system-based approach as a preventive tool to combat the risk of food hazards. Food business operators make use of various food safety management tools throughout the food chain to build a positive reputation amongst the consumer. These food safety management systems or tools offers a structured framework that defines and employs a set of measures to ensure consistent product manufacturing with all the set quality and safety standards (Mayes et al., 1993). Food business operators have been presented with multiple options in terms of different food safety and quality management tools from which they can select or adopt the most appropriate one for its specific activity and must ensure its effective establishment, documentation, and implementation. Amongst various available food safety managements those are available at present includes tools such as GMPs (Good Manufacturing Practices), GHPs (Good Hygiene Practices), GAPs (Good Agricultural Practices) as a prerequisite tool and HACCP (Hazard Analysis Critical Control Points) as a major process control tool (Van der Speigel et al., 2003). Apart from these management systems, such as ISO 9000 deals with the quality management system and ISO 22000 that relates to food safety management systems are also adopted by the food business operators throughout the food chain (Rotaru et al., 2005). These systems form an essential foundation for various activities such as production, handling, etc., with a thorough understanding of all of the basic requirements of GHP and GMP for a particular product or commodity. Effective adoption of these basic practices is an essential pre-requisite for any food business. Food quality and safety management systems are essential to ensure a confident position in the market as well as to build a positive reputation amongst the consumer. Implementation of a system approach to combat food quality and food safety concerns is more beneficial as compared to end-product testing as there may be an enhanced probability of contamination as well as may cause wastage of resources till the time of detection of hazard at the end of the production line. Hence, a system-based approach provides better management over quality and safety aspects at every stage of processing and also during their integration as an integrated approach. GMP and GHP form an important part of pre-requisite programs that deals with codes of good practice that encompass the elemental principles, measures, and resources needed for the production of safe food. They form an essential part and foundation for the successful implementation of quality and safety management programs such as HACCP and ISO 22000.

12.6.1 GMP (GOOD MANUFACTURING PRACTICES) AND GHP (GOOD HYGIENIC PRACTICES)

GMP and GHP are the two key pre-requisite approaches that stand necessary for the effective establishment and implementation of the HACCP system (Gorris, 2005). GHP includes all those necessary hygiene steps that must be critically followed and controlled at every stage of manufacture and distribution to ensure food safety throughout the chain. GHP involves instructions that ensure implementation of hygienic processes in the facility and usually cover the following requirements (Bilska and Kowalski, 2014):

- Maintenance of hygienic conditions in the production area;
- Hygienic construction and maintenance of machinery;
- Adoption of effective cleaning and disinfection protocols (such as pest control);
- Adoption of general hygiene and safety measures at each processing step of the production line to avoid any chance of contamination;
- Training of food-handling personnel about hygiene and safety measures;
- Adoption of hygiene measures by each food-handling personnel.

GMP involves all those measures that need to be taken and must be fulfilled during the production of food or other materials that may be intended to come in contact with food in an appropriate way to ensure food safety (Bilska and Kowalski, 2014). It takes into account necessary actions at each stage of food product manufacturing, beginning with the measures concerned with the maintenance of construction and technical facilities in the production area followed by food-handling personnel and ultimately the major consideration is forced on production process starting from the raw material inception to product distribution. It includes all the principles, procedures, and resources needed to develop a suitable production line for the manufacturing of products with acceptable quality. GMP/GHP measures have been designed by governments along with the Codex Alimentarius Committee on Food Hygiene (FAO/WHO) and the food industry, also in association with other food assessment and control authorities. They should always be effectively applied and well documented.

12.6.2 HACCP (HAZARD ANALYSIS AND CRITICAL CONTROL POINT SYSTEM)

HACCP is most widely adopted as a preventive approach to ensure systematic identification and control of hazards linked with the production, distribution, and use of food products. It is an independent food safety management system specifically adopted in the food domain by various food business operators. It is generally preceded by the adoption of GMP and GHP as a pre-requisite requirement for effective implementation of the HACCP system in food production and distribution facilities. A key aspect of this model is to establish targets or standards to alleviate the risk of any source of food hazards, whether biological, chemical, or physical in origin and concurrently provide a tool for achieving satisfactory levels of food safety (Hulebak and Schlosser, 2002). HACCP is a systematic approach for food control that ensures suitable hazard identification and characterization. Its management at each stage of food processing starts from the inception of raw materials to the distribution of end product to the consumers. It involves a critical evaluation of each step of food processing for any possible hazard that may pose any health risk to the consumer. It employs control and preventive measures wherever necessary to eliminate the entire food production line from health hazards. It mainly deals with (1) hazard identification, (2) identification of control measures, (3) continuous evaluation of HACCP system efficacy (Bilska and Kowalski, 2014). HACCP mainly includes seven key principles that must be sequentially followed critically for effective management of food safety throughout the food chain (Hulebak and Schlosser, 2002) (Figure 12.2).

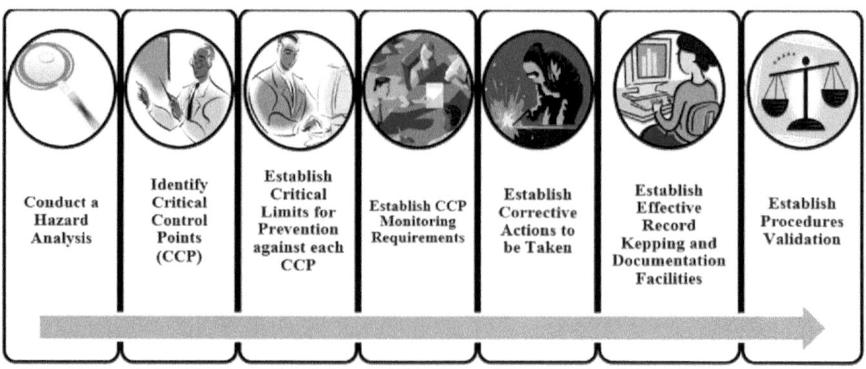

FIGURE 12.2 Key principles of a HACCP system.

Even though GMP and GHP deal with all the generic hygienic measures production of safer food, the HACCP system has an additional advantage of estimating precise determinants exclusive to a particular product and procedure. Coupling of GMP/GHP procedures along with the HACCP system can act synergistically to each other ensuring critical determination of any critical points in the food chain and adoption of appropriate preventive measures to ensure food safety.

12.6.3 ISO 22000 – FOOD SAFETY MANAGEMENT SYSTEM

A variety of advanced food safety management systems have been developed by the government in various countries around the world to facilitate improved food traceability and transparency to promote food safety in both domestic and international trade. Simultaneously, efficient maintenance of the effectiveness of these management systems to ensure food safety has been an intimidating confront. The emergence of various food safety incidences around the world corresponds to the incapability of these food safety management systems in that aspect. To deal with such situations, ISO 22000 has been promulgated as an international standard for the effective adoption of various food safety management systems to assure food safety throughout the food chain, starting from procurement of raw material to distribution of the final product. It involves analysis and mitigation of various hazards that may occur during any stage of production to ensure end-product safety and provides the customer with ultimate safer food.

International Organization of Standardization (ISO) generally reviews its standards after 5 years to monitor the need for any modifications or revisions needed to keep all the standards up-to-date to meet all the legal provisions. ISO 22000 was revised in 2018 to address all the previous shortcomings associated with the previous version of the standard (Ştefan et al., 2018). The previous version of ISO 22000 as the reference norm for a food enterprise does not provide adequate comprehensive information about the pre-requisite programs and was not recognized as an effective reference standard by the Global Food Safety Initiative (GFSI). The revised version of ISO 22000: 2018 was published on June 19, 2018, to ensure that this system meets all the food safety challenges and should be consistent for integration with other ISO standards (Ştefan et al., 2018). It mainly consists of four key elements: (1) the HACCP plan, (2) pre-requisite programs, (3) effective quality management system, and (4) appropriate and adequate sites, facilities, and equipment.

Among these, all the adoption of HACCP plan was the most important element of this system (Chen, 2020). ISO 22000 has been formulated as a tailor-made approach to enable food safety through all the segments of the chain. It does not consider or follow the approach of "one size fits all," as the standards applied to a particular food chain may not be appropriate for another. Implementation of ISO 22000 standard offers various frontiers; some are depicted in Figure 12.3. ISO 22000 is a valuable standard norm as it is an auditable standard with defined necessities, is accepted globally, integrates, and harmonizes with other essential norms such as ISO 9000, Occupation Health and Safety Assessment Series (OHSAH), etc., and contributes to enhanced indulgence and further expansion of HACCP as an essential food safety tool (Færgemand, 2008).

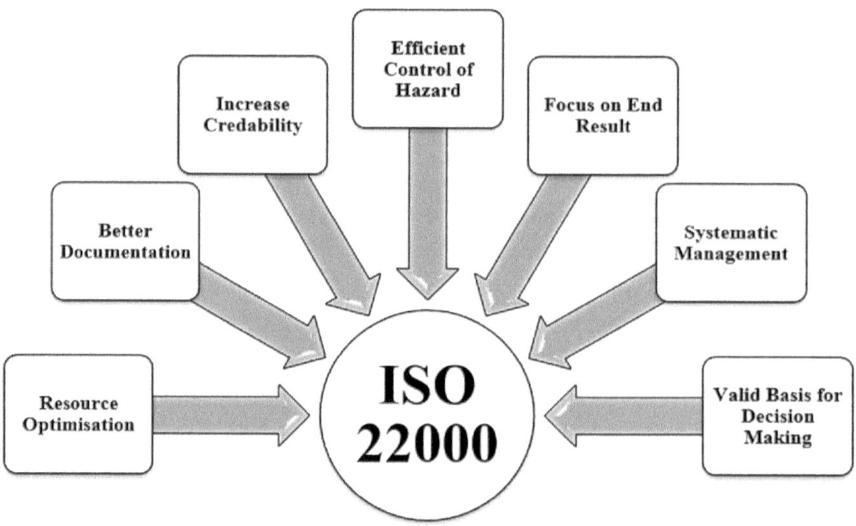

FIGURE 12.3 ISO 22000 system benefits.

ISO 22000 couples HACCP principles with the Codex Alimentarius steps along with pre-requisite programs for the establishment of a valuable system for hazard control. Implementation of HACCP is a fundamental necessity of ISO 22000 norm and hence, they both act complementary to each other with shared corroboration.

12.7 MODERN ANALYTICAL TOOLS

Food safety concerns are incessantly growing and entail the advancement and adoption of newer approaches and technologies for better efficiency. Food quality and safety are amongst the two most critical parameters that have immense economic as well as public health importance. Analysis of food for the properties and composition of constituents stands necessary to meet safety and quality required by the food industry as per consumers want. Food safety deals with assuring that the product is free from any source of toxic compounds, hazardous microorganisms, toxins released by microorganisms, naturally present toxic substances, pesticides, and anti-nutritional compounds. Quality control and food safety assurance serve as two essential elements of food analytical protocols. There have been various analytical techniques such as gas chromatography (GC), high performance liquid chromatography (HPLC), mass spectroscopy (MS) and nuclear magnetic resonance (NMR) are used for the analysis of food safety and quality but have been associated with the constraints of complicated sample preparation procedures. Ongoing analytical procedures used in the food industry for analyzes of food contaminants and toxicants involve protocols that are time taking, require skilled labor, extensive equipment, long separation process and chemicals with high purity. For rapid screening of these contaminants, various complementary techniques have been developed to detect various contaminants. Biosensors are amongst the most suitable choice for this as they offer a more specific and sensitive analysis of an analyte. It monitors the whole food chain in a very organized and lucrative manner. Various biosensors have been effectively employed to assure the safety and quality of foods. Coupling of the high sensitivity of biosensors with HACCP can provide a promising approach for overseeing food safety throughout the chain (Kaur, 2019). Apart from these various molecular approaches have been effectively utilized in combating food safety issues associated with food hazards of biological origin as they are amongst the most prevalent ones.

Foodborne diseases via biological agents like bacteria, fungi, and viruses are amongst the most prevalent causes. Traditional methods used for microbial analysis of food system involve the use of various culturing methods but does not provide a complete characterization of all microbial communities (Sirangelo, 2018). Although various advancements have been made in recent years to mitigate foodborne illness incidences in many parts of the world, but the overall progress has been slow. The role of the various molecular surveillance systems has been emphasized to be beneficial in detecting foodborne outbreaks quickly, their sources and potential transmission routes that can be used to control the spread of foodborne illness (Boxrud et al., 2010; Tauxe, 2010). However, the use of these molecular tools has raised concerns because of the reduced concentration of isolated pathogen that hinders in performing subtyping.

Adoption of omic technologies can serve as a solution for partially decreasing this issue as it allows the rapid detection of pathogens without isolation. Hence, the use of various omic tools have been found effective in reducing foodborne illness cases and outbreaks as it helps in the essential characterization of food quality and safety (Ferri et al., 2015). Omics can be said to have much potential for the detection of foodborne pathogens, as well as monitoring contamination and disease outbreaks, particularly when integrated with emerging technologies (Ellis et al., 2018). Genomics, transcriptomics, and proteomics are rapidly transforming approaches for detection, prevention, and treatment of foodborne pathogens. Microbial genome sequencing, in particular, has evolved from a research tool into an approach that can be used to characterize foodborne pathogen isolates as part of routine surveillance systems. Genome sequencing efforts will not only improve outbreak detection and source tracking but will also create a large amount of foodborne pathogen genome sequence data, which will be available for data mining efforts and could facilitate better source attribution and provide new insights into foodborne pathogen biology and transmission. Practical uses and application of metagenomics, transcriptomics, and proteomics data and associated tools are less prominent, although, these tools are also starting to yield practical food safety solutions (Bergholz et al., 2014). Full consideration to benefits and challenges associated with these tools is essential to ensure the full potential of these tools.

12.8 CONCLUSION

Increased prevalence of the occurrence of foodborne disease outbreaks over recent times has made food quality and safety a booming concern worldwide that stimulates the trade as well as possesses serious public health concerns. Increasing consumer dependence on processed products has led to an increased demand for consistently high quality. However, for consumers, the most important aspect associated with food quality is food safety. Thus, quality, and safety are the two key aspects that have acquired an important position in the life of a food product that governs its success in the market. Contamination via biological agents has been perceived as the most considerate threat to human health that may lead to infections, injuries, deaths, and economic loss. However, chemical contaminants have been mainly associated with chronic illness. The adoption of various voluntary and involuntary food safety and quality management system is the most effective control measure. Also, the development of faster, more sensitive, and specific tools for rapid as well as effective determination will play a critical role in facilitating the use of various modern analytical tools such as "omics" tools, and more importantly, improving the current sequencing technologies can serve well.

KEYWORDS

- **food quality**
- **HACCP**
- **management tools**
- **risk analysis**
- **safety**

REFERENCES

Beier, R. C., & Pillai, S. D., (2007). Future directions in food safety. In: Shabbir, S., (ed.), *Foodborne Diseases* (pp. 511–530). Totowa, NJ: Humana Press.

Bergholz, T. M., Switt, A. I. M., & Wiedmann, M., (2014). Omics approaches in food safety: Fulfilling the promise? *Trends in Microbiology, 22*(5), 275–281.

Bilska, A., & Kowalski, R., (2014). Food quality and safety management. *LogForum, 10*(3), 351–361.

Borchers, A., Teuber, S. S., Keen, C. L., & Gershwin, M. E., (2010). Food safety. *Clinical Reviews in Allergy and Immunology, 39*(2), 95–141.

Boxrud, D., Monson, T., Stiles, T., & Besser, J., (2010). The role, challenges, and support of pulsenet laboratories in detecting foodborne disease outbreaks. *Public Health Reports, 125*, 57–62.

Chaudhuri, D., (2015). Food safety: A public health priority. *Indian Journal of Public Health, 59*(2), 83.

Chen, H., Liu, S., Chen, Y., et al., (2020). Food safety management systems based on ISO 22000: 2018 methodology of hazard analysis compared to ISO 22000: 2005. *Accreditation and Quality Assurance, 25*(1), 23–37.

Chung, E. Y., Kee, D. M. H., Chan, J. W., et al., (2020). Improving food safety and food quality: The case of Nestle. *International Journal of Tourism and Hospitality in Asia Pacific, 3*(1), 57–67.

Ellis, D. I., Muhamadali, H., Chisanga, M., & Goodacre, R., (2018). Omics methods for the detection of foodborne pathogens. *Encyclopedia of Food Chemistry*, 364–370.

Færgemand, J., (2008). The ISO 22000 series global standards for safe food supply chains. *ISO Management Systems, 8*(3), 4–7.

Ferri, E., Galimberti, A., Casiraghi, M., et al., (2015). Towards a universal approach based on omics technologies for the quality control of food. *BioMed Research International*, 1–14.

Giuseppe, E., Monica, S., & GianFranco, G., (2010). Science for food safety, security and quality: A review-part 1. *Quality of Life, 1*(1).

Gorris, L. G., (2005). Food safety objective: An integral part of food chain management. *Food Control, 16*(9), 801–809.

Guérin, A., Rouger, A., Tareb, R., et al., (2019). Advanced omics approaches applied to microbial food safety and quality. *Food Molecular Microbiology,* 1888.

Heggum, C., (2001). Trends in hygiene management–the dairy sector example. *Food Control, 12*(4), 241–246.

Hulebak, K. L., & Schlosser, W., (2002). Hazard analysis and critical control point (HACCP) history and conceptual overview. *Risk Analysis, 22*(3), 547–552.

Jouve, J. L., Stringer, M. F., & Baird-Parker, A. C., (1998). Food safety management tools. *ILSI Europe Report Series, Brussels, Belgium*, 1–23.

Käferstein, F., & Abdussalam, M., (1999). Food safety in the 21st century. *Bulletin of the World Health Organization, 77*(4), 347.

Kaur, K., & Kaushal, P., (2019). Enzymes as analytical tools for the assessment of food quality and food safety. In: *Advances in Enzyme Technology* (pp. 273–292). Elsevier.

Kotsanopoulos, K. V., & Arvanitoyannis, I. S., (2017). The role of auditing, food safety, and food quality standards in the food industry: A review. *Comprehensive Reviews in Food Science and Food Safety, 16*(5), 760–775.

Manning, L., & Soon, J. M., (2016). Food safety, food fraud, and food defense: A fast-evolving literature. *Journal of Food Science, 81*(4), 823–834.

Mayes, T., (1993). The application of management systems to food safety and quality. *Trends in Food Science and Technology, 4*(7), 216–219.

Miyagishima, K., Moy, G., Miyagawa, S., et al., (1995). Food safety and public health. *Food Control, 6*(5), 253–259.

Morkis, G., (2006). The degree of GHP, GMP and HACCP implementation in the food industry. Żywność. *Nauka. Technologia. Jakość, 3*(48), 129–145.

Piskuła, M. K., Strączkowski, M., Żmudzki, J., et al., (2011). The characteristic of consumer safety- and pro health food quality determining factors. *Polish Journal of Agronomy, 7*, 82–91.

Ramos, S., Silva, N., Hébraud, M., et al., (2016). Proteomics for drug resistance on the food chain? Multidrug-resistant *Escherichia coli* proteomes from slaughtered pigs. *Omics: A Journal of Integrative Biology, 20*(6), 362–374.

Rotaru, G., Sava, N., Borda, D., & Stanciu, S., (2005). Food quality and safety management systems: A brief analysis of the individual and integrated approaches. *Scientifical Researches: Agroalimentary Processes and Technologies, 11*(1), 229–236.

Scognamiglio, V., Arduini, F., Palleschi, G., & Rea, G., (2014). Biosensing technology for sustainable food safety. *Trends in Analytical Chemistry, 62*, 1–10.

Sirangelo, T., (2018). Food microbiology and multi-omics approaches. In *International Conference on Food Microbiology and Food Safety*.

Ştefan, Z. P., Raluca, D., & Carmen, V., (2018). Comparative study of certification schemes for food safety management systems in The European Union context. *Amfiteatru Economics, 47*(1), 9–29.

Tauxe, R. V., (2006). Molecular subtyping and the transformation of public health. *Foodborne Pathogens and Disease, 3*, 4.

Trienekens, J., & Zuurbier, P., (2008). Quality and safety standards in the food industry, developments and challenges. *International Journal of Production Economics, 113*(1), 107–122.

Van, D. S. M., Luning, P. A., Yiggers, G. W., & Jongen, W. M. F., (2003). Towards a conceptual model to measure effectiveness of food quality systems. *Trends in Food Science and Technology, 14*, 424–431.

Van, R. W., & Frewer, L. J., (2008). Consumer perceptions of food quality and safety and their relation to traceability. *British Food Journal*, 1034–1046.

Wijtzes, T., Vant, R. K., Int, V. J. H., & Zwietering, M. H., (1998). A decision support system for the prediction of microbial food safety and food quality. *International Journal of Food Microbiology, 42*(1, 2), 79–90.

World Health Organization, (1999). *Food Safety.* (No. EM/RC46/6).

Index

D

F

G

H

I

M

O

T

U

V